Biologists under Hitler

Ute Deichmann
Translated by Thomas Dunlap

Harvard University Press
Cambridge, Massachusetts
London, England
1996

To Milada Ayrton and George Shrut

Library of Congress Cataloging-in-Publication Data

Deichmann, Ute, 1951–
    [Biologen unter Hitler.   English]
    Biologists under Hitler / Ute Deichmann ; translated by Thomas
Dunlap.
      p.   cm.
    Includes bibliographical references and index.
    ISBN 0-674-07404-1 (alk. paper)
    1. Biology—Germany—History—20th century.   2. Biology—Austria—
History—20th century.   3. Biologists—Germany—History—20th
century.   4. Biologists—Austria—History—20th century.
5. National socialism and science.   I. Title.
QH305.2.G3D4513   1996
574'.0943'09043—dc20   95-33338

Biologists under Hitler

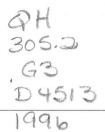

Contents

Foreword

Success in today's scientific world demands that scientists specialize and devote all their energy to solving a *single* problem. Competition is fierce, and the rat race is faster than ever before. All this would be tolerable, but a badly informed public sees dangers from science especially where there are none. Education about what science is, what it can and is permitted to do, is inadequate. In Germany, criticism of science is harsher than elsewhere, which I believe has to do with our unfortunate past. First in 1914, and again in 1933, Germany fell from the heights of scientific success into a maelstrom that destroyed the country in 1945.

What happened in Germany during the years of the Third Reich? A decade ago it struck me as scandalous that there existed no history of human genetics during those years. After all, this had been the very field whose scientists had collaborated most effectively with the Nazis. To remedy this situation I wrote my brief essay *Tödliche Wissenschaft: Die Aussonderung von Juden, Zigeunern und Geisteskranken 1933–1945* (Reinbek, 1984; English translation *Murderous Science: Elimination by Selection of Jews, Gypsies, and Others in Germany, 1933–1945*, Oxford, 1988). No such studies existed for chemistry, the field in which I had obtained my doctorate, or for biology, the field in which I now teach, and I had neither the time nor the energy to write them. And so it was a pleasant surprise when, in the fall of 1987, Ute Deichmann came to see me and asked if I would supervise a dissertation on the history of biology under National Socialism. She told me that she was a biology teacher and was planning to take a leave from teaching for this purpose. It took some time to clarify whether this was possible and whether she was a suitable person for this undertaking. As the present book attests, the answers to both questions were overwhelmingly positive. Once the necessary preliminary work had been completed, the Deutsche Forschungsgemeinschaft (German Research Association, DFG) agreed to support the project, and I am most grateful that it did.

What, then, was the situation of biology in Germany at that time? Anyone who, like myself, as an assistant or a young professor in the 1960s, listened to those who had pursued successful careers during those years, came away with the impression that the Nazis had been a band of ignoramuses hostile to science. Yet at the research institutes of the Kaiser Wilhelm Society for the Advancement of Science (KWG), in particular, despite a terrible lack of funds and constant political interference, a sworn group of idealistically minded scientists had come together and had continued to pursue first-rate science under the most difficult of circumstances. Such was the comfortable legend, but was it the truth?

Before Ute Deichmann set to work, we had to decide what we wanted to know and what we didn't want to know. We were most interested in three questions:

1. How large was the circle of ousted Jewish professors and researchers? In what areas of research did the ousted scientists surpass those who remained at home? What became of the expelled scientists? How great a scientific influence did they gain in the countries in which they sought refuge?

2. How did governmental support for scientific research function in the Third Reich? Was it professional or dilettantish? Did it pertain to the work itself or was it ideological?

3. In what way did scientists cooperate when those in power approached them with proposals to participate in crimes or profit from them?

It was clear to us that in pursuing these questions we had to dispense with the description of two areas. We did not want to devote a special account to the fools and charlatans who pushed themselves to the fore especially during the first six years after the *Machtergreifung* (Nazi seizure of power) and in this way temporarily attained status and influence. We preferred to focus on the defensive efforts by real scientists. Next we left out a description of the sphere of popular science and ideology, including biology as taught in the schools. It turned out to have been the right decision, since Änne Bäumer has since published a study on this topic (*NS-Biologie*, Stuttgart, 1990).

Before reviewing the findings of this book, I should point out that neither Ute Deichmann nor I have any academic background in history. We tried to compensate for this shortcoming by meeting weekly to discuss every newly discovered source, analyze its significance, and define what was still missing. In this way the dissertation took shape in the manner of experimental science—by a process of constant dialogue. Ute Deichmann deserves the credit for writing this book, stimulated by this method.

Support for Biological Research in the Third Reich

All previous statements about the advancement of science in the Third Reich have come from those involved. After 1945, German scientists tried to explain why their work had not yielded very much, even though they were the best scientists. They blamed the decline primarily on a lack of support for scientific research. Foreign scientists who pondered the decline of their German colleagues also believed this explanation.

What really happened? Ute Deichmann was in the fortunate position of being able to collect objective data. She began by drawing up a list of all biologists working in Großdeutschland (Germany, Austria, and the Sudeten region of Czechoslovakia) between 1933 and 1945. For this she used *Kürschner's Deutsches Gelehrtenlexikon*, the course catalogs of all universities, and the annual reports of the Kaiser Wilhelm Society. The list totaled 445 people. Next, she looked up the personal files of all these people in the Document Center in Berlin, which housed the entire record of the Nazi party as an organization, to determine their membership in the party and the SS. Her work showed that 60% of the non-Jewish biologists who remained in Großdeutschland were party members.

Armed with this information she went to the Koblenz Bundesarchiv (Federal Archive) to investigate which of these biologists had been funded in their work during the Third Reich by the DFG (until 1937 called Notgemeinschaft der deutschen Wissenschaft, Emergency Association of German Science). Luckily nearly all the records of the German Research Society pertaining to research funding during these years have been preserved.

Of the 445 biologists, 222 submitted grant applications to the DFG between 1933 and 1945. The DFG's volume of funding for biological research rose *tenfold* between 1932 and 1939, and subsequently remained on that level during the entire war. Until 1937, expert opinions on the projects under consideration were regularly solicited. Later the chief reviewer decided at his discretion. In 1934 party comrades enjoyed a real advantage when it came to receiving support, but in all subsequent years to the end of the war it made statistically no difference whether an applicant was a member of the party or not. Support shifted from the universities to the Kaiser Wilhelm Society, from 10% for the latter in 1932 to 50% in the years 1939 to 1944.

A detailed analysis of the sums given in support does not indicate that any egregiously false decisions were made. The decisions were guided in fact by the quality of the applications. A lack of funds could thus not have been the reason if the results did not satisfy the researchers themselves. What then was the reason? Was it perhaps that with the Jews the elite of scientists had been driven out and the mediocre ones remained?

The Expulsion of Jews from German Institutes of Biology

Ute Deichmann traced the fate of all biologists who were dismissed, first in Germany and later in Austria and Czechoslovakia. She discovered that two-thirds were dismissed for "racial" reasons. One-third were dismissed for other, mostly political, reasons. In total, 45 of the 337 biologists working at the same time were dismissed, 30 of them for "racial" reasons. The dismissed thus accounted for about 13% of all biologists. They had worked in the most diverse fields of biology; no crucial field was entirely depopulated by the dismissals. Among those driven out were some important persons. If we look at the Science Citation Index (which did not exist until 1945, however), we find that after the war the works of the dismissed scientists were cited three times more frequently than those of the scientists who remained in place. Their publications thus received more international attention than did the publications of those in Germany. This pattern may also reflect the fact that after 1933 the dismissed scientists began to publish in international journals in English, while those in Germany for the most part remained loyal to their German journals. It is beyond question that the international influence of the dismissed scientists was greater after 1945 than that of the scientists in Germany.

But is this loss of 13% of the scientists a sufficient explanation for why German molecular biology lagged behind for decades after 1945? I don't think so; I believe that the reason lies deeper. With the expulsion of the Jews and with the attempt to exclude Jews entirely from the life of the sciences, Germany had violated a fundamental principle of science: in science only the experiment and the argument matter, never the person who presents them, unless he or she is a known liar or fraud. The great majority of those who remained behind had silently accepted the anti-Jewish Nazi doctrine, and a minority had profited from it, namely, those who took over the chairs, directorships, and other positions from those who had been expelled. It was not the worst scientists who moved into the open positions: Alfred Kühn replaced Richard Goldschmidt as director of the Kaiser Wilhelm Institute for Biology, and Adolf Butenandt replaced Carl Neuberg as director of the Kaiser Wilhelm Institute for Biochemistry. After the war, none of those who took over the position of an expelled scientist could speak openly with Jewish foreign colleagues.

This isolation probably also explains why not a single German rector's speech after 1945 remembered the expelled Jewish colleagues, and why attempts to win back the dismissed scientists were few and successful in only two cases: the botanist Leo Brauner, in 1933 an associate professor in Jena, returned to Munich in 1955; the botanist Alfred Heilbronn, in 1933 an associate professor in Münster, returned to Münster in 1956. Even without an international boycott, German scholars had a difficult time returning to the community of international science. Science, however, lives on international dialogue, and after the war this dialaogue simply did not develop, all claims to the contrary notwithstanding.

Science thus did proceed during the Third Reich. It was financially supported (at least in the field of biology) as never before. But the expulsion of the Jews had touched it at its essence. True science is cosmopolitan and not nationalistic. Nationalism was the fault from which German science was suffering. Was it the only fault? Or were there others as well? Ute Deichmann uncovered some disturbing facts.

Biology in Proximity to Nazi Crimes

Those who remained behind took over the chairs or other positions of the expelled Jews in a matter-of-fact way. The people who came to occupy these positions often, though not always, had the best of reputations. They thought they were acting in behalf of science itself when they moved into these posts. How far this thinking went is revealed by the example of Fritz von Wettstein. In 1934 he succeeded the recently deceased Carl Correns and became director of the Kaiser Wilhelm Institute for Biology in Berlin. He was apolitical, not a party member, and much too good a scientist to have ever had to consider joining the party. But when it appeared that the German troops would finally overrun the Soviet Union, it was his intention, as part of the General Plan East, to absorb the most important Soviet institutes of agriculture into the Kaiser Wilhelm Society. In a letter to the general secretary of the KWG, dated March 9, 1942, he listed nine target institutes situated in the region between Murmansk and Baku. To these he added provisionally those in Murmansk and Baku, since "they are, after all, not yet under our control at this time." Wettstein's colleague Wilhelm Rudorf, director of the Kaiser Wilhelm Institute for Plant Breeding (after 1938 the Erwin Baur Institute) had moved even more quickly: he had directly approached Alfred Rosenberg, the minister of the occupied eastern territories, with the same request!

Here the issue was not merely a position to be occupied by one person but entire institutes to be taken over, some of which had not even been captured. However one may wish to assess the actions of these men—as possible protection for Soviet colleagues, or as a way of securing their data and projects—we are left with a deep sense of unease. If Wettstein's letter had not survived we would know nothing about this plan, for none of those involved ever mentioned it.

Similar circumstances surround a research project linking the Kaiser Wilhelm Institute for Plant Breeding with Auschwitz. Once again only the records document the result of a meeting on June 25, 1943, attended by, among other people, Wilhelm Rudorf from the institute and four SS officers from the concentration camp at Auschwitz. Like all other participants at the meeting, Rudorf, who after the war became the founding director of the Max Planck Institute for Breeding Research in Cologne-Vogelsang, in later years never mentioned the cooperation with Auschwitz.

Then there is the case of the Nobel laureate Konrad Lorenz. It would appear that the essays he published during the Third Reich, which established his fame also with the Nobel committee in Stockholm, were not closely read by anyone. How could someone who twisted quotes as needed have attained such fame? "A good man, in his dark striving, still knows where lies the moral course" ("Ein guter Mensch in seinem dunklen Drange, ist sich des rechten Weges wohl bewußt"), we read in Goethe. In 1940 Lorenz turned this into: "A good man, in his dark striving, knows quite well whether someone else is a scoundrel." And who is the scoundrel? In 1943 he wrote: "The old biblical saying 'Beware of the marked' has always existed for a good reason." Neither the Old nor the New Testament contains such a saying, which in 1943, two years after the introduction of the yellow star, had to be understood as being directed against the Jews. But all this is well known. What is new is that Ute Deichmann has filled a gap in the biography of Konrad Lorenz. In 1942 in Posen, he participated in a study headed by Dr. Rudolf Hippius on the psychic makeup of 877 children of mixed German-Polish marriages. In 1943 Hippius published the study under his name as a book in Prague, where he had in the meantime become a professor. Konrad Lorenz himself thus participated actively in the racial-genetic analysis of human beings, which formed the basis of selection, the fundamental crime of the Third Reich.

What Were the Outstanding Results of Scientific Research?

If one measures the biological research of the Third Reich by the number of Nobel prizes awarded to those whose work derived from foundations laid in those years, we have to mention the work in behavioral biology by Karl von Frisch with bees and Konrad Lorenz with geese. As a "second-degree *Mischling* [half-breed]," Karl von Frisch had to struggle with great difficulties. What allowed him to survive those difficult years as a scientist and a human being were his superior intellect and his personal integrity. The Austrian Konrad Lorenz initially also had some trouble getting sufficient funding in Germany for his research. He made great efforts to demonstrate to his patrons that the results of his research fit National Socialist racial ideology. It seems to me amazing and remarkable that Max Delbrück, a man of superior intelligence, a Nobel laureate and emigrant by accident, recommended Lorenz for the Nobel prize. Was it so easy to separate the Lorenzian mixture of behavioral biology and National Socialist ideology? Could one forget the ideological part? Was it really only the meaningless, opportunistic blather of an ambitious man, to which one could counterpoise his mature scientific accomplishments?

Max Delbrück was a molecular biologist. Why did he not recom-
mend one of the German molecular biologists for the Nobel prize?
There was Hans Friedrich-Freksa, who in 1940 had suggested a model
of self-complementary genetic substance. At the time this was far-
sighted, but in the years between 1945 and 1953, when scientists were
searching for a structural model of DNA, Friedrich-Freksa contributed
nothing more. There was Alfred Kühn, who together with Adolf Butenandt had analyzed color mutations of the eye of *Ephestia* (flour
moths) in 1940, and had, two years later, published an interpretation
of his findings that came close to the "one gene–one enzyme" hypothesis of George Beadle. All that is true, but while George Beadle successfully proceeded to the next step of analyzing metabolic mutants
in *Neurospora*, Alfred Kühn and Adolf Butenandt did not follow.

There was Gerhard Schramm. He had discovered that the coat of
the tobacco mosaic virus (TMV) was made up of identical protein
subunits and that its RNA was macromolecular. In 1942 he had discovered that TMV could be mutated through nitrite. But at the time
he inexplicably interpreted the nitrite mutagenesis wrongly as a protein catabolism. He made a belated correction in 1958. Moreover,
later in life, in Tübingen, he had been taken in by a fraud in his laboratory. It seems to me that after the war all of the above-mentioned
German molecular biologists no longer had any real contacts, any
discourse, with foreign molecular biologists. They lacked colleagues
to talk to whose ideas would have led them to correct their own thinking. Thus, then and now, it was not the money that was lacking. What
was lacking was discourse with Jewish colleagues, which had been
deliberately broken in 1933 and could be resumed only occasionally
after 1945.

Ute Deichmann concludes her study with a letter Lise Meitner wrote
to Otto Hahn in June of 1945. In it Meitner accused German scientists
as a whole of having "lost the measure of justice and fairness." Some
readers of this book have found fault with this ending. But that's how
it was. Science cannot flourish if it has suffered such a loss, no matter
how much money there might be for it. This simple statement is the
sum of Ute Deichmann's book. Let us hope that many will take note
of it.

Benno Müller-Hill
University of Cologne

Acknowledgments

This book is the revised and enlarged version of the dissertation I wrote between September 1987 and December 1990 at the Institute for Genetics of the University of Cologne under the supervision of Benno Müller-Hill.

I am grateful to Professor Müller-Hill for accepting this unusual topic and thoroughly discussing with me the questions and problems that arose in the process of working with and assessing the material. I wish to thank him for the stimulating intellectual support and the moral encouragement I received from him.

I also thank Eberhard Kolb for his encouragement and suggestions.

My thanks go out to all those who, as contemporary witnesses, were willing to answer questions. They include Hansjochem Autrum, Doris Baumann, Carsten Bresch, Franz Duspiva, Salome Gluecksohn-Waelsch, Viktor Hamburger, François Jacob, Romilde Kramer, Hermann Kuckuck, Hans Marquardt, Renate Mattick, Georg Melchers, Ursula Nürnberg, Werner Rauh, Nikolaus Riehl, Lutz Rosenkötter, Werner Schäfer, Walter Vielmetter, Kurt Wallenfels, Veronika Wieczorek.

My work could not have been done without the help of the directors and staff members at the archives listed in the Sources. It is not possible to thank by name all those who aided my research in the archives. I am especially grateful to the staff of the Bundesarchiv in Koblenz, where I researched a large part of the materials, for their friendly advice and help.

I would like to thank Diane Paul for an intensive exchange of ideas and stimulating discussions; Otto Geudtner, who read the manuscript and contributed much to its improvement through many comments

xvii

and discussions; Brigitte Kisters-Woike and Hans Gubitz, who helped me in my work at the computer; and Rudolf Sack, who encouraged and supported me in many ways. I also thank Mitchell Ash, Jonathan Harwood, Elisabeth Kramer, Helga Jantzen, and Mark Walker for discussions and suggestions.

Finally, I am grateful to the DFG for supporting the research project "The Development of Biology under National Socialism," which provided the framework for the present work, from October 1988 to September 1992 through its focus program "Scientific Emigration" (Az Mu 575).

Biologists under Hitler

Introduction

This study is an attempt to answer the question of how National Socialist politics and ideology influenced the development of biological research at the universities and the Kaiser Wilhelm Institutes in Germany. The findings of previous studies on National Socialist science and research policy led me to expect a noticeable decline in scientific accomplishments in Germany after 1933. One of the earliest analyses came from the American physicist Samuel Goudsmit. Examining the development of nuclear physics in Germany during World War II, he arrived at a scathing verdict: "Despite the German reputation for organization, the failure of their science during the war was due in part to the way they mismanaged the organization of their research. Scientific work did not advance under the administration of men like SS Brigadier General Ministerial Direktor Professor Doktor Rudolf Mentzel, and Ministerial Manager Professor Doktor Erich Schumann" (Goudsmit 1947, p. 239).[1] Many natural scientists and historians share Goudsmit's view that poor administration, meager support for scientific research, and the general science phobia of leading National Socialists strongly impeded the development of science in Germany.

Besides this, the exodus of Jewish scientists is considered the most important reason for the decline of scientific accomplishments during the Third Reich and after. For example, the physicist Werner Heisenberg noted in 1949 that Germany had not participated in the international scientific development during the years of the Third Reich: "A number of outstanding scholars were forced to emigrate. At the universities, the party frequently placed greater value on the ideological orientation of young people than on their scientific training. In part the state even supported the most absurd pseudo-scientific theories, such as the 'cosmic-ice' theory or the so-called 'German physics'" (Heisenberg 1953, p. 216).

The historian Alan Beyerchen (1977) emphasizes the ideological pressure on the universities from the state and the party, which by 1939 led to the suppression of modern physics in favor of "German physics." The historian and former vice-president of the Deutsche Forschungsgemeinschaft (German Research Association, DFG) Kurt Zierold (1968, p. 269) also shares Goudsmit's assessment that scientific research was poorly organized during National Socialism. He believes that the DFG and the National Socialist Reichsforschungsrat (Reich Research Council, RFR) did their job so poorly because the leading men, Bernhard Rust and Rudolf Mentzel, were not at all up to the task. Several historians, for example Beyerchen, Zierold, Helmuth Albrecht, and Armin Hermann, emphasize the antiscientific atmosphere that Hitler created around himself. In their view, Hitler's attitude derived from his failure to recognize the importance of science for the economy, politics, and warfare in the technologized world of the twentieth century.[2] Indeed, certain state and party authorities, particularly Albert Speer, minister for armament and war production, had recognized even during the war the inadequacy of the Reich Research Council as an instrument for the effective promotion of research.[3]

Significant negative repercussions of National Socialist science policy have been described not only for the universities but also for the Kaiser Wilhelm Institutes, that is, research institutes outside the universities that were financially supported by industry. Anthony Michaelis and Roswitha Schmidt (1983, p. 59), in their analysis of the rebuilding of science in Germany after 1945, say this about the Max Planck Society and its predecessor, the Kaiser Wilhelm Society: "The most difficult period in the history of the society was surely the Third Reich, when its basic principle 'ad personam' was suppressed by the

National Socialist motto 'contra personam.'" ("Ad personam" means supporting the best scientists.) In a more recent study of the Kaiser Wilhelm Society in the Third Reich, the National Socialist seizure of power is also seen as a break in science policy. Albrecht and Hermann (1990, pp. 356f.) argue that the economic and military significance of basic research, which had been one of the justifications for establishing the society, no longer counted as an argument after 1933. Hitler, they maintain, lacked all understanding of the importance of science, and he regarded the scientific innovations since World War I for the most part as "humbug put out by the Jews."

By contrast, other authors have questioned for individual fields the thesis that the decline of scientific performance in Germany after World War II was caused by National Socialist science policy and the expulsion of Jewish scientists. Ulfried Geuter has shown that the professionalization of psychology after 1933 progressed through the development and application of theories and methods that were no different from those before that time. However, in the process Geuter makes clear that psychologists, even if they did not let themselves be co-opted professionally by National Socialist racial ideology, did not violate the goals of National Socialism; in fact, they often placed themselves in service to its politics. For example, psychologists were to devise social techniques for the population planning of the occupied eastern territories and develop selection methods for assigning foreign workers to specific jobs (Geuter 1984, pp. 421–427).

Beyerchen (1977, p. 202), too, notes a relaxation of ideological pressure after 1939, which he attributes to the influence of utilitarian concerns and power on the science policy of the regime. He concludes that National Socialism, despite the strong position of "German physics" at the beginning of the Nazi era, did not cause the decline of physics in Germany. The historian Mark Walker (1989) has shown that modern physics under National Socialism, once the connection between academic physics and the industrial production of modern weapons had been recognized, was supported by an alliance of industry, military, and government.

As these examples demonstrate, the evidence for the continuity as well as the discontinuity of scientific research after 1933 has been drawn chiefly from physics. By contrast, biological research during these years has so far been largely ignored.[4] In the 1920s biology in Germany, like the natural sciences as a whole, was internationally acclaimed. After World War II, however, the situation had changed profoundly. For instance, a study of the state of scientific research in Germany, conducted at the request of the German Research Association, noted that many fields still lagged far behind those in other countries; in biology the field of molecular biology, in particular, was considered inferior (Clausen 1964, p. 12). As the primary reason for this, Richard Clausen, too, identified the "bloodletting of German science through the policies of the Third Reich" and the increasing neglect of basic research during these years (p. 18). Marie Luise Zarnitz, in her analysis of molecular biology and genetics in Germany (1968, p. 61), arrived at the following conclusion: "The wave of emigration at the beginning of the thirties was very damaging to scientific genetics as a field in our country. For we lost not only individual researchers but with them also the possibility of forming schools in genetics." By contrast, the reports on research work in biology during the war, drawn up by German scientists shortly after 1945 at the request of the occupying authorities, reveal that this basic research had led to good results even during the period of National Socialism.[5]

To clarify the question about continuity and discontinuity, I shall examine—quantitatively as well as qualitatively—what effect the dismissal and forced emigration of the many Jewish and a few decidedly liberal or leftist non-Jewish biologists between 1933 and 1939 had on biology in Germany. I shall compare the findings with those in other fields. In individual cases I will show the importance émigré biologists attained in their host countries. Then, as now, experimental research at the universities was nearly impossible without outside funding. Therefore, I will analyze—again, quantitatively and qualitatively—the support biological research at the universities and the Kaiser Wilhelm Institutes received during the Nazi period from the German Research Association, the most important funding organization of German science, and from the Reich Research Council (established in 1937). I will investigate whether the careers of the biologists who remained in the German Reich tended to be the result of professional qualification or party membership. Along the way I will examine the thesis that Geuter (1984) and Ash and Geuter (1985) have formulated for the field of psychology, namely, that ideological considerations were in the foreground in appointment policy until 1936, after which time the issue of party membership receded behind other, for example professional, criteria.

I will outline the contents and goals of biological research of Das Ahnenerbe (ancestral heritage), the research and teaching society of the SS, and examine the question of biological warfare in Germany during World War II. Five brief biographical accounts will illuminate the political attitudes and scientific accomplishments of various leading researchers and exemplify characteristic aspects of biological research and teaching during the period of National Socialism.

Let me anticipate some central findings. About 13% of biologists were dismissed between 1933 and 1938, four-fifths for "racial" reasons. About 75% of those who lost their jobs emigrated. Measured by the frequency with which their publications were cited, the expelled biologists were on average considerably more successful than their colleagues who remained in Germany. With respect to the fields of research, however, there were no significant differences between these two groups. Research projects in biology that were purely ideological in orientation and scientifically meaningless were not funded in Nazi Germany. Up until 1944, funding increased in absolute terms. Moreover, the relative share that the Kaiser Wilhelm Society had of the total amount of funding for biological research rose sharply. The reasons for this support lie in the political, economic, military, and ideological spheres. Research projects in genetics and radiology received particularly strong support. But fields other than the established branches of genetics received attention (as Jonathan Harwood [1987] and Jan Sapp [1987] have emphasized); to an increasing degree funding was also given to the then modern fields of virus and mutation research.

The decline of German biological research that became apparent after World War II therefore had causes other than a lack of research funding after 1933; nor can it be explained solely by the expulsion of Jewish scientists. This decline is marked to a special degree by the late onset of research in the molecular genetics of bacteria and phage. The scientific innovations developed abroad, especially in the United States, met a delayed reception in Germany. After 1933, Germany had expelled all Jewish scientists almost without any opposition. Later it pursued a policy of annihilation against all European Jews and conducted an aggressive war against European and non-European countries. It is not surprising that after 1945 for the most part a dialogue did not develop between German scientists and their Jewish as well as other foreign colleagues.

*   *   *

This study is based on a list of 445 biologists (most of whom had attained their *Habilitation*, the qualification to teach at the university level) who worked at universities and the Kaiser Wilhelm Institutes (KWIs) between 1932 and 1945. To determine who was a biologist, I followed institutional affiliations. To begin with, I tried to compile a list of all the names of biologists with *Habilitation* who worked in Germany during this period, in Austria and at the German University in Prague between 1937 and 1945, and at botanical, zoological, and genetic university institutes (383). I took names from the course listings of all German and Austrian universities, the then German universities in Posen, Prague, and Strasbourg (twenty-nine universities), the Landwirtschaftliche Hochschule (Agricultural College) in Berlin, and the Technische Hochschulen (Institutes of Technology, TH) in Braunschweig, Darmstadt, Karlsruhe, and Munich.

To this list I added the names of all biologists with *Habilitation* and a few with doctorates but without *Habilitation*,[6] including a few biochemists and biophysicists, who worked at one of the KWIs relevant to biology between 1932 and 1945.[7] These numbered 62. I drew their names from the Kaiser Wilhelm Society's annual reports of activities from 1932 to 1943.

As I indicated above, very little has been written on the coerced emigration, the funding for and results of research, and the party membership and careers of biologists. I therefore examined these matters almost exclusively on the basis of original sources. For this purpose I analyzed the scientific publications of biologists as well as material in the files of various archives in Germany and abroad. As an additional source I drew on the papers of German, American, and English biologists and interviewed contemporary witnesses.

What follows is an overview of the most important sources and an explanation of how I used the Science Citation Index to assess scientific accomplishments. The following materials allowed me to determine the dismissal and emigration of biologists and the reasons behind them: the *International Biographical Dictionary of Central European Émigrés, 1933–1945;* documents from the archive of the Research Foundation for Jewish Immigration of the Zentrum für Antisemitismusforschung (Center for Research in Anti-Semitism) at the Technical University (TU) in Berlin; lists of the Notgemeinschaft Deutscher Wissenschaftler im Ausland (Emergency Association of German Scientists Abroad); files of the Reichserziehungsministerium (Reich Education

Ministry) in the Bundesarchiv (BAK; Federal Archive) in Koblenz; files of the Rockefeller Archive Center in Tarrytown (RAC); files of the Society for the Protection of Science and Learning (Ms. SPSL) in the Bodleian Library in Oxford; files of the Archiv zur Geschichte der Max-Planck-Gesellschaft (Arch. MPG; Archive for the History of the Max Planck Society); files of various archives of German and Austrian universities (UA); references to possible new positions in the scientific publications of biologists who were probably émigrés.

Information about research content and support for biological research can be found in the *FIAT-Reviews of German Science, 1939–1946* (on the FIAT reviews, see Chapter 3) and in the records of the DFG and the RFR. After the war the occupying powers ordered German scientists to write reports about the research work that had been conducted in Germany during the war. These reports contain detailed information on advances in the various fields and areas of research. Four volumes discuss biology and biochemistry, two volumes biophysics. The documents concerning the research support from the DFG and the RFR are in the Federal Archive in Koblenz. Of the 445 biologists, 222 submitted at least one grant application to the DFG between 1933 and 1945.

Information concerning possible membership in the Nationalsozialistische Deutsche Arbeiterpartei (National Socialist German Workers' Party, NSDAP), the SS, and the Sturmabteilung (Storm Troopers, SA) of 440 biologists was obtained in the Berlin Document Center (BDC). The BDC has preserved the central membership files and other documents of the NSDAP. In addition, many personal records of the SS and, though less complete, of the SA can be found here. The archive, founded in 1945, was under American control and administration until 1994. It was transferred to the custody the German Bundesarchiv on July 1, 1994. Additional information about political activities on the part of individual biologists and political motives behind research funding came from documents in university archives, the various collections of the BAK, and the Institute of Contemporary History in Munich, among them the files from the Rosenberg office and the SS.

On the question of the careers of biologists and the possible background facts, I evaluated faculty, institutional, and personnel files in various university archives, as well as the files of the general administration of the Kaiser Wilhelm Society for the Advancement of Science

(KWG) pertaining to specific institutes. I extracted additional data from the records of the Reich Education Ministry (BAK, Preußisches Geheimes Staatsarchiv Berlin, and Zentrales Staatsarchiv [today Federal Archive] Potsdam und Merseburg) and the KWG in the archive of the Academy of Sciences in Berlin. The files in these archives dealing with appointments to various positions are unfortunately not complete.

On the topic of biological warfare I evaluated above all documents from the American and British secret services in the National Archives in Washington.

One impediment to procuring archival material is the federal law in Germany that establishes a restrictive period for personal files. According to the Bundesarchivgesetz (federal law regarding archives), access to a file can be denied when the person in question has not been dead for at least thirty years. If the death cannot be conclusively established, the file can be closed for a period of one hundred and ten years from the person's birth.

To assess the scientific importance of individual scientists, I identified the cited papers of biologists at universities and Kaiser Wilhelm Institutes and the number of citations in the Science Citation Index (SCI) between 1945 and 1954. The SCI lists all earlier publications in the natural sciences that were cited during the indicated period. The total citations of all papers by a particular scientist is intended to indicate the influence of his or her research. The ratio of the total of the citations a scientist earned to the number of cited papers provided an indicator for the quality of research. The method has limits, for the number of citations reflects not only the scientific importance of a person at the time but depends also on the person's prestige and on the field of research, as well as on the availability of a specific publication to those who might cite it (Weingart and Winterhager 1984). Despite these reservations, a citation analysis seems justified for a comparison of the research among groups of scientists working on similar research projects. Using this method, I have avoided certain technical problems in the use of the SCI. Identifiable self-citations were subtracted from the total number; since some scientists cannot be distinguished in name, I identified these citations on the basis of differences in content. The same goes for errors that appear in the SCI in regard to the listing of names and initials.

One thing that could not be avoided is the problem of multiple authorship of publications. During the period under consideration this was rare in publications from universities and more frequent in those from the Kaiser Wilhelm Institutes. The share of total publications from KWIs in the years 1939 to 1942 ranged from 20% to 30%. Those who coauthored several publications often changed the order of names in the publications. In contrast to current German practice, where directors of institutes usually appear as the last authors in publications with multiple authors, at that time directors and section heads were frequently the first or only authors. Their influence thus appears undiminished. It therefore seems possible, despite repeated multiple authorship in individual cases, to use the SCI as the basis for the assessment of the importance of a scientist. Wherever possible, I have supplemented the citation analysis with an evaluation of a person's scientific accomplishments by colleagues in the field.

1

The Expulsion and Emigration of Scientists, 1933–1939

1. A Brief Summary of Legal Measures

1.1. The Gleichschaltung of the Universities

In Germany all universities and technical colleges have been state institutions. In universities, in contrast to the research institutes of the Kaiser Wilhelm Society (since 1948, Max Planck Society, MPG), science has been pursued according to the principle of the unity of research and teaching, a principle set up by the philosopher and statesman Wilhelm von Humboldt (1767–1835). Humboldt founded the university in Berlin. Before 1933 universities were governed primarily by faculty committees under the guidance of the Ministry of Education, a system called "academic self-government." Traditionally, universities were divided into four faculties (philosophy, which included the natural sciences; law; medicine; and theology), each under a dean of its own election. Only since the 1930s have some universities created faculties in the natural sciences and mathematics. The faculties exercised the three traditional powers with regard to university teach-

10

ers: (1) authorizing a Ph.D. as a lecturer *(Privatdozent)*, after completion of a second dissertation *(Habilitationsschrift)*; (2) recommending the promotion of a lecturer to the rank of associate professor; (3) recommending the offer of a position to a professor from another university. The minister of education had the right to select one of three names presented by the faculty; in most cases the minister chose the first one on the faculty's list. The faculties were thus largely autonomous concerning central decisions of university policy (for further information on the German university system and the impact of Nazi policy on them, see Hartshorne 1937).

The National Socialists deprived the German universities of their independence. This process is called *Gleichschaltung*, literally, putting into the same gear; it refers to the subordination of people and institutions to Nazi policies (see, for example, Stuckart and Globke 1936; Hartshorne 1937; Walk 1981). In the first step, professors and university lecturers who were Jewish or openly liberal or politically left were driven out. From the very beginning, the main goal of Nazi policy was to "purge" the entire civil service and the public sector from Jews and people with liberal or left-leaning sympathies. The laws and decrees concerning the civil service and the public sector applied to the civil servants and employees at the universities. Next, the führer principle was introduced in university administration: the decision-making authority of the faculties in questions relating to *Habilitation*, promotions, and appointments was transferred to the rector; the faculties were left with only an advisory function. The rector became the führer of the university; he was no longer elected but appointed by the Reich education minister, the sole person to whom he was accountable. The führer principle was introduced at German universities in the fall of 1933 and thus formed the second step in *Gleichschaltung*.

The Law for the Restoration of the Professional Civil Service of April 7, 1933 (hereafter referred to as the Civil Service Law) was the first and most important of the laws and decrees that formed the "legal basis" for discriminating against and dismissing persons at universities and higher institutes who were "non-Aryan" or "married to a Jew" *(jüdisch versippt)*, and politically undesirable persons at universities and technical colleges. Soon afterward, a decree extended the law to professors and university teachers without civil service status, that is, lecturers and associate professors without civil service status. Section 3 of the Civil Service Law called for the dismissal of Jews and so-called Jewish *Mischlinge* (half-breeds) as non-Aryans, which included persons who had at least one Jewish grandparent.[1] An exemption issued at the initiative of President Paul von Hindenburg allowed Jewish front-line veterans of World War I and professors and lecturers who had attained civil status prior to 1914 to remain at the universities, at least temporarily. However, these exemptions were eliminated when the Reich Citizenship Law, one of the Nuremberg Laws of September 1935, came into force. Non-Jews who had married a Jew after January 7, 1933, were given the same status as non-Aryans; those who had married earlier were permitted to keep their positions for the time being, though they were frequently driven out on the basis of other sections of the law. For example, they would be accused of political misconduct or dismissed according to section 6 of the Civil Service Law ("to simplify the administration"). In 1937 a decree from the Ministry of the Interior prohibited civil servants related to Jews by marriage *(versippt)* from hoisting the Reich flag. For this reason they had to retire, "except in special cases where their staying on could be justified" (Walk 1981). Section 4 of the Civil Service Law required the dismissal of all civil servants whose prior political attitude put their unconditional loyalty to the state in doubt.

German laws came into force at Austrian universities immediately following the Anschluß on March 13, 1938. That same month all university teachers had to swear an oath to Hitler, but non-Aryans were excluded from doing so. In most cases, Austrian non-Aryans were dismissed in April of 1938 along with politically undesirable persons. Political persecution in Austria was also directed against supporters of the former Dollfuß-Schuschnigg government and against persons with obvious Christian-social attitudes. At the German University in Prague, the dismissals took place in many cases even by the end of 1938, and at the latest immediately following German occupation in March of 1939. In Strasbourg and Posen, the German universities were set up in 1940 and 1941, respectively, with German personnel; members of the previously existing and now closed-down French and Polish universities were not employed.

1.2. The "Voluntary *Gleichschaltung*" of the Kaiser Wilhelm Society

A few remarks on the development of the Kaiser Wilhelm Society will serve to make clear the repercussions of National Socialist

Gleichschaltung. The establishment of research institutes outside universities and academies with the financial support of industry began in many industrialized countries in the late nineteenth century. Examples include the foundation of the Institut Pasteur in Paris (1888), the Rockefeller Institute for Medical Research in New York (1901), and the Carnegie Institution of Washington for Fundamental and Scientific Research (1902). The Kaiser Wilhelm Society was founded in 1911 in Berlin. The Prussian minister of culture and education had become convinced that the universities could no longer meet the demands of modern research in science and technology. Those representatives from science, the government, and industry who supported the plan for the establishment of the KWG considered it necessary that outstanding scientists have at their disposal research institutes with modern equipment and be unimpeded in their work by teaching responsibilities. The best of the university professors were hired as institute directors. As the future director general of the Kaiser Wilhelm Society, Friedrich Glum, put it, the establishment of the society was intended "to restore Germany's competitive abilities in the field of science—indeed, by means of a national effort to secure for Germany the rank at the top of science which it had always claimed for itself and which now seemed in danger" (Glum 1964, p. 199).

The KWG was to be funded primarily from private sources, especially industry, but it would also receive state subsidies. Though a self-governing body, it was according to the Prussian tradition controlled by the state. Thus the managing board of the society also comprised representatives of some ministries.

Unlike most private and economic organizations, the Kaiser Wilhelm Society was not "coordinated" *(gleichgeschaltet)* in 1933. General Director Glum persuaded the relevant state secretary in the Interior Ministry that the introduction of the führer principle would draw strong criticism from abroad, and that for the same reason not all Jewish members of the society's senate should be dismissed (Glum 1964, p. 441). As a result only some of the members of the senate were replaced with persons appointed by the Interior Ministry and the Ministry of Culture.[2] Glum was proud that he had saved the neutral and scientific character of the Kaiser Wilhelm Society with his policy of "voluntary *Gleichschaltung.*" The introduction of the führer principle "was avoided by quick action, though this could not be accomplished without concessions. When *Gleichschaltung* came, we were able to say that we would not be affected by it since we had already carried it out on our own" (Glum 1964, p. 443).

The personnel policy of the society's general administration was characterized during the first years of National Socialism by efforts to retain eminent Jewish scientists—among whom were many institute directors (Richard Goldschmidt, Fritz Haber, Otto Meyerhof, Carl Neuberg, and Ernst Rabel)—for as long as possible, for example, by interpreting the rules rather liberally, while in most other instances the rules were followed to the letter (Albrecht and Hermann 1990). Despite the independent standing of the Kaiser Wilhelm Society, the regulations of the Civil Service Law were to be applied in those institutes that were financed at least 50% with public funds. In a letter of April 27, 1933, Glum ordered all institute directors to dismiss the relevant persons from such institutes.[3] The institutes for anthropology, biochemistry, biology, and medical research were completely financed with public funds, whereas the Institute for Breeding Research was financed with public (the percentage could not be determined) and private funds (Vierhaus and Brocke 1990, p. 619).

The official comments by the president of the Kaiser Wilhelm Society, Max Planck, veil the active role that the society played in the expulsions and reveal the readiness on the part of many scientists of the institutes to subordinate themselves to National Socialist policy without any major objections. Planck wrote the following about the dismissal of the biologists Curt Stern, Viktor Jollos, and Fabius Gross from the Kaiser Wilhelm Institute for Biology: "Dr. Stern accepted an appointment to the University of Rochester. Professor Jollos took up an appointment at the Genetics Institute at the University of Wisconsin in Madison. Dr. Gross went to London as an assistant to Huxley and Fisher."[4] In similar manner he commented on the dismissal of Richard Goldschmidt, the second director of the KWI for Biology, at the end of 1935 as a result of the Nuremberg Laws: "Professor Dr. R. Goldschmidt accepted an appointment at the University of California, Berkeley."[5]

Some institutes had no problems. For example, no Jewish scientists worked at the KWI for Breeding Research. However, a woman gardener, Fanny du Bois-Reymond, was dismissed in 1934.[6] Du Bois-Reymond was the granddaughter of the well-known physiologist Emile du Bois-Reymond. After her dismissal she wrote to Glum:

When I was with you the other day, I needed all my self-control to show the necessary outward composure, therefore I could not express what it means to me to have to leave the community of the Kaiser Wilhelm Society. Of all people, you, Herr Professor, who have done so much for the Kaiser Wilhelm Society, will understand how deeply attached one is to this community. Here one breathes the air that I was used to from the great scientific tradition of my family. Even in my modest position, this atmosphere suited me so well that today I still don't know how I am to exist outside it. I was always aware of the honor of being a part of it, and during the entire six and a half years I have tried to do justice to it by complete devotion to my work. May it be granted to you to keep the tradition of the Kaiser Wilhelm Society nice and clean in the Third Reich.[7]

The *Gleichschaltung* of the society was officially carried out in 1937, on the occasion of the appointment of the successor to the retired president Max Planck. Glum himself was dismissed. His successor, Ernst Telschow, was a member of the NSDAP. The new president was Carl Bosch, chairman of the board of directors of IG Farben.

2. *"Non-Aryan" Dismissals and Emigrations*

In 1933, 254 biologists had *Habilitation* at German universities and 24 assistants or DFG fellows without this qualification worked at Kaiser Wilhelm Institutes; in 1938, 55 persons had *Habilitation* at Austrian universities and at the German University in Prague. According to present information, 30 (8.9%) of this total of 337 persons were dismissed between 1933 and 1939 as non-Aryans or because they had a non-Aryan wife (Table 1.1). Of the 30 dismissed biologists, 26 (86%) emigrated. Let me stress once again that because this analysis is limited to certain institutions, it does not include physicists who became molecular biologists after being expelled, biologists working in institutes other than the ones mentioned (among them the physiologist and later Nobel prize winner Bernhard Katz, who emigrated in 1935), and biochemists. Furthermore, Jewish assistants at the universities who did not have *Habilitation,* such as Ernst Caspari and Hans Grüneberg, as well as biologists without a permanent position, for example Salome Gluecksohn-Waelsch, are not part of this study. Among the Jewish students also not considered was Charlotte Auerbach, who was working at the KWI for Biology and had not yet fin-

Table 1.1 Dismissals and emigration of biologists who were of Jewish descent or married to a Jew

Name	First name	Position	Institution	Field	Year of dismissal	Successor	Year of emigration	Countries	First permanent position	Year of return	Institution
Bodenstein	Dietrich	sc. collaborator	KWI Biology	zoology	1933	—	1933	Italy, U.S.	Stanford Univ.	—	—
Brauner	Leo	associate prof.	Univ. of Jena	botany	1933	—	1933	Turkey	Univ. Istanbul	1955	Univ. Munich
Bresslau	Ernst	full professor	Univ. of Cologne	zoology	1933	O. Kuhn	1934	Brazil	Univ. São Paulo	—	—
Brieger	Friedrich	lecturer	Univ. of Berlin	bot./gen.	1934	—	1934	England, Brazil	Univ. São Paulo	—	—
Fraenkel	Gottfried	lecturer	Univ. Frankfurt	zoology	1933	—	1933	England, U.S.	Imperial College London	—	—
Freund	Ludwig	associate prof.	Ger. U. Prague	zoology	1938	—	[1]	—	—	1949	Univ. Halle
Gams	Helmut	associate prof.	Univ. Innsbruck	botany	1938	—	[2]	—	—	1945	Univ. Innsbruck
Goldschmidt	Richard B.	director	KWI Biology	zool./gen.	1935	A. Kühn	1935	U.S.	Univ. of Calif. Berkeley	—	—
Groß	Fabius	assistant	KWI Biology	zool./gen.	1933	—	1933	England	Marine Biol. Ass.	—	—
Hamburger	Viktor	lecturer	Univ. Freiburg	zoology	1933	—	1934	U.S.	Washington Univ.	—	—
Heilbronn	Alfred	associate prof.	Univ. Münster	botany	1933	—	1933	Turkey	Univ. Istanbul	1956	Univ. Münster
Heitz	Emil	associate prof.	Univ. Hamburg	bot./gen.	1937	—	1937	Switzerland	Univ. Basel	—	—
Hermann	Siegwart	lecturer	Ger. U. Prague	bacteriology	1938	—	1939	France, U.S.	pharmac. comp.	—	—
Hertz	Mathilde	lecturer	KWI Biology	zoology	[3]	—	1935	England	—	—	—
Hirmer	Max	associate prof.	Univ. Munich	botany	1936[4]	—	—	—	—	—	—
Japha	Arnold	associate prof.	Univ. of Halle	zoology	1936	—	[5]	—	—	—	—
Jollos	Viktor	associate prof.	KWI Biology	zool./gen.	1933	—	1934	U.S.	—	—	—
Kalmus	Hans	lecturer	Ger. U. Prague	zool./gen.	1938	—	1938	Yugosl., Engl.	Univ. of London	—	—
Marcus	Ernst-G.	associate prof.	Univ. of Berlin	zoology	1935	—	1936	Brazil	Univ. São Paulo	—	—
Marton	Hugo	associate prof.	Univ. Heidelberg	zoology	1935		1936	Scotland			

Peterfi	Tibor	guest	KWI Biology	cytology	1935[6]	—	1935	Engl., Denmark, Turkey	U. Copenhagen	—
Philip	Ursula	collaborator	KWI Biology	zool./gen.	1933[7]	—	1938	England	—	—
Pringsheim	Ernst-G.	full professor	Ger. U. Prague	botany	1938	V. Czurda	1939	England	Cambridge Univ.	—
Przibram	Hans	associate prof.	Univ. of Vienna	zoology	1938	—	[5]	—	—	—
Schwarz[8]	Walter	lecturer	TH Darmstadt	botany	1933	—	1933	Palestine	Hebr. Univ. Jer.	—
Steinitz	Walter	lecturer	Univ. Breslau	zoology	1935	—	1935	Palestine	—	—
Stern	Curt	lecturer	KWI Biology	zool./gen.	1933	—	1933	U.S.	Univ. Rochester	—
v. Ubisch	Gerta	associate prof.	Univ. Heidelberg	bot./gen.	1933[9]	—	1934	Brazil	Butantan, São Paulo	—
v. Ubisch	Leopold	full professor	Univ. Münster	zoology	1935	H. Weber	1935	Norway	—	—
Wolf	Ernst	lecturer	Univ. Heidelberg	zoology	1933	—	1933	U.S.	Am. Optical Co.	—

Notes: The following abbreviations and terms are used: sc. collaborator = scientific collaborator with Ph.D.; bot. = botany; zool. = zoology; gen. = genetics; return = return of émigré and dismissed but non-émigré university teachers to a position at a German or Austrian university. Successors are indicated only for dismissed full professors and directors.

1. Ludwig Freund was arrested in Prague in 1943 and taken to Theresienstadt. He was freed in May of 1945, but as a German (German Jew) he was not allowed to resume his teaching in Prague. In 1949 he was appointed full professor in Halle.

2. Helmut Gams was dismissed because he had a Jewish grandfather. He did not emigrate. According to information supplied by the university archives in Innsbruck in January 1990, he was presumably of interest to the General Staff because he spoke Russian and Scandinavian languages.

3. Mathilde Hertz was the daughter of the physicist Heinrich Hertz, the inventor of wireless waves. In 1933 she was dismissed from her position as lecturer at the University of Berlin, even though her grandparents had been Protestants. The KWG, in recognition of her father's accomplishments, made it possible for her to retain her position as assistant at the KWI for Biology. Owing to the growing anti-Semitism in Germany, she decided in 1935 to emigrate to England; in all likelihood she would have been dismissed once the Reich Citizenship Law came into effect (Ms. SPSL, file M. Hertz).

4. Max Hirmer was dismissed in accord with section 6 of the Civil Service Law. He did not emigrate.

5. See main text.

6. The Hungarian professor Tibor Peterfi was the editor of *Berichte der wissenschaftlichen Biologie,* published by Springer Verlag, and had been working since 1922 as a guest at the KWI for Biology. In this position he could not be dismissed, but because the situation became increasingly unbearable for him, he emigrated via Cambridge to Denmark. In 1944 he became professor at the University of Istanbul; in 1946 he returned to Hungary.

7. Since she had only a temporary position, Ursula Philip was not officially dismissed, but she had no prospects for a renewal of her position.

8. After emigrating to Palestine, Walter Schwarz changed his name to Michael Evenari.

9. See Section 10 of this chapter.

ished her doctorate when she emigrated to Scotland, where she later discovered the mutagenic effect of chemicals.

Of the dismissed biologists, 5 (16%) held the title of full professor *(ordentlicher Professor)*; 3 of those 5 held chairs, that is, headed a university institute; 12 (40%) were associate professors *(außerordentliche* or *außerplanmäßige Professoren)*; 10 (33%) were lecturers *(Privatdozent)*; and 3 (10%) were assistants or DFG fellows without *Habilitation* at Kaiser Wilhelm Institutes. Eight (26%) of the dismissed biologists were working at one of the above-mentioned Kaiser Wilhelm Institutes at the time of their dismissal.

A few examples will elucidate the consequences of dismissal and emigration for those affected. Hans Przibram, born 1874, was, according to information presently available, the only biologist unable to emigrate in time. In 1903 he became lecturer and in 1921 associate professor of zoology at the University of Vienna. Together with L. Porges von Portheim and Wilhelm Figdor, he had founded the biological research institute, the so-called Vivarium, at the Wiener Prater in 1903, donating it to the Academy of Sciences in Vienna in 1914. Many Jewish, liberal, and socialist scientists worked at the Vivarium. Only a few of them, like Przibram, who headed the Zoology Section of the institute, also had positions at the university.

The enforcement of the Civil Service Law in April 1938 led to a large number of dismissals at the research institute. The hopes that this aroused in other Austrian scientists are reflected in a letter by the new rector of the University of Vienna, Professor Fritz Knoll: "Immediately after the revolution, all kinds of people gave me names of persons they wanted to see appointed as station and department heads in place of the departed Jews. Also, various people have already come to me in person to introduce themselves for this purpose as universally competent candidates."[8]

The new plenipotentiary of the research institute installed by the National Socialists described the "Aryanization" as follows: "Until the revolution of 1938, the leadership and organization of the biological research institute was largely in Jewish hands. The Jews were removed after the revolution, and in so far as funds of the biological research institute (private and governmental) were being used for the personal work of the Jews, Aryans were paid with these funds."[9]

As already mentioned, this research institute had been a gift from several of these now "removed Jews." Portheim emigrated to London in 1938. For a number of reasons Przibram and his wife did not arrive in Amsterdam until December of 1939. They had planned to emigrate to America via England, where the London aid organization the Society for the Protection of Science and Learning (SPSL) had procured a position for him.[10] However, with the outbreak of the war their British visas became useless. Until 1941 Przibram and his wife waited in vain for an American visa.[11] They were eventually deported to the concentration camp Theresienstadt. Hans Przibram died there of exhaustion on May 20, 1944; a day later his wife committed suicide.

The Viennese professor of zoology Heinrich Joseph, born in 1875 and retired in 1934, also committed suicide in 1941 together with his wife. Arnold Japha, born 1877, was associate professor of zoology in Halle and after World War I also city physician at the city's health office. He came from a Protestant family, but his parents were of Jewish descent. Effective May 1, 1933, he was stripped of his post as *Magistratsmedizinalrat* in accord with the Civil Service Law. In December of 1935 he was dismissed as professor, in accord with the Reich Citizenship Law. Because of his military time as a front-line soldier and his years of service in the communal health service, he was given a small pension. He could not bear the idleness to which he had been condemned. A string of insults, defamations, and harassments by the Gestapo drove him to commit suicide on May 16, 1943 (Heindorf and Schwabe 1968, pp. 125–142).

About a third of the biologists who emigrated were unable to secure a permanent position at a university, in a research institute, or in industry. The best known example of the failure of a German scientist abroad is Viktor Jollos. He had become known for his discovery of environmental changes in protozoa, which under asexual propagation were retained over hundreds of generations and which, since they did not alter the genetic material, he called permanent modifications. Documents of the Rockefeller Archive Center show that in 1933 Jollos emigrated with his family to Edinburgh, where he received an invitation to continue his genetic research of protozoa and *Drosophila* at the University of Wisconsin in Madison.[12]

From February 1934 to August 1935, he was supported by the Emergency Committee in Aid of Displaced Scholars and by the Rockefeller Foundation. As was the case with nearly all such aid for emigrants, this support was intended to be transitional and was therefore limited in time. However, neither then nor later was Jollos able to obtain a permanent position. The University of Wisconsin did not grant him tenure, and other universities, too, declined to offer him an appointment.

The aid organizations and Jollos's colleagues attributed his failure to his difficulty adjusting to conditions in America and the serious mistakes he had made in the period right after his arrival. For example, his high financial demands and his authoritarian conduct toward students were criticized. He had failed to take into consideration that his position was unclear and that his status was no longer what it had been prior to his expulsion.[13] In addition, shortly before his expulsion, he published a paper claiming the existence of directed mutations, which soon turned out to be a mistake.[14] The error complicated his situation, even though his scientific accomplishments were generally recognized and appreciated in the United States. For some years he lived with his family in very difficult material conditions and subsisted primarily on private donations. He died on January 15, 1941, after suffering a heart attack.

Gerta von Ubisch, appointed associate professor of botany in Heidelberg in 1929, emigrated to Brazil in 1934. Her fate is described below in Section 10 of this chapter. After her emigration she, too, for a number of reasons never again obtained a permanent position and the opportunity to do scientific work. Her brother Leopold von Ubisch, until 1935 professor of zoology in Münster, was materially much better off, but after his emigration to Norway he also failed to get another permanent post. Ernst Bresslau, dismissed in 1933 from his position as full professor of zoology in Cologne, died in 1935 in Brazil after suffering an attack of angina pectoris. With great effort he had participated in setting up the institutes of the University of São Paulo, which had been founded in 1933, and a few days before his death he had delivered his inaugural lecture as professor of zoology.[15] Hugo Merton, until 1935 associate professor of zoology in Heidelberg, also died a few years after his emigration in England from an ailment brought on by his internment in the concentration camp in Dachau (see Section 9 of this chapter).

Other biologists had no great difficulties obtaining new permanent positions after emigration. Richard Goldschmidt, director of the Kaiser Wilhelm Institute for Biology, tells us that he had been preparing his emigration to the United States since 1933. Before he could leave "voluntarily," he was dismissed effective January 1, 1936. The same year he was given the chair for cytology and genetics at the University of California at Berkeley. The later career of Viktor Hamburger, until 1933 assistant and lecturer in Freiburg, will be discussed in Section 5 of this chapter. His rapid rise in the United States, a rarity among émigré biologists, resulted from his extraordinary scientific accomplishments. Curt Stern had similarly favorable conditions. Assistant at the Kaiser Wilhelm Institute for Biology until 1933, he was a recognized scientist through his work in the area of the genetics of *Drosophila*. He had spent a year in the United States on a fellowship from the Rockefeller Foundation and then resigned from his position at the Kaiser Wilhelm Society shortly before he would have been dismissed. At that point he had not yet found a new position. In a letter to Max Hartmann, a copy of which he sent to several zoologists and geneticists in Germany (among them Alfred Kühn and Fritz von Wettstein), he explained this step and made clear at the same time that the events in Germany were utterly contrary to the principle of scientific universalism:

> Today, through Professor Dr. Goldschmidt, I have requested my dismissal from the Kaiser Wilhelm Society, if that step has not already been taken in keeping with the Civil Service Law . . . However, the new laws exclude from scientific work not particular persons who have demonstrated that they are unworthy of participating, but all "non-Aryans" without regard for their accomplishments and efforts. Therefore, even if it were possible, as an exception, to preserve my position, I am still being excluded not only from the teaching of our science—needless to say, the offer by Frisch to the post of conservator in Munich has been voided—but in general terms from any activity as a citizen . . . It is terribly difficult for my wife and me to separate ourselves externally from Germany. You know that I have always considered myself fully German.[16]

After his emigration in 1933 he became scientific collaborator, in 1935 assistant, in 1937 associate professor, and in 1941 full professor of zoology at the University of Rochester; in 1947 he was appointed professor of zoology and genetics at the University of California at Berkeley.

As Stern's letter indicates, many Jewish émigrés had deep emotional bonds to Germany; even nationalistic feelings were not rare. Expulsion therefore represented a great humiliation and meant more than the loss of a job and the difficulty of finding a new position in another country. An excerpt from a letter by the zoologist Ernst Marcus to the Reich education minister, dated June 1, 1936, illustrates these feelings: "After growing up the son of a Prussian judge, allowed to serve at the front line four years in the Prussian Guard, decorated as a sergeant with the Iron Cross First Class, and having done all my work at Prussian departments . . . it is unworthy of me to beg for a placement with foreign charitable institutions. If I . . . am expelled from the fatherland and made into a homeless beggar, I will see this as the greatest injustice and the deepest insult that could be inflicted on me. This I have not deserved, if people acted in accordance with the Prussian principle 'suum cuique.'"[17]

The acceptance by their German colleagues of the expulsion of Jewish scientists, almost without any objections, let alone significant protest, was strongly criticized abroad and by some of those affected. Among the sternest critics was Albert Einstein, who wrote from Oxford on May 30, 1933, to the expelled Max Born: "I think you know that I have never had a particularly favorable opinion about the Germans (in moral and political respects). I must admit, however, that they have surprised me not a little by the degree of their brutality and cowardice."[18] The positions that became vacant through the expulsion of scientists were—at least in biology—newly filled after a short period. Lecturers became full professors, research assistants became lecturers. This rapid professional advancement secured the National Socialist regime the loyalty of a large group of young academics and led, together with strong political and social pressure to toe the line politically, to a growing moral unscrupulousness.

I know of no biologist who declined to accept a position that had been opened up by the expulsion of a colleague. But I would like to point out one notable exception from the field of medicine. The physiologist Otto Krayer turned down the offer of chair of pharmacology at the University of Düsseldorf—the chair from which the Jew Philipp Ellinger had just been expelled. Krayer explained his decision in a letter of June 15, 1933, to the minister of science: "The primary reason for my reluctance is that I feel the exclusion of Jewish scientists to be an injustice, the necessity of which I cannot understand, since it has been justified by reasons that lie outside the domain of science. The feeling of injustice is an ethical phenomenon. It is innate to the structure of my personality, and not something imposed from the outside."[19] This was not an open, political protest, but an individual one.

Krayer acted out of loyalty to his ethical convictions. His decision led to his immediate dismissal and the loss of all academic privileges, such as the right to use public libraries. At the end of 1933, Krayer left Germany in response to an invitation from the Department of Pharmacology at University College London. In 1934 he emigrated to Beirut and in 1937 to the United States, where he became a professor at Harvard Medical School.

3. *Political Dismissals and Emigrations*

Nine biologists were dismissed for political reasons and two for reasons that are still unknown (Table 1.2).[20] Four of these eleven biologists emigrated. The fate of Erich Daumann and Friedrich Eckert after their dismissal is not known. Four of the nine biologists dismissed for political reasons held positions at Austrian universities and two at the German University in Prague. Hans Gaffron emigrated for other reasons; the reasons for the emigration of Karl Arens, Ernst Matthes, and Jakob Seiler are unknown. After the war, six of the dismissed scientists—among them two émigrés—returned to a position at a German or Austrian university.

Julius Schaxel was dismissed in 1933 for Communist sympathies and persecuted by the police and the SA. He fled to Switzerland and then emigrated to the Soviet Union at the invitation of the Academy of Sciences in Leningrad. There he not only launched vigorous attacks against the racial doctrine of the Nazis but also criticized Lysenko's teachings in the Soviet Union (Strauß and Röder 1983, p. 1026; on Lysenko, see also Chapter 4, Section 7). He and his wife were arrested in 1937 during the Stalinist purges, and he died in Moscow on July 15, 1943, under unclear circumstances.

Gustav Gassner, politically very conservative and German-national in outlook, rejected the policy of the National Socialists from the beginning. He prohibited the Hitler salute and any political activity within the Institute of Technology in Braunschweig, where he was professor of botany and rector. Student unrest and student boycotts eventually led to his dismissal.[21] He went to Turkey, where he founded the Central State Institute for Plant Protection, in Ankara, and worked as an expert on plant protection in the Turkish Agriculture Ministry from 1934 to 1939. In 1939 he returned to Germany, where he headed a private plant protection institute until 1945.

Table 1.2 The dismissal and emigration of non-Jewish biologists for political or unknown reasons

Name	First name	Position	Institution	Field	Year dismissed	Reason for dismissal or emigration	Successor	Year of emigration	Countries	First permanent position	Year of return	Institution
Arens	Karl	lecturer	U. Cologne	botany	—	unknown	—	—	Brazil	Univ. of Rio de Janeiro	—	—
Daumann	Erich	lecturer	German Univ. of Prague	botany	1939?	political	—	—	Tasmania?	—	—	—
Eckert	Friedrich	lecturer	German Univ. of Prague	zoology	1938	unknown	—	[1]	—	—	—	—
Gaffron	Hans	guest	KWI Biology	biochemistry		[2]	—	1937	U.S.	Univ. of Chicago	—	—
Gassner	Gustav	full professor	TH Braunschweig	botany	1933	political	Jaretzky	1934	Turkey	[3]	1945	TH Braunschweig
Holtfreter	Johannes	lecturer	Univ. Munich	zoology	1938	unknown	—	1939	England, Canada, U.S.	Univ. of Rochester	—	—
Kisser	Josef	lecturer	Univ. Vienna	botany	1938	political	—	—	—	—	1945	Univ. of Vienna
Kosswig	Curt	associate prof.	TH Braunschweig	zool./gen.	1937	political	—	1937	Turkey	Univ. of Istanbul	1955	Univ. of Hamburg
Mainx	Felix	associate prof.	German Univ. of Prague	botany	1938	political	—	—	—	—	1947	Univ. of Vienna
Matthes	Ernst	full professor	Univ. of Greifswald	zoology	?	unknown	Heidermanns	1937	Portugal	Univ. of Coimbra	—	—
Penners	Andreas	associate prof.	Univ. Vienna	zoology	1938	political	—	—	—	—	—	—
Schaxel	Julius	associate prof.	Univ. Jena	zoology	1933	political	—	1933	USSR	—	—	—
Seiler	Jakob	associate prof.	Univ. Munich	zoology		unknown	—	1933	Switzerland	Univ. of Zurich	—	—
Storch	Otto	full professor	Univ. Graz	zoology	1938	political	J. Meixner	—	—	—	1945	Univ. of Vienna
Strouhal	Hans	lecturer	Univ. Vienna	zoology	1938	political	—	—	—	—	1945	Univ. of Vienna

Note: For abbreviations and general comments see Table 1.1.
1. Friedrich Eckert lost his position as lecturer in Prague in 1938 and became a teacher. His subsequent fate is unknown.
2. See main text.

The dismissal of Curt Kosswig from the Technical College in Braun-schweig in 1937 is a special case, for he was at first quite willing to work with the NSDAP after 1933 in the area of the genetic, that is racial, education of the population. He had even joined the SS in 1933. Kosswig had become known as a geneticist through publication in 1927 of his *Das Gen in fremder Erbmasse* (The gene in foreign gen-otype), where, through genetic experiments with cyprinodonts, he had anticipated the concept we now know as gene transfer in carcinogen-esis (Anders 1991). Kosswig became a political target because of his refusal to advocate racist positions—what he, as a member of the SS, was supposed to do—because of his public support for nonconformist colleagues, and because of his friendly relations with Jewish col-leagues. His situation deteriorated further when he quit the SS in 1936. With the help of the émigré biologist Alfred Heilbronn he was ap-pointed to the university in Istanbul. In 1937 he just barely got out, avoiding arrest by the Gestapo and the loss of his passport.[22]

The experimental embryologist Johannes Holtfreter was dismissed in 1938 from the University of Munich. He was considered a liberal, but the exact reason for his dismissal is unknown. At the beginning of 1939 he emigrated to England. When war broke out he became an "enemy alien" and was taken from England to an internment camp in Canada. Released in 1941, he was able to continue his research at McGill University, in Montreal, until 1946, when he was offered a position at the University of Rochester (see Section 5 of this chapter).

4. The Impact of the Expulsion of Biologists on Research in Germany

As already noted, my analysis of the repercussions of the expulsions of biologists it does not include university assistants without *Habili-tation* or students of biology. All in all, 45 (13%) of the 337 biologists under consideration were dismissed and/or emigrated between 1933 and 1939. We know for certain that 34 (10%) of them emigrated, and 2 others are presumed to have done so. Of the 45 biologists who were dismissed and/or emigrated, 15 (33%) were botanists, 28 (62%) zo-ologists, 1 was a biochemist, and 1 a cytologist. Four of the botanists and 8 of the zoologists were geneticists; 5 of the zoologists were de-velopmental physiologists. After World War II, 9 of the 45 biologists were again offered and accepted a position at a German or Austrian university, among them 4 émigrés. This list does not include those biologists who were able to pursue biological research in Germany after 1933 in private firms or independent research institutes, in spite of their dismissal (see below).

The outstanding scientific accomplishments of the émigrés cannot be acknowledged in detail here. Instead, I shall try to assess the impact of the forced emigration on biology in Germany through a comparison with losses in other fields and with the help of the Science Citation Index. The number of émigrés shows that losses in biology were fewer than those in physics and mathematics, where, according to the data of Alan Beyerchen (1977, p. 44), 25% and 20%, respectively, of the scientists emigrated. However, more recent studies by Klaus Fischer (1988) suggest that these numbers are too high, at least in physics; according to his data, 15.5% of the university teachers in physics emigrated. The impact was particularly great at the University of Göttingen. It lost most of its professors of physics and mathematics, as a result of which teaching and research came to a halt for some time (Becker, Dahms, and Wegeler 1987). Many of the dismissed physicists and mathematicians were leading scientists of the time, among them a number of Nobel laureates. Biochemists, too, in particular physiologists of metabolism, lost many who were forced to emigrate.[23] But biochemistry, as indicated above, is not included in this analysis. In biology, the dismissals did not on the whole lead to scientific changes in Germany that were comparable in severity to what occurred in other fields. The greatest number of dismissals was recorded at the German University in Prague, where the full professor of plant physiology Ernst Pringsheim and six other lecturers or associate professors of biology were driven out.

An analysis of the loss with the help of the citations reveals that the qualitative impact from dismissals and emigration was far greater than one would expect based on the numbers alone. If we take the number of citations in the SCI between 1945 and 1954 as a measure of the scientific influence of a scientist, we find that the émigrés, with an average of 130 citations per person, were more successful than the non-émigré biologists, who averaged 42 citations per person, by the remarkable factor of three.

The following émigrés clearly stand out from the others with respect to the ratio of citations to papers (Table 1.3): the developmental physiologists Viktor Hamburger, Johannes Holtfreter, and, to a lesser de-

Table 1.3 The ten émigré biologists with the highest number of citations in the SCI 1945–1954

Name	Field	Institution	Number of citations	Number of articles	Citations per article (avg.)
Holtfreter, Johannes	zoology	Univ. of Munich	559	100	5.6
Goldschmidt, Richard	genetics	KWI for Biology	456	219	2.1
Fraenkel, Gottfried	zoology	Univ. of Frankfurt	447	219	4.7
Gaffron, Hans	biochemistry	KWI for Biology	375	71	5.3
Hamburger, Viktor	zoology	Univ. of Freiburg	331	53	6.2
Stern, Curt	genetics	KWI for Biology	289	86	3.4
Heitz, Emil	botany	Univ. of Hamburg	257	59	4.4
Pringsheim, Ernst-G.	botany	German Univ. of Prague	203	102	2.0
Bodenstein, Dietrich	zoology	KWI for Biology	177	57	3.1
Brauner, Leo	botany	Univ. of Jena	151	49	3.1

Notes: Institution = place of last position held in Germany. For a comparison with biologists who remained in Germany, see Table 2.7.

gree, Dietrich Bodenstein; the insect physiologist Gottfried Fraenkel; the biochemist Hans Gaffron; the geneticists Emil Heitz and Curt Stern; and the plant physiologist Leo Brauner. The focus of the research of these scientists, as of the other émigrés, was no different from that of their colleagues in Germany in corresponding positions. For example, all three developmental physiologists continued research in the tradition of the by then classic field of development physiology (experimental embryology) of the school of Hans Spemann, which included questions of embryological regulation, induction, morphogenetic fields and their auto-organization. (The careers and scientific development of Hamburger and Holtfreter will be discussed in the next section.)

None of the émigré geneticists later worked in the field of the genetics of viruses, bacteria, or bacteriophage. In 1931 Stern, working on *Drosophila* at the Kaiser Wilhelm Institute for Biology, came up with the cytological proof of crossing over (simultaneously with Barbara McClintock, who was working on corn). In the United States, he continued to work successfully in the field of cytogenetics and later became a human geneticist. Richard Goldschmidt, an internationally recognized geneticist since the twenties, had discovered a phenomenon he called *Phänokopie* in 1935 through experiments with *Drosophila*. He used the word to describe environmentally caused, nonheritable changes of the phenotype, which outwardly corresponded to the phenotype of certain mutants. Beginning even in the thirties his publications were controversial.[24] For example, he maintained to his death that genes had the characteristics of enzymes. Moreover, he drew a sharp distinction between micro- and macroevolution, positing macromutations ("hopeful monsters") as the basis for the latter. This view, too, proved to be untenable.

Hans Gaffron, who as a biochemist is included here because of his work under Wettstein, worked on the photochemical processes of photosynthesis. Gaffron lost his position as assistant at the KWI for Biology when Adolf Butenandt became director of the institute following the dismissal of Carl Neuberg. With the help of Wettstein, Gaffron was able to continue his work for two years as guest at the KWI for Biology. Wettstein tried unsuccessfully to get him a fellowship or a permanent position. Gaffron therefore decided not to return to Germany while working at the Hopkins Marine Station on a Rockefeller Foundation fellowship.[25]

Gottfried Fraenkel's work on the physiology of nutrition and hormones is regarded as internationally preeminent (see, for example, Bhaskaran, Friedman, and Rodriguez 1982). The discovery of the hormonal regulation of insect metamorphosis (Fraenkel 1934) was one of the fundamental works in insect endocrinology. An experiment described by Fraenkel (1935) was used by Peter Karlson and his working group, which clarified the structure of ecdysone in 1965, as a biological test for ecdysone (Karlson 1990, pp. 196ff.). In 1962, Fraenkel reported on another insect hormone, which he characterized further in 1965 and called bursicon (Bhaskaran, Friedmann, and Rodriguez 1982, p. 4).

Leo Brauner's research focused on the field of the physiology of stimulus and movement in plants. In Turkey he built up the teaching and research of plant physiology virtually from nothing, so that most of the plant physiologists working in Turkey after the war came from his school.[26]

In summary we can say that the most outstanding successes of the émigrés in biology came in the "classic fields" of experimental embryology, genetics, and animal and plant physiology. Here, too, we note a difference from physics, where the majority of the scientifically influential émigrés were theoretical physicists or nuclear physicists, that is, they represented modern directions of the field. This circumstance contributed to the establishment of a "German physics" as a reaction to a "Jewish science." However, of the biologists, at least Hamburger, Holtfreter, and Fraenkel were among the internationally leading scientists in their respective fields. Experimental embryology in Germany had lost its most important representatives after Spemann.

As a final example of the importance an émigré biologist attained in his host country I shall mention the work of the Jewish botanist Walter Schwarz, who was dismissed in 1933 from his post as lecturer at the Technical College in Darmstadt. (His importance is not revealed by a citation analysis alone.) Schwarz began with experimental ecological studies of outdoor plants under Bruno Huber at the Technical College in Darmstadt. Experimental ecology had been introduced as a field of botanical research at the end of the 1920s. Of fundamental importance was Otto Stocker's monograph of 1928, *Der Wasserhaushalt ägyptischer Wüsten- und Salzpflanzen vom Standpunkt einer experimentellen und vergleichenden Pflanzengeographie* (The water balance of Egyptian desert and salt plants from the perspective of an experimental and comparative plant geography) (Lange 1989). It contained the first quantitative data on the water metabolism of plants under extremely arid conditions. Schwarz emigrated to Palestine in 1933 and changed his name to Michael Evenari. He taught plant physiology and ecology at the Hebrew University in Jerusalem. After the war, he and colleagues built up the Botany Department of the university, which he headed for twenty years. Evenari is today considered the founder of modern ecophysiological research in Israel, and he made desert research into one of its main fields. In addition, through his study and reconstruction of historical "run-off farms" in the Negev desert he made a contribution to applied botany. "Run-off" agriculture can be traced back two thousand years to the ancient culture of the Nabateans, who, by using this method, were able to settle extreme desert regions without irrigation. Today the research at the experimental farms in the Negev is conducted jointly by the Hebrew University and the Ben Gurion University. Evenari published his autobiography in 1987. Apart from an impressive description of his various activities after his emigration, it contains reflections worth reading about his time as assistant and lecturer at German universities.

5. Viktor Hamburger and Johannes Holtfreter: The Expulsion of Two Eminent Experimental Embryologists

This section traces the scientific path of two of the most eminent émigrés: Viktor Hamburger and Johannes Holtfreter. What were their outstanding achievements? What role did the emigration and the scientific influences of the host country, the United States, play in their careers? To understand the context or their work we must first look at some aspects of the history of classical experimental embryology (also discussed in Chapter 3, Section 2.5).[27]

Both Hamburger and Holtfreter were students of Hans Spemann. During the first half of this century, Spemann was the most important exponent of classical experimental embryology in Germany, a field of zoological research founded by Wilhelm Roux in 1885. Spemann worked primarily with embryos of amphibians. In the 1920s, experimental embryology focused on the timing and course of the determination of the subsequent developmental direction of an embryo part in early amphibian development.

Spemann, through his transplantation experiments, in which he moved small pieces of tissue from one embryo into another to analyze the state of determination, demonstrated that a certain part of the amphibian embryo—the dorsal lip of the blastopore—behaved differently than other parts. It was capable of self-differentiation, whereas the regions located farther away still depended on inductive influences from their environment. These observations led to the famous organizer experiment that Spemann and his student Hilde Mangold published in 1924. Mangold transplanted the dorsal lip of the blastopore of one embryo into the ventral region of a second, differently colored embryo. The result was remarkable: even with transplantation involving embryos of different species, the development of a second embryo was induced in this region; the embryo comprised cells from the donor and recipient embryo. For this experiment Spemann was awarded the Nobel prize in 1935, the only classical experimental embryologist to be so honored. It continues to be seen as the outstanding achievement of experimental embryology, and it triggered numerous experiments aimed at discovering the nature of induction, to find an inductive substance or process.[28]

Ross Harrison took experimental embryology in a different direction, an approach important for the discussion that follows. Harrison was one of the most important classical developmental physiologists in the United States; at the beginning of the century he established experimental neuroembryology as a separate field of research. Among other things, Harrison examined the question of how nerves innervating amphibian limbs formed a normal pattern. With the help of a special method of culturing tissue, which he had developed himself, he recognized that nerve fibers are an extension of nerve cells themselves. In this way he refuted a widely held notion according to which the innervation pattern arose from other cell types of the limbs that were transformed into nerve fibers.

Owing to its methodological limitations, experimental embryology had stagnated by the 1940s. Research on embryonal induction, in particular, was affected by this stagnation, since the search for "wonder molecules" that possessed organizing and inductive abilities proved fruitless. After World War II interest shifted from induction of entire embryonal parts to the study of developmental processes on the cellular, subcellular, and—since the 1950s—molecular level (Hamburger 1988, p. vii). Against the background of the stagnation and later transformation of the field of experimental embryology, the question concerning the scientific development of Viktor Hamburger and Johannes Holtfreter in connection with their emigration takes on special interest. According to T. J. Horder and Paul J. Weindling (1985, p. 229), all four émigré experimental embryologists of the Spemann school who went to the United States—Hamburger, Oscar Schotté, Salome Gluecksohn-Waelsch, and Holtfreter—aligned the content of their research with the orientations predominating in the United States. Was their emigration perhaps one prerequisite for the necessary change in their methods and research results and thus for their success?

Holtfreter was born in 1901. After completing his studies in Rostock, Leipzig, and Freiburg, and taking his doctorate under Spemann in 1925, he became assistant to Otto Mangold at the Kaiser Wilhelm Institute for Biology in 1928 and lecturer in zoology at the University of Munich in 1933. Although he was already "one of the most promising researchers . . . among the exponents of experimental embryology,"[29] he was never offered a professorship or associate professorship in Germany. In September of 1935, he went to the United States for one year as a fellow of the Rockefeller Foundation. Some of that time he worked under Ross Harrison at Yale University. In 1938 he ran into problems with the Nazi authorities, the exact reason being unclear. Holtfreter was not Jewish, he was liberal but nonpolitical and had never joined any political organizations. After a brief imprisonment, he decided at the beginning of 1939 to leave the country and to accept an invitation—issued at the initiative of, among other people, Joseph Needham—from Cambridge University to a position as (unpaid) guest researcher. He was by then an internationally recognized scientist. When war broke out he became, like many émigrés, an "enemy alien" and was sent to an internment camp in Canada in 1940. He was released in November of 1941, following the intervention on his behalf by American scientists, among them Ross Harrison and the German émigré Richard Goldschmidt, who had been teaching at Berkeley for four years. He was able to continue his research in developmental physiology as a guest at McGill University, in Montreal; financial support came from fellowships from the Rockefeller and Guggenheim Foundations. In 1946 he was given a position in the Biology Department of the University of Rochester, where he became professor in 1948.

Holtfreter's work stood in the Spemann tradition. At first he studied the developmental potential of the earliest embryonal states *in vivo* with the help of transplantation experiments involving live organisms. In 1931 he succeeded in developing a new inorganic culture medium and thus moved on to *in vitro* transplantation experiments with amphibian embryos ("Holtfreter's solution"; Holtfreter 1931). In 1932 he demonstrated the inductive ability of destroyed organizer tissue and other destroyed embryonal parts (Bautzmann, Holtfreter, Spemann, and Mangold 1932).[30] In 1933, with the help of the culture medium, he was able to prove the chemical nature of the inductive effect. Spemann himself had wondered about the possible chemical nature of the organizer as early as the 1920s, but he never carried out experiments with completely destroyed organizer tissue (Hamburger 1988, p. 95).[31]

The necessity of creating a method of isolation that made possible the culturing of embryonal parts completely separate from the organism seemed increasingly important to Holtfreter not only for experiments to clarify the inductive effect but also for studies of the capacity of embryonal tissue to self-differentiate. In the 1930s, his interests shifted more and more to the activities of individual embryonal cells or tissue, for he was convinced that the basic question of experimental embryology could not be solved by studying larger embryonal parts. He maintained that "early attempts to interpret embryogenetic processes in physico-chemical terms and to trace the behavior of tissues back to the basic properties of the single cell (Roux 1984, Rhumbler 1899–1927) have found little interest among later students of development. The problem of induction with its chemical implications overshadowed everything else" (Holtfreter 1943b, p. 251).

In the early 1930s he described the differentiation of small, isolated embryonal cell clusters *in vitro* (the results were published later, Holtfreter 1938).[32] According to this study, ectodermal tissue proved to be incapable of self-differentiation and regulation. In fusion, isolation, and defect experiments with various amphibian species in the stages of cleavage, gastrulation, and neurulation, he discovered the significance of positive and negative affinities of individual embryonal parts to each other. These tissue affinities were tissue specific but not species specific and led to the attraction or rejection of tissues as an important shaping factor in development (Holtfreter 1939a, 1939b).

After his emigration, Holtfreter continued these isolation experiments on the cellular level in Cambridge and at McGill. He successfully produced *in vitro* aggregation of embryonal cells, and he showed that a change in the surface properties of cells constituted one precondition for morphogenetic movements, that is, mass movements of cells, such as occurred in gastrulation (Holtfreter 1943a, 1943b, 1944; see also Trinkaus 1966). As these examples reveal, Holtfreter made a decisive contribution to the gradual transition of analysis in developmental biology from entire organisms to smaller and smaller units, a change that had already begun in Germany. Because he made the complex processes in embryonal development analyzable at the level of the cells, he is today credited with the first decisive advance in the understanding of morphogenesis since the 1920s (Oppenheimer 1970; Browder 1986, preface). To this day his experiments and methods have influenced research in developmental biology in fields dealing with the properties of interacting cells (Oppenheimer and Willier 1974, pp. 187f.).

Hamburger, born in 1900, was dismissed from his position as assistant and lecturer in zoology at the University of Freiburg in 1933 for being a non-Aryan. He received the letter with news of his dismissal in accord with the Civil Service Law while at the University of Chicago, where he was working for a year as a fellow of the Rockefeller Foundation. Having been informed of his dismissal, he remained in Chicago with Professor Frank Lillie after his fellowship had expired in October of 1933. The Rockefeller Foundation, which had started a project of support for displaced scholars, supported him until he found a position at Washington University in St. Louis in 1935. He advanced rapidly there: in 1935 he was named assistant professor, in 1938 associate professor, and in 1941 full professor and director of the Zoological Institute.

In less than ten years, Hamburger succeeded in transforming Washington University into a center of research in embryology. In the process he concentrated on questions of neuroembryology, which he analyzed in limbs of chick embryos.

In contrast to Holtfreter, Hamburger became internationally known for his contributions in the field of neuroembryology only after his emigration. Did emigration prompt his transition from experimental embryology involving amphibians in the tradition of Spemann to neuroembryology in the tradition of Harrison?

Interestingly enough, Hamburger had already begun with neuroembryological research under Spemann in Freiburg, where he received his doctorate in 1925. He later recalled that the experiment Spemann gave him, namely, to study limb innervation in amphibians, "came as a surprise, since it had nothing to do with his own line of work."[33] All of Spemann's other doctoral candidates were working on early amphibian development. Hamburger later said "the *bon mot* circulated that too many people were already hanging on the upper blastoporal lip."[34] But neither Spemann himself nor any other researcher in Germany took any greater interest in questions concerning the development of the nervous system. Hamburger, however, was advised by Ross Harrison, a frequent summertime visitor at the institute of his friend Spemann: "Since the interest of Spemann in the nervous system ended with the closure of the neural tube and that of Harrison began at that stage, I got more help in my embryological work from him than from Spemann" (Hamburger 1979/1980, p. 612). But for the most part Hamburger was left to his own devices. Since both Spemann and Harrison were classical experimental embryologists, Hamburger applied the techniques (extirpation and transplantation) and the questions (determination and morphogenesis) of this discipline to the nervous system.

When Hamburger came to Chicago to Frank Lillie on his fellowship, he was given the task of using the microsurgical instruments developed by Spemann (for example, glass needles and scalpels) to conduct corresponding studies of chick embryos. He developed an extirpation and transplantation technique for the chick embryos, but when it came to his research on the nervous system he was left to himself. In Chicago, too, nobody was interested in this topic, and the situation did not improve when he came to St. Louis. As before, he had to rely on his own energy and initiative.

In Chicago he replaced amphibian larvae as research objects with chick embryos and analyzed the effects of wing bud extirpation on spinal motor neurons and sensory nerve cells (Hamburger 1934). He showed that the growth and differentiation of limb buds *(Anlagen)* in chick embryos, too, did not take place under the control of the nervous system, and that the pattern of nerve paths in the limbs was determined by structures in the differentiating limb and not by factors in the nerves themselves (Hamburger 1939).[35] Starting from the hypothesis that peripheral organs such as muscles and sensory organs quantitatively influenced the development of their nerve centers, he searched for factors that controlled the size and differentiation of the nervous system. This paradigm of the existence of specific stimuli that were formed in the peripheral organs and transported to the respective nerve centers, whose quantitative growth and differentiation they regulated, became the basis for all subsequent studies in this area.

The Italian scientist Rita Levi-Montalcini and her mentor at the time, Guiseppe Levi, stimulated by Hamburger's experiment of 1934, worked on similar questions about the effect of limb bud extirpation on the development of specific nerve centers.[36] With respect to the control mechanism of the developing nerve centers by peripheral tissues, they arrived at a different result, which involved a massive cell death as part of the differentiation process. When Hamburger found out about these studies after the war, he invited Levi-Montalcini to examine this problem once again with him. She accepted his invitation, and from 1947 she spent the next thirty years in St. Louis. Her arrival ended Hamburger's isolation and was the beginning of a fruitful collaboration at Washington University.

From the moment when the search for the control factors of the nervous system became purely biochemical, Levi-Montalcini continued it with the biochemist Stanley Cohen. This research eventually led to the discovery and characterization of nerve growth factor, a diffusible factor that alters the differentiation and growth properties of sensory ganglion cells. For this work Levi-Montalcini and Cohen were awarded the Nobel prize in 1986. Hamburger participated in the research until it became purely biochemical, when, in his words, "I realized that I could be of no further help, and I turned to other interests" (Hamburger 1990, p. 124). He himself never did any work in biochemistry or molecular biology.

The discovery of nerve growth factor is one of the greatest successes of biology in the twentieth century. It reveals the existence of an additional integration system of the vertebrate organism, one that is presumably older than the neurons or hormonal systems (Oppenheimer and Willier 1974, p. 227). Here I should emphasize once more that the fundamental research on which it was based was the direct result of the neuroembryological research Hamburger had initiated. His great influence on experimental neuroembryology cannot be attributed to the scientific influences in the United States. Nor was there a specific research direction that he, like Holtfreter, continued in a logical progression. Rather, Hamburger's influence can be attributed to his creativity, to the tenacity with which he tried to solve problems that had nothing to do with contemporary research, and to many coincidences. As an example one could mention his change of research objects, which came about rather by chance, and the choice of the sensory ganglion by Levi-Montalcini (to this day no comparable results could be produced with motor ganglions).[37] Nobody could foresee that neuroembryology, then a small discipline, would acquire such importance in research in developmental physiology. Hamburger later stated (1985, p. 387): "Experimental neuroembryology started as a modest offshoot of experimental embryology. Its founder was Ross Harrison. As fate would have it, experimental embryologists nowadays try to redefine the old problems, while developmental neurobiology flourishes as never before."

We can only speculate whether Hamburger would have established experimental neuroembryology in Germany if he had not been forced to emigrate. We do not know whether he would have been given the necessary institutional and financial means. Klaus Sander told me that the influence of the school of Spemann and of the conservative tradition in Germany might have prevented Hamburger from striking out in a different direction.[38]

An opportunity to bring Hamburger back to Germany was missed in 1950. After the early death of Eckhard Rotmann, a student of Spemann's and since 1948 full professor of developmental biology at the University of Cologne, Hamburger was offered this position as an associate professor. His department would have belonged to the Institute of Zoology, whose director, Otto Kuhn, a former member of the NSDAP and the SS and an incompetent scientist, had just been restored as full professor after having been dismissed by the British occupational authorities. Under these circumstances, Hamburger, who had already been full professor in St. Louis for nine years, declined the offer.[39]

I should add that Salome Glueckson-Waelsch, also a student of Spemann's, became a pioneer in the developmental genetics of small mammals after her emigration to the United States. After years of working as a research associate at Columbia University, she was given a position at the Albert Einstein College of Medicine in New York, where she became full professor in 1958. According to her own statements, her switch to genetics was not due to scientific influences in the United States. She had already tried to pursue genetic research after obtaining her doctorate in 1932 and had always felt that the lack of any genetics in her training was a shortcoming.[40] In her view, "in science, European émigrés introduced their methods and fields of research into the U.S. laboratories rather than be affected by Americans. This was certainly the case in biochemistry, e.g. with Fritz Lipmann and Rudolf Schoenheimer, and also in developmental biology. However, on the level of social and political interactions, the émigrés became adapted to the American system."[41]

It remains to be noted that after the expulsion of Spemann's most important students, experimental embryology of vertebrates in Germany declined to the point where it was for many years insignificant at the international level. Spemann's successor, Otto Mangold, remained throughout his life rooted in classical experimental embryology, that is, he followed in Spemann's footsteps. After 1945, he was suspended from his post for being a National Socialist and thereafter headed a small private institute in Schloß Heiligenberg at Lake Constance.[42] What remained in Germany was the developmental biology of insects, represented by Friedrich Seidel and his students.[43]

6. Dismissed Biologists Able to Continue Their Work in Germany

A number of biologists dismissed for political reasons were given the opportunity to continue their research somewhere else in Germany, for example at one of the Kaiser Wilhelm Institutes or in private plant-breeding institutes. Elisabeth Schiemann lost her position as associate professor at the University of Berlin in 1940. Unlike other colleagues who also did not belong to the National Socialist party, she refused to submit to certain regulations. These harmless transgressions—she did not attend the obligatory meetings of the Nationalsozialistischer Dozentenbund (National Socialist League of University Lecturers, NSDB), did not join the May 1 marches, and cited Jewish and Russian authors in her lectures—eventually led to her dismissal. To illustrate Schiemann's attitude, Anton Lang (1987, p. 25) has described an episode that occurred in Berlin. A meeting had been organized by the section of the National Socialist Student Organization (NSDStB) at the Botanical Museum on the topic "Heredity and Racial Aspects of Culture and Science":

The main speaker had been imported from the central Nazi party organization of Berlin; if my memory is right he had the truly Germanic name of Hadamitzky. He started his speech with the declaration that there were three human races: good races, poor races, and the Jews. The following speakers either sought to provide support for this classification by citing various examples, or avoided the "Jewish question" altogether. Only Elisabeth Schiemann got up and stated, with a clear although slightly breaking voice—as was her wont in times of great stress or passion—that we should acknowledge the great contributions of different people to German culture and science—French, Italian, "and yes, let us say it clearly, the Jews." She went on to name names, too, including scientists and writers who at that time—it was 1937 or 1938—were harassed by the Nazi authorities or had already been forced to leave the country. (Interestingly, nobody challenged her; it is true that Party Comrade Hadamitzky had already left the meeting).

The immediate reason for her dismissal was presumably her refusal to accept the dissertation of a woman who was a party member.[44] She was able to continue her research with a DFG fellowship that Fritz von Wettstein obtained for her. In 1943 she was appointed departmental head at the newly established Kaiser Wilhelm Institute for Cultivated Plant Research near Vienna.

The three assistants at the KWI for Breeding Research—Hermann Kuckuck, Rudolf Schick, and Hans Stubbe—were dismissed in 1937. On the occasion of an internal quarrel, they were accused of having disrupted the peace at the institute. That was the pretext. Their dismissal was in fact due to strong pressure from a group at the institute that was composed of members of the NSDAP and the SS; it determined the political climate at this institute and wanted to get rid of the three scientists, who were not National Socialists.[45] Stubbe was able to continue his work at the KWI for Biology under Wettstein. In 1943 he was appointed director of the KWI for Cultivated Plant Research. Kuckuck and Schick carried out research in plant breeding at private firms or on their own farms. After the war both were given chairs at German universities.

Three full professors were retired against their will: Hans Driesch (philosopher and biologist in Leipzig) and Richard Hesse (zoologist in Berlin) for political reasons, and Carl Zimmer (zoologist in Berlin) for unknown reasons.

7. Karl von Frisch, the Mischling, and the Solidarity of His Colleagues

Karl von Frisch was not dismissed, but his fate during the Nazi period shows the continuing importance of the Civil Service Law in the later years of National Socialist rule. The campaign launched against Frisch at the University of Munich with the goal of getting him dismissed is another urgent reminder that the Nazi terror at the universities was at its core a racist terror. Jewish ancestry was sufficient reason for expulsion, no matter what the person's accomplishments and political views. Moreover, the fate of Frisch provides insight into the political atmosphere that existed at some universities as a result of the racial laws.

Karl von Frisch was born in Vienna on November 20, 1886. After studying medicine and zoology in Vienna and Munich, he worked at the Zoological Institute at the University of Munich as an assistant to Richard Hertwig from 1910 on, and after 1912 also as a lecturer. He was offered a professorship in zoology at the University of Rostock in 1921 and at the University of Breslau in 1923. In 1925 he was appointed to succeed Hertwig as chair of zoology at the University of Munich. There, in 1932–33, he set up a new institute with funds from the Rockefeller Foundation.

Frisch became known through his work on the sensory physiology of insects and vertebrates. Beginning in 1912, he conducted experiments on color vision in fish and bees and on bee communication. These experiments led him to the discovery of the "language of bees," that is to say, the use of certain types of dance as a means of communication. It had been known for centuries that bees have some form of communication about sources of pollen and nectar. Frisch was the first to succeed in uncovering the nature of this communication. The necessary observations and experiments extended over a period of some twenty years. In 1923 he and his collaborators arrived at the conclusion that bees informed their hive mates about food sources through dances on the honeycomb. A round dance indicated a source of nectar, while a tail-wagging dance was described as characteristic of pollen gatherers.

Frisch stood by this interpretation for many years, until his long-time collaborator Ruth Beutler came to suspect, on the basis of her experiments, that bees also communicated about the distance of a food source. Beutler's observations led Frisch to return to the study of bee communication. Experiments conducted between 1944 and 1946 on his country estate near St. Gilgen (Salzburg) compelled him to revise his original view on the meaning of the dances. His results showed that the two dances indicated food sources located at various distances, regardless of the type of food involved. In the waggle dance, which indicated distances of between 50 and 100 meters, the orientation of the dance in the hive contained additional information about the direction of the food in relation to the position of the sun (Frisch 1946). After World War II, other zoologists largely confirmed Frisch's results. Consequently, today there is little doubt that the dance language represents the most effective means of communication among bees.[46]

Frisch also did research in the field of sensory physiology and communication among fish. While analyzing the sense of smell in fish, he demonstrated in the 1930s that carp released a defensive scent in the presence of enemies, and that certain species of fish (such as minnows) possess a sense of smell comparable to that of dogs. In 1973 Frisch was awarded the Nobel prize in medicine (jointly with Niko Tinbergen and Konrad Lorenz) for his work on bee communication.

The Berlin Document Center has a file on Karl von Frisch, which sheds light on many of the events at the Zoological Institute in Munich between 1933 and 1945. The Civil Service Law, passed in April of 1933, put Frisch's position in jeopardy for the first time. It required all civil servants to present "proof of Aryan ancestry" *(Ariernachweis),* and Frisch was unable to furnish certificates of ancestry for his grandmother on his mother's side. He stated that she was possibly of non-Aryan descent. The office of the rector let him remain at his post with the classification "one-eighth Jew."

But within the student body and among the lecturers there were active groups who did not agree with this decision and worked to have Frisch dismissed. In December of 1934, the *Deutsche Studentenzeitung* (German student paper) in Munich described him as a cold specialist with no interest in Germany and its revival.[47] The spokesman for the lecturers, Johannes Scharnke, tried unsuccessfully to get him dismissed. A group of younger members of the NSDB—among them Wilhelm Führer, head of the organization's local district (Gaudozentenführer), and Ernst Bergdolt, lecturer in botany and a member of the NSDAP since 1922—wanted Frisch replaced with a committed National Socialist. They produced the following political evaluation:

It is high time that the most modern and best equipped Zoological Institute in Germany, which is at this time still ruled by a petty-minded, narrow specialist who has no understanding of the new era and is extremely hostile toward it, be given a leader who will put an end to this situation and who is worthy of the capital of the movement [*Hauptstadt der Bewegung*] in terms of his science, his character, and his politics.[48]

The NSDAP engaged in its lengthy campaign against Frisch because he employed Jewish assistants, even helping them after the National Socialist seizure of power, and disapproved of National Socialism. The lecturer Bergdolt made these motives clear in a letter of May 15, 1936, to the Reich Education Ministry. In it he criticized Frisch's support for non-Aryans, on the grounds that, among other things, he had given the Jewish zoologist Otto Löwenstein a position at the institute and in 1932 had considered Curt Stern for a post as curator.[49] Bergdolt also denounced Frisch for allowing a Jewish scientist, Dora Ilse, to work at his institute after 1933, when she had failed to obtain a fellowship abroad. Prior to that Ilse had been employed at the institute as a private research assistant. Bergdolt charged that after the Nazi seizure of power the "institute, under the leadership of Frisch, has turned into a reservoir of reactionary and subversive forces." Among the forty-seven members of the institutes there were only five "who are committed to National Socialism." Not only reactionary Germans were employed there, but many foreigners, among them "elements who cannot be expected to contribute to the understanding of National Socialism later on." Moreover, Bergdolt noted, Frisch's students included an extraordinarily large number of women.

Bergdolt, a National Socialist from the very beginning, was offended that, as he wrote in the same letter, it would be held against a person at the Zoological Institute if he used the "German salute" or "professed his loyalty to National Socialism." Old Professor Ludwig Döderlein, who had been "driven out of Strasbourg" and was still working at the institute even though he was already eighty, had undoubtedly been given an inferior room only because he had joined the NSDAP. Zoologists who were members of the party had no chance of obtaining jobs as assistants with Frisch. On the strength of these arguments, and with reference to Frisch's probable Jewish descent, Bergdolt—in the name of the NSDB and the Dozentenschaft (Association of University Teachers) of the University of Munich—submitted a request to the Bavarian Ministry of Culture and to the Reich Education Ministry asking that Frisch be transferred by the end of the summer semester of 1936. He requested that someone be put in charge of the Zoological Institute "who is beyond reproach both scientifically and politically."

This campaign by the NSDB paralleled the attempt by the followers in Munich of what was called "Aryan physics" to fill the chair of theoretical physics with a member of the party after it became vacant in 1937 upon the retirement of Arnold Sommerfeld. Wilhelm Führer, the regional head of the NSDB, led the campaign against Werner Heisenberg, whom the faculty had recommended as Sommerfeld's successor (Beyerchen 1977, p. 164). The campaign ended with the appointment of Wilhelm Müller, a member of the party but not a theoretical physicist. Bergdolt coedited the *Zeitschrift für die gesamte Naturwissenschaft* (Journal for all natural sciences), which became the unofficial organ of the followers of "Aryan physics." The rejection by Führer and his circle of theoretical physics as a "Jewish science" played a decisive role in the campaign against Heisenberg. In the case of Frisch, as well, these circles tried to associate him with the cliché of "Jewish science" by accusing him of narrow specialization. For example, we read the following in Bergdolt's letter to the Education Ministry: "Professor v. Frisch has an unusual ability to make propagandistic use of the results of his research, the sort of ability we know from Jewish scientists. In contrast, he lacks entirely the ability to survey his work from a broader point of view, let alone to find connections to the natural establishment of a volkish polity, something that seems so self-evident and would be so easy given his areas of expertise, bees."

The real threat to Frisch's position did not come, however, from accusations concerning his "Jewish" or un-German science, his political misconduct, and his preferential treatment of Jews; it came from the written testimonies Bergdolt had submitted to the ministry substantiating Frisch's Jewish descent. According to these documents, Frisch appeared to be not a "one-eighth Jew" but actually a "one-quarter Jew." In the following years, Frisch was therefore repeatedly asked to furnish his grandmother's still missing certificate of ancestry

(Frisch 1973, p.118). In January of 1941, the Reich Genealogy Office (Reichssippenamt) finally declared him to be a "*Mischling* of the second degree" and thus a "one-quarter Jew." Consequently, on January 9, 1941, Frisch received the following letter from the Bavarian Ministry of Culture through the office of the rector: "According to the determination of the Reich Education Ministry, Dr. Karl von Frisch, full professor at the University of Munich, is a *Mischling* of the second degree. The Reich [education] minister . . . thus intends to retire him in accord with Section 72 of the German Civil Servant Law. I request that you inform Dr. Karl von Frisch of this intent."

The diverse reactions of his colleagues to the impending dismissal reveal, on the one hand, the great popularity Frisch enjoyed, and, on the other, the varying extent of conformism to National Socialist policy among German biologists. The first support for Frisch came from Hans Spemann, professor emeritus of zoology in Freiburg and winner of the Nobel prize in 1935. In a letter to Reich Education Minister Bernhard Rust, dated January 21, 1941, he came out strongly in favor of keeping Frisch in his present position. His letter was passed on through the rector of the University of Freiburg, who indirectly supported Spemann's cause by recommending him to Rust as one of the most eminent German scientists. The directors of the Kaiser Wilhelm Institute for Biology, Alfred Kühn and Fritz von Wettstein, also wrote to Rust as early as January of 1941, and on February 6, 1941, they called on Max Demmel, a senior official in the Reich Education Ministry in person to ask that Frisch be given an exemption.

An unknown supporter arranged for Frisch to publish an article in *Das Reich* at the end of January of 1941, which gave him an opportunity to present to a broader public a detailed account of the variety of research being carried out at the Zoological Institute.[50] What is remarkable about the article are the repeated references to the importance of the work at the institute for economic and food policy. For example, it included a photograph of a coworker of Frisch's in an officer's uniform with the comment that he had been given leave from the military to carry on his scientific work. Another photograph showed a colleague who was finishing up "an important task for the Four-Year Plan during her leave from the Air Raid Warning Service."

There followed additional petitions on behalf of Frisch. On March 29, 1941, Professor Zange of Jena, in his capacity as chairman of the

Society of German Ear, Nose, and Throat Doctors, wrote to Rust concerning the impending dismissal of Frisch. He justified his supportive action by pointing out that Frisch, quite apart from his accomplishments in the field of zoology, had so substantially advanced the knowledge of hearing in fish and other animals, that he had been named an honorary member of the society. Word of Frisch's imminent dismissal also reached Hungary: Professor J. von Gelei, a member of the Hungarian Academy of Sciences, turned to the Chancellery of the Reich with the urgent plea to refrain from dismissing Frisch. He spoke of the "irreplaceable loss" to the German people if Frisch were no longer able to continue his teaching, for after the death of Spemann he was "the best educator of the next generation of German scientists."

It cannot be ruled out that these initiatives contributed to the postponement of the announced dismissal. Still, the party chancellery—headed by Martin Bormann—informed the Reich Education Ministry on November 31, 1941, that no exception could be made for Frisch.

On January 22, 1942, the following professors and lecturers at the University of Munich expressed their support for Frisch in a brief petition to Rust: Ruth Beutler, Reinhard Demoll, (Max?) Dingler, Karl Escherich, Werner Jacobs, Hermann Kahmann, and two others whose names cannot be deciphered. The letter closed without a "Heil Hitler." Karl Escherich was the first Nazi-appointed rector of the University of Munich. Dingler was probably the zoologist Max Dingler.

The decisive help came from a politically influential person who held a high position in the field of food policy and worked for Frisch behind the scenes. His name is not mentioned in the documents. According to Frisch this man was a vigorous supporter of him and his institute. This person may have also been behind the appearance of Frisch's article in *Das Reich*. His efforts eventually established that Frisch's work was important for food policy and secured a research contract for Frisch from the Reich Food Ministry to combat nosema, a bee disease (Frisch 1973, p. 116).

Beginning in 1940, nosema, a single-cell intestinal parasite of bees, had spread in Germany and neighboring countries. In the worst year, 1941, it killed several hundred thousand bee colonies. This outbreak not only caused a decrease in the honey harvest; owing to the role of the bees in pollinating useful plants, it also damaged agriculture and fruit growing. A nosema committee was formed to combat the infestation, with the above-mentioned unknown person as its chairman.

In 1941 Frisch and his collaborator, Beutler, began working on the problem on a broad basis. They showed that all conventional remedies were ineffective. The study of the connection between the spread of the disease and external conditions was not completed by the end of the war. The research contract, however, was soon broadened. Frisch was asked to study, verify, and expand upon experiments—known from Russian beekeeping journals—to encourage bees to visit certain kinds of blossoms. These experiments were carried out in 1942 on orders from the Reichsfachgruppe Imker (Reich Beekeeping Organization) and produced some preliminary results. According to Frisch's own account, the experiments with scent guiding led him to return to the "language of the bees," and this produced the new findings on the communicative meaning of the round dance and the tail-wagging dance.

This research aimed at combating the nosema infestation led the party chancellery to change its original decision to dismiss Frisch. The first decision (January 31, 1942) gave him the opportunity to continue his work despite his forced retirement. On April 27, 1942, the decision changed again. Bormann wrote to Rust: "It has now been brought to my attention after the fact that in this case, as well, retirement would be a considerable impediment to his work, which I am told is very important also in terms of food policy. For these reasons it would therefore be expedient to postpone Professor von Frisch's retirement until after the war." In response, Rust, on July 27, 1942, suspended Frisch's retirement until after the war.

One last petition on behalf of Frisch merits special attention. The executive board of the German Zoological Society—Carl Apstein and the professors Otto Mangold, Adolf Remane, Hermann Weber, and Hans-Jürgen Stammer—did not neglect to request (on May 11, 1942) that the Reich Chancellery conduct a friendly review of Frisch's case. In contrast to all other petitions, the authors began by expressing their support for National Socialist goals in clear terms with repulsive anti-Semitic statements before delivering a plea on behalf of pure science and Frisch. The board, they wrote, "does this with full appreciation of 1.) the incredible fierceness of the war waged against the German people by Jewry; 2.) the fundamental importance of all measures by our National Socialist state for the preservation and promotion of our

Volk; and 3.) the need for the active National Socialist attitude on the part of the German professor. But it also does it with the full awareness that it is necessary to protect German science from an irreplaceable loss." This letter was the only one that surely made no difference for Frisch, since it was written at a time when the decision by the party chancellery to let him stay on had already been made. But by writing it the representatives of zoology in Germany could later avoid being accused of not having done anything to support Frisch.

The threat to his own position did not stop Frisch from helping Polish colleagues who were threatened by the National Socialist racial and population policies. On November 6, 1939, shortly after the German invasion of Poland, the Gestapo arrested 183 professors from all faculties at the University of Cracow, among them 22 life scientists. They were taken to the Sachsenhausen concentration camp and later to Dachau. Fifteen of them died in the camps, while others died later from the effects of imprisonment. This "special action" by the Gestapo was the beginning of the German campaign to liquidate the Polish intelligentsia. Thus by 1940 *Einsatzgruppen* (action groups) of the SS had killed several thousand Polish intellectuals. In a unique act, most of the Cracow professors—provided they were not Jewish—were released after international protests. Karl von Frisch—presumably as the only German colleague in the field—was significantly involved in the efforts to effect the release from Dachau of Roman J. Wojtusiak, who had worked at his institute on a fellowship in 1932 (Seyfarth and Pierzchala 1992).

In conclusion I turn briefly to the role that the faculty of natural sciences at the University of Munich played in the disputes involving Frisch. During the visit of Kühn and Wettstein to Oberregierungsrat Demmel in February 1941, Demmel emphasized that, with respect to the planned retirement of Frisch, a statement by the faculty had to be on hand and that it would play a decisive role in determining the opinion in the ministry. The faculty, however, never met for this purpose. After Frisch had received the announcement of his impending dismissal, the dean, Friedrich-Carl von Faber, professor of botany, refused to call a faculty meeting on the issue. Various members of the faculty, among them the physicists Arnold Sommerfeld and Walther Gerlach, had asked him to do so. Faber was among the professors in Munich who supported "Aryan physics." His response to the criticism by Gerlach regarding the appointment of Müller to the chair of theoretical physics would suggest that, to him—head of one of the largest botanical institutes in Germany—scientific truth no longer mattered: "If you only understand theoretical physics to mean so-called modern dogmatic theoretical physics of the Einstein-Sommerfeld stamp, I must inform you that this will indeed no longer be taught in Munich. The appointment of Professor Müller's has been achieved precisely to bring about a definitive change."[51] After such a statement it becomes understandable that even Frisch's outstanding scientific achievements could not alter Faber's negative attitude toward him.

The conduct of the faculty had a small epilogue. In 1946 Ruth Beutler, in a memorandum on the conduct of the faculty, stated: "One could have expected that now, at least to a small degree, amends would be made for the injustice the university committed under the previous regime by ordering the dismissal of Professor von Frisch or at least not objecting to it."[52]

Klaus Clusius, since 1936 full professor of physical chemistry in Munich and after the war dean of the faculty of natural sciences, contradicted this statement emphatically. The following remarkable statement is from his letter to the Bavarian Ministry of Culture: "By questioning various members of the faculty, I have ascertained that until today they had never heard anything about the danger at that time of seeing Professor Dr. von Frisch dismissed on the basis of the racial laws."[53]

The Zoological Institute was substantially destroyed on July 13, 1944. In 1946 Frisch accepted an appointment to the University of Graz; he returned to the University of Munich in 1950. He died on June 12, 1982.

8. The Return of Émigré Biologists to Scientific Institutes in Germany after 1945

Of all the émigré biologists, only the four university teachers who had gone to Turkey returned to universities in Germany. One of them, Gustav Gassner, who emigrated to Turkey following his dismissal in 1934 for political reasons, returned to Germany as early as 1939 (as mentioned above) and headed a private institute of plant protection from 1940 to 1945. In 1945 he returned to his former position as full professor and director of the Botanical Institute at the Technical College in Braunschweig. He also became rector of the university, president of the Biologische Zentralanstalt (today the Biologische Bundesanstalt, the Federal Biological Institute), and was for two years temporary president of the Forschungsanstalt für Landwirtschaft (Research Institute for Agriculture). Thus after the war he had a decisive influence on the scientific life in Braunschweig.[54] The three other émigrés to Turkey had been given positions at the university in Istanbul. Their decision to return to Germany resulted primarily from the policy of the Turkish government to replace foreign university teachers, including German refugees, as quickly as possible with Turkish scientists. For that reason foreigners received only uncertain contracts and extremely low pensions.

Curt Kosswig, who emigrated to Turkey in 1937 for political reasons, became director of the Zoological Institute of the University of Hamburg in 1955. Of the émigré Jewish professors, Leo Brauner and Alfred Heilbronn accepted posts at universities in the Federal Republic, Brauner in 1955, Heilbronn in 1956. Prior to his emigration to Turkey in 1933, Brauner had been associate professor and assistant to Otto Renner in Jena. In 1955, following Renner's retirement, he was appointed chair in botany at the University of Munich. In 1948 Renner had succeeded the full professor Friedrich von Faber, who had been dismissed because of his National Socialist background. Although Brauner was an internationally recognized plant physiologist, his appointment was not greeted with universal approval. Renner expressed doubts about Brauner's qualifications as head of the institute, justifying them—as the following portion of a letter by Friedrich Oehlkers shows—by charging that Brauner was excessively specialized. Oehlkers, who had been asked for an opinion on Brauner, wrote to Renner on November 12, 1954:

> To describe Brauner as a second-rate botanist is totally incomprehensible to me. Once you survey the wealth of his publications . . . there is no doubt that he is not in any way inferior to those so far mentioned for the position in Munich, and that he undoubtedly stands above his contemporaries in uprightness and in the meticulousness of his way of working and thinking. The fact that he has worked primarily in the field of physiology should actually make it all the more clear that he is concerned with the "plant as a whole," for there is no longer a field today in which physiological methods and questions have not been introduced.[55]

In a letter to Erwin Bünning, Oehlkers made clear what he believed impelled the professional criticism of Brauner: "I hope it will prove possible to overcome this accursed Munich anti-Semitism, for that is all it is."[56]

Some émigré biologists turned down appointments to German universities. Leopold von Ubisch did not accept the offer in 1945 to return to his former chair in Münster. Hans Grüneberg, who prior to his dismissal and emigration to England in 1933 had been assistant at the Zoological Institute of the University of Freiburg, declined the appointment in 1954 to a chair at the Free University of Berlin. He wrote the following to Hans Nachtsheim concerning his decision: "I do not believe that I would be able to banish the shadows of the past and to regain my freedom from bias in dealing with people unknown to me. By now I have put down roots here."[57] He also turned down in 1960 the offer of a position as head of a planned department for animal and human genetics at the Max Planck Institute for Genetics and Genetic Pathology in Berlin.[58] According to Grüneberg, the chair in Berlin was also declined by Ernst Caspari, who had been assistant at the Zoological Institute of the University of Göttingen and who was then working at a small university in Connecticut.[59] In 1950 Viktor Hamburger was offered an associate professorship in developmental physiology at the Institute of Zoology at the University of Cologne, which he declined (as described above, in Chapter 1, Section 5).

Ernst Pringsheim, expelled as a Jew from his position as full professor of botany at the German University in Prague in 1938, returned to Germany after his active work as a researcher in Cambridge. He lived in Göttingen, where he received a pension from the German government. Gerta von Ubisch also returned to Germany late in her life. Her unpleasant experiences trying to obtain compensation are described below (in Section 10 of this chapter).

9. Wiedergutmachung *in Public and Civil Service*

The term *Wiedergutmachung*, literally, to make good again, refers to a material and emotional restoration of the old state. It implies a kind of restitution that relativizes guilt. This, of course, is impossible, the crimes of humiliation and murder can never be undone. In Israel the term used for *Wiedergutmachung* is *shilumim* (payment). It refers only to material restitution.

The laws on restitution and their implementation in practice have been analyzed in many different publications (among others Herbst and Goschler 1989). Most of the monies for restitution went to compensation payments. The categories of damage included damage to life, limb, and health; to property and possessions; and to professional and economic advancement. Into this last category fell compensations for disadvantages suffered under the Civil Service Law of April 1933. Until 1953 compensation was handled exclusively according to laws passed by the various *Länder* (states). The Supplementary Federal Law for Victims of National Socialist Persecution of October 1, 1953, constituted the first regulation on the level of federal law; it formed the basis for the Federal Law for Compensation to victims of National Socialist Persecution of June 29, 1956 (Herbst and Goschler 1989, p. 26). *Wiedergutmachung* in civil service was often criticized because of the payment of large sums to people who already had a good income.[60]

As an example of the restitution regulations of the state of Baden-Württemberg I will briefly introduce the compensation laws for members of the civil service. The state's legislature passed the first compensation law, the Law for the Restitution of National Socialist Injustice, on August 16, 1949; it provided for compensation for damage done to a person's economic advancement. According to the Baden-Württemberg Law Regulating Compensation for Members of the Civil Service, ongoing support was provided only for those persons who, as members of the civil service, would have a claim to retirement pay without the "persecutory measures."[61] Assistants and lecturers, as well as associate professors without civil service status who had been dismissed between 1933 and 1935, were not included in this group. The majority of the expelled Jewish university teachers were thus, on legal grounds alone, excluded from the possibility of receiving ongoing support. Only in 1955 did a federal law allow former lecturers to claim compensation, provided the "regular course of their career" might have led to a position with civil service status. On December 23, 1955, Baden-Württemberg also made this provision part of an act amending the law regulating compensation (Mussgnug 1988, p. 197).

The response to the application by the widow of the zoologist Hugo Merton shows how the State Office for Compensation for Northern Baden, in Karlsruhe, interpreted the compensation law. Merton, associate professor in Heidelberg and of Jewish descent, lost his right to teach zoology in 1935 through the Reich Citizenship Law. In 1937 he emigrated with his family to Edinburgh. However, in 1938, when he returned to Germany for a brief period, he was arrested and taken to the concentration camp in Dachau. He fell ill at Dachau, and though he was able to get out in 1939, he died in Edinburgh in 1940, probably as a result of this illness.[62] When his widow requested restitution, the amount of the compensation payment was calculated with great precision by the Karlsruhe office:

> The period from January 1, 1936 to March 23, 1940, is recognized as the period of compensation . . . The loss of income thus comes to . . . 2668.83 [Reich marks]. In accordance with §3 Sect. I Suppl. Law, this amount is converted into *deutsch marks* at the rate of 10:2, which comes to 533.76. According to §25 Sect. II Suppl. Law, the claims of the deceased pass on to his heirs only to the extent to which it compensates for a loss which the family of the deceased suffered as a result of the loss of his income. In accordance with these regulations the compensation for the widow is set at one third, that is 177.92 [deutsch marks].[63]

This decision was changed only in 1956 following a protest by the rectorate lodged in response to pressure from the faculty. The state of Baden-Württemberg had to grant Mrs. Merton a maintenance allowance for the remainder of her life in the amount of the legal provision for surviving dependents.[64]

The Baden-Württemberg Law Regulating Compensation for Members of the Civil Service provided restitution essentially only for former civil servants. Those who had been prevented from embarking on a university career before 1933 were thus punished a second time after the war, as the following account of the fate of Gerta von Ubisch illustrates.

10. Gerta von Ubisch: The Emigration and Return of a Professor

Gerta von Ubisch was born in Metz on October 3, 1882. She came from a protestant family, but her grandparents on her mother's side were Jewish.[65] First she studied physics and then, after receiving her doctorate, botany with Ludwig Jost in Strasbourg. Erwin Baur, who had obtained the full professorship of genetics at the Landwirtschaftliche Hochschule (Agricultural College) in Berlin in 1911, procured a fellowship for her to work on grains. She began with the analysis of genetic factors of barley at a time when genetic research had not yet found any recognition in Germany and was dismissed more or less as a trifle.[66]

Since a position as assistant at the Agricultural College was reserved for agriculturalists, she went in 1914 to the newly founded Kaiser Wilhelm Institute for Biology as an assistant to Carl Correns. Shortly after the outbreak of the war she left the institute and worked for several years in private plant-breeding companies until she was given, on May 1, 1921, a position as assistant with Ludwig Jost at the Institute of Botany in Heidelberg, where she worked for twelve years. She could accept the position only on the condition that in addition to her genetic research she also work in the area of physiology. At Jost's suggestion she obtained *Habilitation* in the field of botany in 1923.[67] In 1929 she was given the title of associate professor. She was the first woman to obtain *Habilitation* in the state of Baden, and one of the first women to do so in all of Germany.[68] Prior to 1919 it had not been possible for women to become lecturers.

Her *Habilitation* thesis, "Experiments of Hereditary Transmission and Fertility with Regard to Heterostyly and Doubling," drew the following comments from Jost in his report: "It is clear that one must be particularly critical in approaching the application for *Habilitation* by a woman. The decision is made easier in this case by the fact that there are already so many accomplishments. It is therefore not necessary to build on *hopes*, as is usually the case with *Habilitationen*, rather one can say that the scientific development has reached a high point."[69]

He called her *Habilitation* thesis "superb" and had no qualms about recommending Ubisch for an associate professorship. However, he did not think that she would become full professor, and he could also "not imagine that at the moment any other woman is conceivable as the director of a botanical institute and garden."[70] In fact during her ten years of work as a lecturer, Ubisch did not receive an appointment to a chair as associate professor from one of the larger universities. She herself thought the reasons for this were twofold: first, it was very difficult, in fact nearly impossible, for a woman to obtain a majority vote in a faculty; second, few positions in the field of heredity were provided for in university budgets, since this field was represented by the professors of botany and zoology themselves. Until 1935 there was only a single genetic institute at a university in Germany, namely the Institute for Heredity in Berlin, which Baur had founded.

Ubisch, the only lecturer teaching genetics in Heidelberg, was financially dependent on her lecturer salary and teaching assignments. In the course of her genetic research on factor analysis in barley, she came up with a Mendelian interpretation for phenomena that seemed to deviate from Mendel's Laws (Ubisch 1919). She studied linkage relationships in barley and discovered a genetic explanation for heterostyly (styles of varying lengths) and the irregular sex distribution of Antennaria (Ubisch 1934). Although she did not stand out internationally or nationally, the Science Citation Index shows that her work was cited. She also received positive recommendations from internationally recognized scientists such as Correns, Jost, and Friedrich Went when she applied for a professorship in South Africa in 1933.[71]

After the Civil Service Law—which applied also to university employees without civil servant status, such as assistants and lecturers—came into force, Ubisch was terminated effective April 20, 1933, on the grounds that she was of non-Aryan descent. Various professors pleaded that she be allowed to remain at the university. Jost, who in Ubisch's view was a democrat and remained so even after 1933 (he was retired in 1934 for reasons of age), pointed out in several letters to the rector and the minister of culture that her father had been an officer and front-line veteran and in this way secured her *venia legendi* (right to teach). The National Socialist Student Organization thereupon called for a boycott of her lectures; the students responded to the call and the classes were cancelled for one semester.[72] At the time such actions were not unusual among students for whom the official measures against Jewish and liberal university teachers did not go far enough. In Heidelberg the professors Arthur Rosenthal and Heinrich Liebmann in the faculty of natural sciences, and Ernst Levy, Walter Jellinek, and Max Gutzwiller in the faculty of law, were affected in 1935 by boycott actions from the National Socialist Student Organization or the professional organizations *(Fachschaften)* (Mussgnug 1988, p. 60). Ubisch was also officially prohibited from examining doctoral candidates or candidates for the *Physikum* (an intermediary preclinical examination for students of medicine). In 1935 the university, in accord with the Reich Citizenship Law, officially deprived her of the right to teach.

In the summer of 1933 Ubisch left Heidelberg to seek a position abroad, officially taking a leave of absence to conduct scientific work. She accepted an invitation from the Dutch Association of Female Academics to spend one semester at Utrecht and received a fellowship in Switzerland for the summer semester of 1934. She had great difficulty finding a permanent position. She was not chosen for a professorship in plant physiology in South Africa, despite a recommendation from Jost, who as one of the leading plant physiologists in Germany had been asked to suggest someone, and the positive reviews from others—presumably because of her age (she was fifty-one at the time).[73] Her efforts to be placed via the Academic Assistance Council in London, an aid organization for dismissed academics, were unsuccessful. Her case did not appear to be urgent, since her brother still held a full professorship in Münster.

Through the mediation of Ernst Bresslau, who had been dismissed as professor of zoology in Cologne because of his Jewish descent and had emigrated to Brazil, she was offered, in December of 1934, a position as head of a department of genetic science at the "snake serum institute" Butantan in São Paulo. Bresslau, a friend of the von Ubisch family, died shortly after her arrival in Brazil (see Chapter 1, Section 2), leaving her to manage without his help. The work that awaited her at the institute, completely different from what had previously been described to her, was complicated by many intrigues. The director had in mind for her to breed horses for the production of serum against snake bites; to arrive at basic genetic findings in the serum question, she decided to begin by breeding guinea pigs, in which she crossed wild strains with various inbred strains for several generations. As a result of internal power struggles at the institute, the director was fired in 1938 and the Department of Genetics was dissolved. In 1939 Ubisch succeeded in obtaining a one-year position at the Ministry of Agriculture, but after 1940 she no longer had a permanent position. Among other things she worked on an agricultural estate and taught biology in São Paulo.

After the war she immediately sought a visa for Norway, where her brother Leopold von Ubisch was living with his family. It took a year and a half before she received permission to enter, the reason being that Norway was so impoverished as a result of the German occupation that, according to her own statements, foreigners were at first not allowed in. She was given a visa for one year, but living with her brother, who had not found a permanent position in Norway, and with her sister-in-law proved difficult. She therefore decided to look once again for possibilities in Brazil, and she accepted the offer from a shipowner for free passage in return for translating services. This time, too, she did not find a post—she decided to do housework and childcare for people she knew—and so after a few months she returned to Norway and then to Germany.

In May of 1952 she arrived in Heidelberg; she was almost seventy years old and completely penniless. Because of her financial hardship she applied for ongoing support by way of compensation to the Ministry of Culture in Stuttgart on May 28, 1952. The rector at the university, Professor Eberhard Schmidt, supported her application by writing repeatedly to the ministry. Walter Jellinek, who had been dismissed as professor of public law in 1935 and had been given back his chair in 1945, wrote an expert opinion for her. Still, her application was turned down on March 12, 1953, on the grounds that she did not have civil servant status at the University of Heidelberg either as a lecturer or as an assistant and was thus not legally entitled to a pension.

Ubisch was incensed at this decision with its slavish adherence to the letter of the law. She wrote to the ministry: "Since you have demonstrated it to me paragraph by paragraph, I have no doubt that I am not entitled to compensation according to the letter of the law. But should it not be possible to put forward a human factor that ameliorates the injustices that arise with every law?"[74]

Schmidt advised her to appeal the decision to the Compensation Court of the Regional Court in Karlsruhe and drafted the plaint for her. The complaint maintained that a *Privatdozent* did not simply "lecture privately" and therefore belong to those who worked as freelancers; on the basis of the Reich Habilitation Law of 1939 it was highly likely that Ubisch would have been given a position as a supernumerary university lecturer or a corresponding civil servant position if she had not been expelled from the university because of racial persecution.

A court hearing was held on October 20—the only one in this case. Landgerichtsdirektor (Chief Judge of the State Court) Dr. Goltzen presided; Rössinger, the regierungsassessor, was the lawyer representing the Ministry of Culture. No official protocol was taken, but excerpts from the protocol drawn up by Ubisch and sent to the rector on March 24, 1954, show that the court was determined from the outset to reject her complaint. Ubisch was first asked why she had no money even though her grandfather had been a wealthy man, and she had to clarify that her grandfather died in 1881 and that his wealth was lost to inflation and currency reform. In the end she was given the blame for her own situation: "The presiding judge Dr. Goltzen charged that I left *voluntarily* in 1933. The *venia legendi* had been restored to me in the fall of 1933, hence I should have been teaching until 1935. Then came the Nuremberg Laws which said: whoever is employed remains. In that case I would still have been lecturer in 1939 when the new Reich Habilitation Law was issued, and I would have become a civil servant. By leaving voluntarily and without cause I deprived myself of this chance." When, in her response, she pointed to the student boycott of her classes, among other factors, Rössinger dismissed her assertion, declaring that at the time other Jewish professors were also kept on. (Only after the hearing did she find out that the Nuremberg Laws had the opposite effect of what had been claimed in court, namely, that previously protected non-Aryan university teachers were now expelled, and that the professors mentioned by name in the hearing also had to leave by 1935, at the latest.)

The court proposed a settlement that included a monthly payment of 150 Deutsch Marks (DM), subject to termination at any time. (At the special request of the then Minister of Culture Gotthilf A. Schenkel, Ubisch had been given, starting in 1953, DM 120 a month for three years without a legal entitlement, the settlement thus meant an increase of this payment by thirty marks under the same conditions.) She was advised by all sides, including the university, to accept the settlement, which she eventually did. However, she felt deeply humiliated by the hearing, in which officials had not shrunk from making false statements, in which her persecution had been denied, and in which she, the victim, had been blamed for her own situation. In 1956 she wrote to the rector: "Twenty years are a long time, but it should not be so long that all crimes of the Nazis should be forgotten. I never would have thought that I would ever attend a hearing—and it would seem not as the plaintiff but the defendant—in which the Nazi methods would be denied. The Ministry of Culture must be criticized for the fact that it sent as the representative of the government, and thus as my opponent, [an official] who was so young that he was ignorant of the provisions of the infamous Aryan-paragraph—or believed he could ignore them."[75] The small financial support awarded to her was an insult, since "it is so flagrantly different from the treatment which my male colleagues received from the various universities."[76]

Only after the compensation law of Baden-Württemberg was amended on December 23, 1955, extending entitlement to compensation claims to former lecturers and associate professors without civil servant status, did she request a review of her case, with the help of the rectorate. This time she was successful. The compensation decision of the Ministry of Culture on June 1, 1956, granted to her—four years after her return to Germany and at the age seventy-three—the legal status of a retired associate professor with civil servant status and awarded her the payment of a pension retroactive to January 1, 1954. At the request of the rector, the faculty considered the question of whether to admit Ubisch into the register of employees of the university, and on June 25, 1956, the faculty voted the motion down (for a comparison, see the case of Ernst Lehmann in Chapter 2). Gerta von Ubisch died in Heidelberg on March 31, 1965.

2

NSDAP Membership, Careers, and Research Funding

It is appropriate to begin an analysis of the relationship between science and National Socialism with a look at membership of scientists in National Socialist organizations. Therefore I shall first present statistical data on the party membership of biologists before examining the political preconditions for scientific careers and the awarding of research funds.

The Nationalsozialistische Deutsche Arbeiterpartei (National Socialist German Workers' Party) succeeded the small Workers' Party in 1920, and Hitler become its leader in 1921. The party suffered a crisis in 1924, when Hitler was imprisoned for nine months because of his attempt to overthrow the Bavarian government. It was, however, reorganized in 1925 and, after the world economic crisis of the late 1920s, attracted an increasing number of followers, not only among farmers, workers, and students, but also among industrialists, conservative nationalists, and active members of the army. In the general elections of 1930, the number of votes for the NSDAP increased from 0.8 to 6.4 million. The NSDAP was the party that received the most votes in the elections of July 1932 (nearly 14 million, or 37%). With the assistance of conservative politicians, Hitler was appointed Reich chancellor by President Hindenburg on January 30, 1933. In March 1933 the promulgation of the Ermächtigungsgesetz, which led to the liquidation of the parliament, constituted a decisive step toward a totalitarian one-party state. By July 1933 all other parties had dissolved themselves under the pressure of the NSDAP. After the death of President Hindenburg on August 2, 1934, Hitler became officially "Führer and Reichskanzler" of Germany.

Hitler allowed National Socialist leaders like Hermann Göring, Heinrich Himmler, and Joseph Goebbels to control their own spheres of power, though he always remained the person to make the final decisions. With respect to science policy, a variety of authorities in party and government competed with one another. In addition, the old authorities in the universities and the Kaiser Wilhelm Society remained in many cases influential, in spite of the institutional changes mandated by the National Socialist "leadership principle." On the level of science policy, the National Socialist system seems to have been polycratic rather than monolithic. By making this statement, I do not mean to question the overall power of Hitler. I address the various authorities deciding on appointments at universities and decisions of research grants below (in Sections 2 and 5 of this chapter). Here I want to briefly mention only two of the most important subdivisions of the NSDAP, the SA, and the SS. The SA was a powerful paramilitary organization of the NSDAP until 1934. After the "Röhm Putsch" of June 30, 1934, however, in which Hitler liquidated Chief of Staff Ernst Röhm and eighty-four of his followers, the SA had only little importance.

The SS was led by Himmler beginning in 1929. After 1934, it grew into a state within the state. Its branches included the armed Waffen-SS and the general SS, an élite organization in terms of racial purity. Only men could become members, and they had to undergo a racial and racial-hygienic assessment beforehand. If they wanted to get married, they had to ask permission of the Central office for Race and Resettlement (Rasse- und Siedlungshauptamt) of the SS. Brides-to-be also had to be racially and race-hygienically acceptable. The SS provided the guards and staff of the concentration camps. Chapter 5, Section 1, provides a more detailed analysis of the SS and Himmler's scientific aims.

1. NSDAP Membership

I examined membership in the NSDAP, the SA, and the SS for 440 biologists who worked at universities and Kaiser Wilhelm Institutes in Germany between 1933 and 1945 and at universities in Austria and the German University in Prague between 1938 and 1945. The data for this analysis come for the most part from the Document Center in Berlin. While records can be considered largely complete as far as membership in the NSDAP is concerned, the information on SA and SS membership is not, as I have discovered in a number of cases through a comparison with information in the DFG files or in personal files. Wherever possible the information has been supplemented with data in the personal record sheets of the Reich Education Ministry.[1]

Of these 440 biologists, 234 (53.2%) joined the NSDAP, of the 404 biologists who did not emigrate, 233 (57.6%) joined. (Karl Arens, who emigrated in 1936, was a member of the party.) This figure is somewhat higher than that for physicians (44.8%, according to Kater 1989, p. 56)[2] and corresponds roughly to the percentage for psychologists who did not emigrate (54.7%) as calculated by Mitchell Ash and Ulfried Geuter (1985, pp. 263–278). Of the 440 biologists, 92 (20.9%)— 91 of the 404 (22.5%)—were at least for a time members of the SA;[3] 23 of the 440 (5.2%)—22 of the 404 (5.4%)—were members of the SS. (As noted in Chapter 1, Section 3, Curt Kosswig was a member of the SS; he left it prior to emigrating in 1937.) Only 14 of the SA members and 2 of the SS members were not members of the NSDAP. At 5.6%, SS membership of the 393 male biologists who did not emigrate was somewhat lower than among the male physicians registered with the Reich Physician Chamber after 1935 (about 7%, Kater 1989, p. 70). Another 14 (3.2%) of the biologists were "supporting members" of the SS. Of the 440 biologists, 14 (3.0%) were women; among the 404 who did not emigrate the number was 11 (2.7%). Of those, 4 (28.5% and 36.3%, respectively) joined the NSDAP.

Clusters of biologists entering the party occurred in the years 1933, 1937, 1938, and 1940 (Table 2.1). As was the case with physicians (Kater 1989, p. 55), most joined the party in 1937. Presumably the timing had something to do with the fact that the party, fearful of opportunists because of the rush of new members, closed the membership rolls after 1933 and did not reopen them to the public—with

Table 2.1   Biologists who became members of the NSDAP

Year	Number of biologists who joined the party
1922	1
1930	3
1931	3
1932	7
1933	68
1934	3
1935	2
1936	1
1937	88
1938	25
1939	8
1940	18
1941	3
1942	1
1943	0
1944	2
1945	0
?	1
Total	234

few exceptions—until 1937. Twenty of the 25 biologists who joined in 1938 were Austrian university teachers, who could legally become members of the NSDAP after the Anschluß. In 1940 it was probably the German victories in the war that led so many to join.

An analysis of NSDAP membership from the perspective of members' age reveals that the readiness to join the party declined noticeably with a person's age. Thus fewer than a third of the biologists who were between fifty and sixty in 1933 joined the party; of those who were thirty or younger in 1933, 70% did so (Table 2.2). It is likely that this age distribution of party members is at least in part connected with hopes for better chances of advancement through correct political behavior. However, a relatively large percentage (26%) of biologists who were past sixty in 1933 also became members of the NSDAP.

With the exception of applied botany, there was no clustering of party membership in a specific field: the percentage in applied botany (65.8%) was above the average for biologists, in genetics it was below (ca. 40%).[4] Among botanists the percentage was somewhat higher than among zoologists (Table 2.3): 121 of 212 botanists (57.0%) were

Table 2.2 Age structure and NSDAP membership among biologists

Age in 1933	Number	% of all biologists	No. of NSDAP members	% NSDAP membership	% of age group among all NSDAP members
61 +	50	11.3	13	26.0	5.6
51–60	72	16.4	23	32.0	9.8
41–50	99	22.5	52	52.5	22.2
31–40	114	25.9	72	63.2	30.8
30 and younger	105	23.9	74	70.5	31.6
Total	440	100.0	234	—	100.0

Table 2.3 NSDAP membership and area of specialization and academic positions of biologists

Area	Number	Number of NSDAP members	% NSDAP membership
Zoology	221	111	50.2
Botany	212	121	57.0
Applied zoology	18	10	55.6
Applied botany	38	25	65.8
Genetics	58	23	39.7
Others	7	2	28.5
Full professors	132	58	43.9
Associate professors with civil servant status	17	11	64.7
Associate professors without civil servant status	144	84	58.3
Lecturers	89	56	62.9
Fellows	58	25	43.1
Total	440	234	53.2

Note: The highest position held prior to 1945 was counted.

Others = biochemists, biophysicists, and virologists. Applied botany and applied zoology = biologists who specialized in applied research. Genetics = biologists who specialized in genetics. Fellows = biologists without teaching credentials, for example DFG fellows at Kaiser Wilhelm Institutes.

party members, as were 111 of 221 zoologists (50.2%). Of the 202 botanists who did not emigrate, 120 (59.0%) belonged to the party, as did 199 of the non-émigré zoologists (55.0%). The analysis of the connection between party membership and quality of research (see Section 8 of this chapter) shows that the scientific influence of party members was only about half that of nonmembers: 234 party members had 7,538 citations in the Science Citation Index from 1945 to 1954 (32.2 per person), while 206 nonmembers had 14,066 citations (68.3 per person).

2. The Significance of NSDAP Membership for Habilitation *and Appointments*

According to the Reich Habilitation Law, after 1933 the procedure for obtaining *Habilitation* involved three steps:[5]

1. determination of the teaching qualifications by the faculty;
2. participation in the Reich Camp for Civil Servants;
3. granting of the qualification by the Reich education minister.

The faculty could decline for objective reasons to admit someone to the first step, the demonstration lesson *(Lehrprobe)*. These objective reasons included misgivings about a person's character and politics.[6]

Prior to 1933, the procedure for appointments at universities called for the faculty to submit a list of three candidates to the appropriate minister of culture when a position became vacant. In most cases, as Ulfried Geuter (1984, pp. 105ff.) has shown for psychology, the person at the top of the list was appointed by the minister. With the introduction of the führer principle at the universities in the fall of 1933, the right of the faculty to decide appointments was transferred to the rector. Faculties were given only an advisory function, but their suggestions continued to remain the starting point for the relevant processes.

According to Geuter, when it came to appointments in psychology, faculty recommendations were for the most part followed also after 1933. A faculty passed its recommendation on to the rector, who attached his own statement and submitted it to the relevant Ministry of Culture, after May 1934 to the newly established Reich Ministry for Science, Education, and People's Education (Reich Education Ministry, REM), where the final decision was made by Minister Bernhard Rust. The REM requested political evaluations from the head of the League of University Teachers. Membership in this organization was compulsory for all midlevel academics and was open to those who held chairs. The head of the association of lecturers was appointed by the Reich education minister, after consultation with the rector and the leader of the Regional National Socialist League of University Lecturers (Gaudozentenführer), and was directly responsible to the rector (Hartshorne 1937, p. 51).

The party influenced appointments officially through the "deputy of the führer," Rudolf Heß, and after 1940 through the party chancellery under Martin Bormann. The REM had to consult Heß and later Bormann before it recommended an appointment or the offer of a chair, which was then made by the Reich Chancellery. Heß and Bormann drew for their opinions on statements from the Reich leadership of the NSDB, which got its information from the local leaders of that organization. The NSDB developed out of the National Socialist Teachers' League (NSLB) and, like the National Socialist Student Organization, incorporated into the party on January 1, 1936. All party members who were university teachers were grouped together into a regional NSDB. The NSDB was the arm of the party that was to carry out the National Socialist revolution at the universities. This revolution was to contribute to the "reformation of the volkish life, that is to say, to the creation of the new German," not "through regulations and laws" but by "finding, educating, and leading people."[7] The NSDB advocated that academic positions be filled by Nazis true to the party line. In its eyes, the transformation of the universities by the REM according to political criteria was proceeding much too slowly.[8] Beginning in 1938 the faculty had to inform the local leader of the NSDB directly about its appointment recommendation, and the local leader passed it and the rector's statement on to the deputy of the führer.[9]

Additional political evaluations could be requested from the "Rosenberg office." Alfred Rosenberg, on the basis of a decree by Hitler, called himself the "representative of the führer for supervising the entire spiritual and ideological schooling and education of the NSDAP." The title "Rosenberg office" incorporated his staff. The Central Office for Science in the Rosenberg office developed into a separate agency in 1937. Its task was, among other things, "to observe, assess, and promote scientific life at German universities and outside from an ideological perspective."[10] In 1938 it was given a separate budget, and it became an increasingly active rival of the NSDB when it came to the political evaluation of university teachers (Beyerchen 1977, p. 191).

Despite the party's possibilities of exerting influence, as described above, membership in the NSDAP or in one of its branch organizations never became an official requirement for *Habilitation* or appointment to a university post. But to what extent was it a prerequisite in actual practice? The extant NSDB evaluations of individual biologists show that while membership was very important, it alone was not decisive for a positive evaluation.[11] A few examples: in 1938 the leader of the Regional League of University Teachers in Berlin rejected the granting of the position of lecturer to Hans Breider, member of the NSDAP and the SA. Among the reasons Breider was rejected were his activities on behalf of the Catholic Center Party (Zentrum), his membership in a Catholic student fraternity during the Weimar Republic, and his support for Professor Leopold von Ubisch, who was to be dismissed for "racial reasons" (see below).[12]

The director of the Genetics Department of the Kaiser Wilhelm Institute for Brain Research, Nikolai V. Timoféeff-Ressovsky, was not a party member and not even a German citizen. In 1938 he was given a positive evaluation by the NSDB, which emphasized above all his world renown as a geneticist and his anti-Communism.[13] Hans Nachtsheim, not a party member and at the time associate professor in Berlin, also received a positive evaluation in 1936, in this case from the Regional League of University Teachers in Berlin: he championed the national state at all times and was fully recognized as far as his science was concerned (genetics).[14] By contrast, Fritz von Wettstein, also no party member, received a decidedly bad evaluation from the NSDB in Berlin: he had tried to prevent the *Gleichschaltung* of the Kaiser Wilhelm Society, had clearly discriminated against party comrades in his institute, and had invested in considerable new facilities for Hans Stubbe, who had been dismissed for political reasons (see Chapter 1, Section 6).[15] However, this evaluation had no consequences for Wettstein, especially since he was already director of a Kaiser Wilhelm Institute.

To clarify statistically the importance of membership in the party or its branch organizations, I have compared the percentage of membership among all university lecturers and holders of chairs with the percentage among biologists who acquired *Habilitation* after 1933 or were appointed to a chair. All appointments considered here were connected with appointments to associate professorships with civil servant status or to full professorships.

The results show that the percentages of NSDAP membership among university lecturers (62.9%) and associate professors (64.7%) were noticeably higher than average party membership among all categories. Among full professors, however, it was below the average (43.9%) (Table 2.3). The data take into account the highest position a person held by 1945. Among the biologists who acquired *Habilitation* or were appointed to a chair in 1933 or later, the percentage with party membership is clearly higher than in the comparison group. Of 92 biologists who obtained *Habilitation* in Germany between 1933 and in Austria between 1938 and 1945, 74 (80.4%) were members of the NSDAP. Of the 92 biologists, 84 (91.02%) were members either of the NSDAP or, at least for a time, of the SA or SS. But only 55 (59.7%) of them were party members already prior to their *Habilitation;* 60 (65.0%) were members of the NSDAP, the SA, or the SS. Between 1933 (Austria 1938) and 1945, 44 biologists were given full professorships, while 17 received associate professorships with civil servant status and chairs. Of full professors, 72.7% (prior to their appointment, 61.3%) were party members; 64.7% of associate professors (all of them prior to their appointment) were party members (Table 2.4). Prior to their appointment, 65.0% of the full professors and 70.0% of the associate professors had been members of the NSDAP or one of its branch organizations.

These figures show that party members were strongly favored for appointments. In addition, pressure was often exerted on nonmembers.[16] Over and above this, however, the figures make clear that party membership was also in practice not a necessary prerequisite for *Habilitation* or appointment after 1933. Leaving aside the small number of biologists who never joined the party or one of its branch organizations, some became party members only after their *Habilitation* or appointment. This also applies to biologists who held chairs prior to 1933: of the 74 full professors who had been appointed and called to a chair by 1932, 20 (27.0%) later joined the NSDAP.

According to Geuter (1984), political considerations played a prominent role in appointments above all in the first years of Nazi rule,

Table 2.4 Appointments to full professorships and associate professorships with civil servant status in three periods between 1933 and 1945

	1933–1936 Full profs.	1937–1940 Full profs.	1937–1940 Assoc. profs.	1941–1945 Full profs.	1941–1945 Assoc. profs.	Total 1933–1945 Full profs.	Total 1933–1945 Assoc. profs.
Number	20	12	2	12	15	44	17
Number of NSDAP members	14	9	1	9	10	32	11
% NSDAP members	70.0	75.0	50.0	75.0	66.7	72.7	64.7
Number who were NSDAP members at time of appointment	9	9	1	9	10	27	11
% NSDAP members at time of appointment	45.0	75.0	50.0	75.0	66.7	61.3	64.7
Number of citations per person (SCI 1945–1954)	62.5	34.7	202.5	72.0	48.9	57.5	67.0

Note: The high number of citations among the associate professorships (202.5) is due to S. Struger (TH Hannover), who was cited 390 times.
Assoc. profs. = associate professors with civil servant status.

while from 1937 on, professional considerations and the question of the practical relevance of scientific work were given greater importance. This thesis was confirmed by Ash and Geuter (1985), who determined that party membership had considerable significance in appointments to associate and full professorships in psychology up until 1937, after which time it mattered much less. To test the extent to which a change in appointment policy can be documented in biology, I divided biologists who were appointed full professors between 1933 and 1945 into three groups based on the date of their appointment (1933–1936, 1937–1940, 1941–1945). I then compared these groups in regard to their average scientific achievements (measured by the number of citations in the SCI 1945–1954) and the percentage of party membership (Table 2.4).

Owing to the small number of appointments to full professorship in the various groups, statistically secure statements are not possible. Nevertheless, we can note as a trend that the results do not speak for a change in appointment policy as described above. Thus scientific performance was weakest not in the group of biologists appointed between 1933 and 1936 (an average of 62.5 citations per person), but in the group appointed between 1937 and 1940 (34.7 citations). Its subsequent rise among those appointed after 1939 (72.0 citations), could be explained—leaving aside the possibility of random fluctuations—by the greater importance attached to scientific qualifications for appointments after 1940. This period saw appointments to the universities in Posen and Strasbourg.

The data on the percentages of party membership in the three groups also do not indicate that appointments before 1936 were based more strongly on political considerations than was the case after this time. If we take into account only the joining of the party prior to the time of appointment, we find that 45.0% of biologists appointed full professors by 1936 were party members as compared to 75.0% of those appointed from 1937 on. The larger percentage of party membership after 1936 probably reflects the effect of the moratorium on new memberships prior to 1937: far fewer party members with appropriate qualifications were available than was the case after this date. If we disregard the timing of membership, the share of party members in the three groups is about equal (Table 2.4).

Political activity did not necessarily stand in opposition to professional quality. Otto Mangold, for example, full professor in Erlangen in 1933 and Hans Spemann's successor in Freiburg in 1937, was a recognized developmental physiologist. A member of the NSDAP since 1935, Mangold became rector of the University of Freiburg in 1938 (a post he held until 1940), and in this capacity he implemented National Socialist university policies. Hansjochem Autrum went to Berlin in 1937 as an assistant to Friedrich Seidel, taking over the position of the recently expelled Jewish assistant Ernst Marcus. Autrum, like Seidel a member of the NSDAP and the SA since 1933, became one of the outstanding sensory physiologists of insects. Among the eminent biologists who got involved with the ideology and politics of National Socialism and supported them in essential points were Konrad Lorenz and Kurt Mothes, who are discussed in greater detail below (in Chapters 3 and 2, respectively).

Especially during the first years of National Socialist rule, "old fighters" (people who had joined the party before January 1933) tried to influence questions of appointment or to advance themselves. Bergdolt and the NSDB's unsuccessful attempt to drive Karl von Frisch from the University of Munich is only one example. The efforts by Ernst Lehmann (Tübingen) to be allowed to set up and head an institute for "German biology" also failed to bear fruit (see Section 4 of this chapter). By contrast, the following examples show that in Austria the party took care of the old fighters who had been politically persecuted after 1933.

The NSDAP was outlawed in Austria in 1933. In 1934 Chancellor Engelbert Dollfuß set up an authoritarian Christian corporate state *(Ständestaat),* and a number of openly National Socialist as well as Social Democratic university teachers were dismissed (the SPÖ, the Socialist Party of Austria, was outlawed in 1934). The first group included full professors of biology: Paul Krüger, since 1929 professor of zoology at the University of Vienna, was found guilty in 1935 of "National Socialist activities and sympathies" and was dismissed on March 25, 1935.[17] He had been a member of the Austrian NSDAP since 1933. On July 1, 1935, he was called to the chair of zoology in Heidelberg, which had become vacant with the retirement of Curt Herbst. Friedrich-Carl von Faber had been full professor of botany since 1931, also at the University of Vienna. He joined the Austrian NSDAP in 1934,[18] and on December 1, 1934, he was appointed to the chair of botany at the University of Munich, which had opened up with the appointment of Fritz von Wettstein as director of the Kaiser Wilhelm Institute for Biology.

Othenio Abel, professor of paleobiology at the University of Vienna, was sent into "early retirement" in 1934, also for political reasons. In 1944, the dean of the faculty of philosophy at the University of Vienna wrote about Abel: "As rector and dean at the University of Vienna he was always to be found in the front line of the battle against the threatening Jewification and foreignization. He always worked vigorously and purposefully for the joining of Austria to the German Reich."[19] In 1935 Abel was called to Göttingen as professor of paleobiology.

In Graz, the associate professor Bruno Kubart was sent into temporary retirement in 1936. The Ministry for Internal and Cultural Affairs responded favorably to the application he submitted in 1938 for "compensation for the professional damages he has suffered in the struggle for the National Socialist awakening of Austria."[20] Kubart received retroactive payment of his salary for the period from March 1936 to February 1938 and was immediately returned to his old position.[21]

Cases in which the circumstances of appointments were analyzed through an evaluation of relevant documents show that the faculties were also after 1933 in most instances the key decision-making bodies when it came to appointments. To be sure, political considerations were entertained in the faculty meetings, but professional factors remained an important criterion, one that was in most cases accepted by the rector and the REM. In a few cases compromises between science and politics were already made when it came to the faculty's recommendation. While evaluations from the NSDB could block individual associate professorships and appointments,[22] as far as we know they could not force through the appointment of professionally unqualified university teachers against the will of the faculty or the REM. The SS, too, influenced appointments. In 1941 Hans Stubbe was to take over the chair of plant genetics at the University of Posen. The REM, in agreement with the rector of the University of Posen, wanted to appoint him on the basis of his qualifications in the field of genetics, even though his democratic past was known and the party chancellery had already rejected his appointment. In the end, only a negative evaluation by the security service of the SS prevented Stubbe's appointment in 1943.[23]

The following case illustrates the typical career hopes of active National Socialists, which often went hand in hand with the expulsion of Jewish scientists, as well as the considerations and criteria weighed by universities, the REM, and the party in making appointments.

3. The Chair in Zoology in Münster, 1935–1937

Following the dismissal of Leopold von Ubisch in 1935 for being a non-Aryan (his mother was from a Jewish family), a successor had to be found for the chair of zoology in Münster. Ubisch had been full professor of zoology in Münster since 1927. As a front-line veteran he was not dismissed in 1933, but he was the target of strong criticism, especially from students. In December the Prussian Ministry of Education prohibited him from giving any more lectures on heredity *(Vererbungslehre)*.[24] According to his sister, he was offered the chance to be "Aryanized."[25] When he declined, he was, in accordance with the Nuremberg Laws, sent into retirement at the age of fifty "for reasons of age" and emigrated to Norway.[26]

One of the leading enemies of Ubisch at the University of Münster was Heinrich-Jacob Feuerborn, who had been assistant at the zoological institute and associate professor without civil servant status since 1927. Feuerborn had engaged in little political activity before 1933. A former member of a Catholic fraternity, he was considered a supporter of the Catholic Center Party. After 1933 he worked actively on behalf of the goals of National Socialism. He became a member of the NSDAP in 1933 and joined the Regional Education Office (Gauschulungsamt) of Westphalia-North and the NSDAP's Office of Race Policy.[27] Every year he conducted training courses and lectures on heredity and racial biology in the work of the "Gauführer," the NSLB, the National Socialist Physicians League, and other organizations.[28] He was also active in the area of nature conservation; he headed the League for Nature and Heimat for the districts Westphalia-North and Westphalia-South. In the winter semester of 1935–36 he gave a lecture on the "biological foundations" of National Socialism.

Unless otherwise indicated, the following account is based on information in files concerning the appointment to the chair in zoology at the University of Münster from the Prussian State Archive (Preußische Geheime Staatsarchiv).[29] Beginning in 1933, Feuerborn tried to get rid of Ubisch and take his place as head of the institute. He was supported especially by leading party members of the Westphalia-North district, who called attention to Feuerborn's outstanding work in the field of local history and geography and local biology (*Heimat-kunde* and *Heimatbiologie*). Feuerborn considered his attempt to drive out Ubisch not a personal matter but a struggle for German honor, which National Socialism had once again elevated to a central concern. For the same reason he also informed the students about matters relating to the institute.[30] The biologists in Münster were divided, some supported Ubisch, some stood behind Feuerborn. Ubisch threatened to leave unless Feuerborn were transferred; in 1935 Feuerborn was sent to Braunschweig for three months.

Following Ubisch's forced "retirement," his former student Curt Kosswig, associate professor at the Technical College in Braunschweig, was asked to fill the chair temporarily. Kosswig did not belong to the NSDAP; he had joined the SS in 1933 but had left again after a short time. In Münster he took the side of his former teacher. After the head of the NSDB had informed the REM that Kosswig had gotten jobs in Braunschweig for Ubisch's assistants Altrogge and Hans Breider, who had been transferred away from Münster because of their support of Ubisch, he was no longer considered as a successor to Ubisch.[31] The rector had intended for Feuerborn to succeed Ubisch. The faculty's list of candidates, however, contained the following names: Professor Otto Koehler (Königsberg), Professor Wilhelm J. Schmidt (Gießen), lecturer Karl Henke (Göttingen), associate professor without civil servant status Friedrich Seidel (Königsberg), associate professor without civil servant status Hermann Giersberg (Breslau), and associate professor with civil servant status Curt Kosswig (TH Braunschweig). At the instigation of the expert *(Referent)* in the REM, Konrad Meyer, who had the full support of Walter Mevius, professor of botany in Münster, the faculty added Hermann Weber, associate professor without civil servant status, to its list.

The faculty's letter to the REM, which accompanied the appointment list, closed with the statement: "The faculty could not bring itself to place the local associate professor without civil servant status Feuerborn onto the list. His scientific accomplishments are not sufficient."[32] In a letter to Meyer, Mevius elaborated on this decision by saying that Feuerborn was scientifically "utterly rejected" by all important zoologists (such as Wolfgang Buddenbrock-Hettersdorf, Richard Hesse, and Max Hartmann).[33] More than twenty evaluations had been negative, "and accomplishments should, after all, be the first principle in our state."[34] Moreover, Feuerborn had forfeited the sympathies of "nearly all respectable colleagues in the faculty" because of the way in he had tried to impose himself on the faculty as successor to Ubisch. As a third argument against Feuerborn, Mevius explained that because Feuerborn was an old Catholic Center Party supporter, his attitude to National Socialism was endangered by the "many black [Catholic] secret channels" in Münster. "One must not bring any Catholics to Münster. Only old enemies of the Catholic Center Party belong here."[35]

The rector reversed himself and endorsed the faculty's position with regard to the recommended candidates. One thing that mattered to him was the fact that "good National Socialists" had also voted against Feuerborn in the unanimous faculty evaluation. Similarly, the leader of the NSDB in Münster withdrew his initial support for Feuerborn. In 1936 the REM appointed Hermann Weber, member of the party since 1933 but a recognized zoologist, as professor of zoology in Münster. The same year Feuerborn was asked to temporarily occupy the chair of forest zoology in Freiburg; in 1939 he became associate professor of zoology without civil servant status at the University of Berlin. He was dismissed in 1946.

The following discussion of Ernst Lehmann also deals with a biologist who, on the basis of his efforts on behalf of Volkish and National Socialist goals, nursed hopes for a special kind of career. Lehmann presented himself as the pioneer of "German biology." The account of his professional and political career will therefore center on the question of what kind of grip ideology came to have on biology as compared with "Aryan physics."

4. "German Biology": The Example of Ernst Lehmann

Owing to the central importance the concept of race had for National Socialist ideology and politics, one might easily be led to assume that the research and teaching of biology had a strong ideological component. A look at biological-ideological journals like *Der Biologe* as well as some monographs by prominent biologists could reinforce this notion, for here we do find numerous statements from biology professors in various fields who seek to establish a connection between National Socialist ideology and biology. Biological ideas of holism (the whole is more than the sum of its parts) were seen as a model for ideas of volkish unity: the volkish community or the race was regarded as a biological entity in which the individual was of no value and from which "alien" *(Artfremde)*, genetically diseased, and "parasitic" members had to be removed, eradicated. I shall mention only two examples. The zoologist and holist Friedrich Alverdes tried "to expand and develop the biological perspective of holism on the basis of what we have learned in our fatherland through the rise of the idea of totality" (1935, p. vii). The theoretical biologist Ludwig von Bertalanffy saw his "organismic biology" as a counterpoint to the "analytic-summationary and machinistic-theoretical attitude in biology." He believed that an "organismic age" had begun to replace the technological age, and that the "hope that the atomizing conception of state and society might be followed by a biological conception that recognizes the wholeness of life and of the Volk" had now (in 1941!) been fulfilled (Bertalanffy 1941, pp. 248, 343).

However, within biology no initiative on a larger scale attempted to separate a "German biology" from a "Jewish" or international biology with regard to content and to establish it at universities. The following account reviews the only attempt—by Ernst Lehmann—to found a "German biology" and analyzes the reasons for its failure.

4.1. Plans for an Institute for "German Biology"

Ernst Lehmann, appointed in 1922 professor of botany at the University of Tübingen with a focus on genetics and in 1931 chairman of the German Association of Biologists, wrote in 1934:

> It is truly admirable what the biological will has accomplished since January 30, 1933. To be sure, some positive things had already been done earlier. We can think, for example, of selection by the SS, marriage permits for genetically healthy, superior, Aryan people . . . Soon after the *Machtergreifung* [seizure of power] there began the successful fight against unemployment and its catastrophic consequences for peoples' lives. Then the government started with the practical cleansing of the Volk from people of foreign races. The laws for the restoration of the professional civil service were passed. Despite the most careful implementation, wounds were undoubtedly inflicted in the process, some of them surely unnecessary. This was the last chance to save Germany, which lies so close to the compact

settlement areas of the Jews in the East, from becoming utterly Jewified; hence the individual had to take second place behind the totality . . . We have freed our Volk from foreign races. And if we work
on building up anew its racial composition, the next task, which
arises inevitably from our basic biological knowledge, is to compensate within the ranks of our own Volk for the struggle for survival
that operates freely in nature and thus to take the eugenic measure
of sterilization. (Lehmann 1934a, pp. 37–38)

Lehmann, echoing a statement by Bavarian Minister of Culture Hans
Schemm, to the effect that National Socialism was applied biology,
demanded: "Today, in view of the extreme existential affliction of our
Volk, there arises before the German biologists as his most pressing
and greatest task that of sharply formulating the laws of life from
which our German Volk lives and from which the National Socialist
Weltanschauung is born" (Lehmann 1937, p. 341).

It was not only enthusiasm and opportunism during the first years
of National Socialism that prompted Lehmann to make such statements. They were the result also of years of committed work in the
volkish movement.[36] He had been active in the Volkish Protection and
Defense League (Völkischer Schutz- und Trutzbund) since 1920 and
was also a member of the Pan-German League (Alldeutscher Ver-
band).[37] According to his own statements, "at a time when biological
work in Dahlem was still largely in the hands of Jews—I call to mind
the names Goldschmidt, Stern, Jollos, and Brieger alongside Mr.
Muckermann—. . . [I established] the Association of Biologists in Tübingen in 1931 on a purely volkish basis with its journal *Der Biologe*
in the volkish publishing house J. F. Lehmann."[38] With what were
surely not unjustified expectations he was hoping for a special career
after 1933: "In 1930 the *N.S. Monatshefte* called for the establishment
of a professorship in national biology. It seems to me that I have laid
the groundwork for such a position."[39]

In the winter semester of 1932–33, he gave a lecture entitled "Biology in Contemporary Life" for students of all faculties. In it he propagated his notion of the importance of biology as the foundation of
National Socialist ideology. He spoke of the task of promoting the
Nordic race because it was, owing to its constitution, more strongly
threatened by race mixing than other races (Lehmann 1933a, p. 225).
Lehmann also signed as the author of the declaration of Tübingen

university teachers in support of Hitler and Hindenburg. At the beginning of 1933 he applied for membership in the NSDAP.

From 1933 to 1937 he was dean of the faculty of natural sciences, and in this capacity he requested in 1934 that the publisher and bookseller Julius F. Lehmann (no relation) be awarded an honorary doctorate by the faculty, which was done. He justified his request by pointing out that Lehmann's national-volkish publishing house had issued numerous publications on the topic of racial science *(Rassenkunde)*, especially the "racial works" of Hans F. K. Günther.[40] In 1935 he wrote about his "fierce struggle" in the twenties to prevent the only Jewish lecturer in Tübingen—"for Tübingen has always known how to keep away Jewish professors without saying much about it"—from becoming head of the local Association of Associate Professors (Nichtordinarienverband) (Lehmann 1935).

Lehmann supported the political demands and measures of the National Socialists beyond the first years of the Third Reich. In the forties he was still celebrating the "achievements" of National Socialism—compulsory sterilization and the Nuremberg Laws—as biological necessities. And he did not shrink from making a connection between these measures and the humanist Gregor Mendel. For example, on the one hundred and twentieth anniversary of Mendel's birthday, he wrote:

Anybody who attempted today to survey the impact of what has been done in this country based on the laws of heredity and insights about the human race, would stand quietly in awe before the greatness of what has been achieved and is being aimed for. Work has begun in countless fields, always based on the laws formulated by Mendel ... What the wider population could relate to most easily in the beginning were the eugenic efforts, which expanded into racial hygiene and were aimed at protecting human society from the burden and misery of hereditary diseases ... The study of human constitution with respect to the given hereditary base and the recording of the genetic foundation of the traits of different human races were in equal measures a result of these insights. Individuals have worked and are working tirelessly on unraveling the innumerable problems, and where necessary and possible the results that arise from their efforts are being put on a legal foundation [*Judengesetze*, anti-Jewish laws]. (Lehmann 1942a)

Lehmann's great hopes to rise in the Reich as the representative of "German biology" ended with his being suspended from his professorship in 1937. Moreover, the party whose favor he had curried for many years did not accept him. In July of 1945, Lehmann described himself to the De-Nazification Committee as a victim of National Socialism. How can we explain the sudden end of Ernst Lehmann's promising career and with it that of "German biology"?

Lehmann ran into his first problems with the new state when he tried to become a member of the NSDAP. His first application for party membership was turned down on May 13, 1933, with the explanation that he was of Jewish descent and had also belonged to a (Masonic) lodge.[41] His non-Jewish descent was quickly verified. A second application on April 22, 1937, was turned down with reference to his former membership in a lodge.[42] After a decree from Hitler on April 27, 1938, permitted former lodge members to join the party, Lehmann continued his vigorous efforts for admission. His admission in 1938 by the local group for Württemberg—Hohenzollern was shortly afterward declared invalid by higher party authorities.[43] In letters to party and SS offices, Lehmann tried in vain to clarify possible motives for this rejection.[44] He wrote petitions to Rudolf Heß,[45] Himmler, and Hitler,[46] but had no success with these men, either. It is likely that the reason for his rejection after 1938 had something to do with the disciplinary proceedings that had been brought against him.

Lehmann's plans for building an institute for "German biology" also led nowhere. As he saw it, one of the essential tasks of the planned institute would be to bring "biological ideas to the German people."[47] In his view, a large number of biological questions "with regard to race" were already being taken care of by branches of the Office of Race Policy, for example by Karl Astel's institute at the University of Jena. For that reason he believed it was necessary that "biological thinking" be promoted in other areas—such as the field of evolutionary doctrine (in opposition to the teachings of ecclesiastical circles, which rejected the notion of the evolution of organism)—with regard to the "attitude to many essential questions of the life of the Volk, for example the Volk's will to live and the related question of children, the movement for the protection of nature and animals." He intended to reach the German people in the schools, through his work as head of the Biology Section in the NSLB and through contacts with the student body and with National Socialist organizations such as the Hitler Youth and Kraft durch Freude (Strength through Joy). He also believed it was necessary to oppose the "circles that are once again paving the way for a purely mechanistic biology . . . ignoring National Socialist concerns and supported by Jewish world science."

It becomes apparent that Lehmann's ideas of a "German biology" were not in the least concrete. They contained some of the common prejudices of the time, such as the rejection of a mechanistic science, which was often used to characterize so-called Jewish science. We can assume that he was also hinting at the director of the Kaiser Wilhelm Institute for Biology, Fritz von Wettstein. Although known not to be a National Socialist, Wettstein had, to Lehmann's great annoyance, influence on the decisions of the Reich Education Ministry, in such matters as appointments in biology, one reason being that he was considered an internationally recognized scientist.

From the political quarter Lehmann got a lot of support from Hans Schemm. In 1929 Schemm had founded the radical, ideologically oriented NSLB, before becoming Bavarian minister of culture in 1933. According to Schemm, who based his conceptions largely on Lehmann's plans, a "uniform biological basis" was to be established in Munich for all faculties at the university and the TH. For this purpose the Institute of Botany was to be remodeled, since it was temporarily without a director after Wettstein's appointment as director of the Kaiser Wilhelm Institute for Biology.[48] Wettstein reported that Lehmann, who was not on the faculty's list of recommended candidates, visited the institute in 1934, made far-reaching suggestions for changes, and voiced the opinion that he, as the führer of the Association of Biologists, should be given one of the major professorships.[49] In Lehmann's view, the takeover of the institute in Munich and its transformation into the center of "German biology" failed because Schemm died at the beginning of 1935.[50] There are legitimate doubts about this version, since Wettstein's successor in Munich, Friedrich-Carl von Faber, had already been appointed on December 12, 1934. Moreover, Schemm's influence on scientific affairs at Bavarian universities began to decline sharply as early as May 1, 1934, a result of the centralization of scientific administration in the new Reich Education Ministry, where the man of growing influence was Bernhard Rust.

Lehmann did not abandon his plans right away. But he failed to establish an institute for "German biology" even at the University of Tübingen, where he had set up "workplaces" *(Arbeitsstellen)* for the German Association of Biologists as well as for the Biology Section of in the NSLB.[51] Evidently those in charge at the Reich Education Ministry took no interest in his plans. Moreover, it is very likely that his plans were also thwarted by the head of the Dozentenschaft (Association of University Teachers), Robert Wetzel, a man who had similar biological-political ambitions as Lehmann and who, as a member of the SS and the Sicherheitsdienst (Security Service, SD), had the power to eliminate him as a rival (see Section 4.3 of this chapter).

4.2. "German Biology" and "German Physics"— A Comparison

Lehmann's plans for an institute constituted the only serious attempt to establish a "German biology" within existing university structures. A few comments about "German physics" will provide a basis for a comparison with biology.[52] The controversy about the subject matter of physics, which passed into "German physics," began in the 1920s. Both Philipp Lenard, the founder of "German physics," and Johannes Stark, one of its main proponents under National Socialism, were eminent scientists who had won the Nobel prize in 1905 and 1920, respectively. They equated "German physics" with "Aryan physics" and divided "Jewish" from "Aryan" science by subject matter and methods. In the traditional fields, "German physics" could not be distinguished from generally accepted physics. "German physics" emerged from the hostility of its supporters to the theory of relativity (linked to Albert Einstein) and, in a lesser degree, to quantum mechanics.[53] Along with their shared political and ideological views, the advocates of "German physics" were also united by their rejection of what were at the time the modern currents in physics. Some of the objections to the theory of relativity and quantum mechanics, such as those about the high level of mathematics and abstraction they involved,[54] corresponded to common clichés about "Jewish thinking." This division of physics happened despite the increased role of mathematics in other areas of physics and the fact that many proponents of "Jewish" physics were not Jews.

In contrast to physics, the field of biology produced no movement in which prominent representatives, on the basis of their professional opposition to a so-called Jewish science, brought forth a "German science" of any appreciable influence. There were biologists who published scientifically marginal articles in the *Zeitschrift für die gesamte Naturwissenschaft* (Journal for all natural sciences), which was considered the unofficial organ of "Aryan" physics. This journal had been founded in 1934, with the following goal: "To overcome the positivist and specialist scientific thinking, and in so doing eventually to rebuild the natural sciences from the perspective of a worldview that a German would see as truly alive and that is more than the sum of specific individual facts."[55] Publications by biologists in this journal tended to be general in nature: for example, Wilhelm Troll, "Die Wiedergeburt der Morphologie aus dem Geiste deutscher Wissenschaft" (The rebirth of morphology from the spirit of German science, 1935/1936), Hermann Weber, "Lage und Aufgabe der Biologie in der deutschen Gegenwart" (The situation and task of biology in the present time in Germany, 1935/1936), Hans André, "Der verhaltensgegensätzliche Aufbau der Pflanze im Lichte der biologischen Feldtheorie" (The counterbehavioral structure of the plant in light of biological field theory, 1935/1936), and Eduard May, "Dingler und die Überwindung des Relativismus" (Dingler and the overcoming of relativism, 1941). While the morphologist Troll and the zoologist Weber were recognized biologists, André and May were scientifically insignificant. However, leaving aside May and Ernst Bergdolt, who, as mentioned above, belonged to the circle of anti-Semitic university teachers in Munich,[56] and was one of the main proponents of "German physics," these biologists were not organizationally united.

Despite his many efforts to attain national prominence with "German biology," Lehmann, too, remained an insignificant and isolated case. We can draw superficial parallels between him and Stark. Both had an exaggerated need for recognition, and both tried to build a position of power in the National Socialist scientific organization through close ties with strongly ideologically oriented supporters of National Socialism. But Lehmann's "German biology" was not characterized by an opposition to the content of what was at the time modern biology, and his effort to establish an institute for "German biology" failed in its early stages.

Lehmann was a recognized plant geneticist,[57] though in the area of cytoplasmic inheritance he advocated anomalous positions: his thesis that genes were to be found exclusively in the cell nucleus contradicted nearly all important plant geneticists in Germany, who considered it a proven fact that genes were also located in the cytoplasm. Lehmann tried to explain differences in reciprocal crossbreeds not through plasmatic genes but by assuming the existence of "blocking genes" *(Hemmungsgene)* in the nucleus. Their activity supposedly led to the formation of specific growth hormones that were passed on by the cytoplasm, and thereby to the observed differences of growth retardation of the *Epilobium* hybrids in cases of reciprocal crossbreeds (see Chapter 3, Section 1.2, where I discuss Lehmann et al. 1936 in greater detail). As described in Chapter 3, he believed he had proved the existence of a pleiotropically acting retardation factor in the nucleus (Lehmann 1942b). This belief, as well as his clinging to the "monopoly of the nucleus" in hereditary transmission proved to be untenable. His theories were unequivocally refuted by Peter Michaelis (1938) and Heinz Brücher (1938), who was Lehmann's doctoral student.

Since Lehmann did not accept the existence of cytoplasmic inheritance, he stood for many years in conflict with Otto Renner, the professor of botany in Jena, and he felt slighted by what he believed to be the domination of German plant genetics by Wettstein, whose plasmon theory he rejected.[58] Wettstein was publisher of the *Zeitschrift für induktive Abstammungs-und Vererbungslehre* (Journal for inductive evolution and genetics), where, in 1938, Brücher was allowed to publish work whose findings diverged from Lehmann's results.[59] In contrast to the quarrels among physicists, the scientific disagreements between Lehmann and his colleagues were of no importance for his "German biology" in terms of subject matter.

A stronger movement for "German biology" probably failed to develop because of the state of biology at the time. Unlike physics, biology had no modern subfield in the twenties and thirties that could have been set apart from "German" biology in terms of content and defamed as "Jewish" in practice. The field of biology in which mathematics played the biggest role was undoubtedly the genetics of Mendel and the school of Thomas Hunt Morgan. Many important biologists, not only morphologists but also physiologists and even geneticists, initially looked upon it with a great deal of skepticism. Gerta von Ubisch wrote of the widespread attitude of physiologists toward genetics before World War I: "The fact that results could be attained only with statistical methods, that is, the fact that large numbers were necessary, that only probability calculations yielded unimpeachable results, was unpalatable to proper plant and animal physiologists, indeed it was uncanny—it was inexact. They wanted to get their results without mathematics."[60] According to Hans Kappert, who in 1931 succeeded Baur as professor of genetics in Berlin, the reception of Morgan's theses in Germany in the twenties was often "admiration mixed with a certain unease. 'Such crassly materialistic ideas could only be developed in America'—this was a remark at the time in an otherwise quite positive lecture which, I think, was given by Renner. It was simply the case that among the German biologists, vitalism had emotionally not quite been overcome even after the First World War" (Kappert 1978, p. 46).

Genetics as a separate subject was not established at the universities until the mid-1930s, with the exception of Baur's Institute of Genetics in Berlin (founded in 1911). But since the representatives of National Socialism in the party and the state legitimated their racial doctrine with the help of scientific insights of Mendelian genetics, this field could never have been rejected as "un-German" or as a "Jewish science."

In making the comparison with physics, we must bear in mind one fundamental difference between the two sciences: unlike physics, biology does not possess a firmly established body of theories that would have been affected by decisive innovations in terms of content and method in one particular area of biology. By contrast, the theory of relativity and quantum mechanics profoundly unsettled all of physics as then known and led to a new, comprehensive structure of theories. Despite this difference, developments in biology had a strong impact on the content and the methods of research that was being carried on at the time. For example, molecular biology, which was developing in the forties, led to a profound change in the objects and questions in genetic research. These innovations affected also other areas of experimental biology, among them developmental physiology. The hypothetical question arises: Would there have been an "anti-molecular German biology" on a scale comparable to "Aryan physics" if Jewish scientists had played a substantial role in the development of molecular biology in the twenties?

4.3. The Disciplinary Proceedings against Lehmann

Lehmann was accused of a breach of his official duties and had to submit to an inquiry in 1938, which ended in disciplinary proceedings.[61] As a result he was suspended from his position as university teacher in 1938. Presumably Robert Wetzel headed the campaign against Lehmann, trying in this way to eliminate him as a rival in the field of political biology. A former student of Lehmann's and evidently an influential member of the party commented: "It was Professor Wetzel to whom Lehmann, as the only truly biologically oriented natural scientist in Tübingen, was a thorn in the side. It was he who on several occasions passed off Lehmann's ideas as his own and who later announced Lehmann's plans for the creation of an Institute for German Biology in Tübingen as his own. And it was he who, in order to be the preeminent biologist in Tübingen, tried to bring about Lehmann's downfall with every conceivable means. After all, decisive for all the authorities were his evaluations of Lehmann, which he passed directly to the Reichssicherheitshauptamt [Reich Security Main Office]."[62] This background also explains why the disciplinary proceedings dragged on for so long (until 1943). At the end of November 1939, Lehmann was denounced to the Gestapo. And while the Gestapo established that Lehmann had engaged in "activities that undermined the Volk," it did not consider it necessary to initiate further measures against him.[63] At the beginning of 1943, Lehmann was fully restored to his offices through a decree of the REM. However, the rector, in agreement with the pro-rector, the head of the Dozentenschaft (Wetzel's successor), and the dean, declared that he was unable to comply with the decree. In response, the REM withdrew the decree; Lehmann's suspension was lifted, but he was prohibited from heading the institute and administering examinations.[64]

4.4. De-Nazification

We note that Lehmann's difficulties after 1933 had nothing to do with a rejection of or even opposition to National Socialist ideology and practice. On the contrary, Lehmann was strongly devoted to this ideology and publicly advocated and supported the measures of National Socialist policy. His problems resulted from his ambition to establish himself as the chief representative of "German biology," his poor choice of a political ally (Hans Schemm), and his threat to the career of a rival colleague (Wetzel) with similar ambitions but greater means of political power. In addition, he face opposition to his standing as an academic teacher and a director. All this led to disciplinary proceedings against him and to his temporary suspension, which contributed to the failure of his ambitious goals.

Lehmann was dismissed in 1945 on orders from the French military government. He resisted this move, believing that the new government should make restitution for his suspension and his political persecution under National Socialism.[65] His attempts to present himself to the de-Nazification proceedings as someone who had been politically persecuted are quite remarkable. To throw some light onto his line of argumentation, I will contrast excerpts from his publications and letters after the collapse of the Third Reich with earlier statements.

Lehmann in 1938 on Friedrich Oehlkers, whom he had brought to Tübingen as an assistant: "Shortly afterward he married a woman who was fully Jewish [*Volljüdin*] . . . Since I had not known anything about this relationship, and since Oehlkers knew my anti-Semitic attitude, after this a relationship between me and Oehlkers never developed, and even less so between me and Mrs. Oehlkers."[66] (Oehlkers left the institute in 1928.)

In 1945: "One thing that may have contributed to [the charge of his being of Jewish descent] was the fact that I earlier made Professor Oehlkers—in spite of his Jewish wife—my assistant and an associate professor."[67]

In 1934: "But the German biologist demands emphatically that when it comes to shaping our Volk, we use the weapons that biology offers . . . The work is about the preservation and, if necessary, the advancement of the great races of our people, so that they—by eliminating everything foreign—shall lead in noble harmony to a Volks-biological wholeness" (1934b, p. 142).

In 1937: "National Socialism has put into effect the insights of the laws of life" (1937, p. 340).

In 1945: "However, since I had already been prominent as a *biologist* for decades, and since biology—although in a distorted manner—was being claimed by National Socialism as its preserve, I had *to conduct a struggle on all fronts*."[68]

Circa 1940: "If anybody wanted to reproach me for something, it could only be that I have tried passionately, to the very limits of my strength, to put my science in the service of the volkish idea, and then, after Hitler had shown the way, of the National Socialist idea and the National Socialist state."[69]

In 1945: "Now, I did not understand why, *despite these bitter struggles against the party as a nonmember,* I was again suspended on the basis of the regulations passed by the military government. After what I have described, the reasons could not be political. My biological efforts have nothing to do with politics."[70]

In 1935: "However, we German biologists, in addition to agreement with the greater picture, feel a harmony in a more particular way. 'National Socialism is politically applied biology,' Schemm said when the paths were being charted on which our German Association of Biologists could be affiliated with the NSLB. These are words that enable us to realize how Schemm conceived of biology as the core of National Socialist education. And so biological thinking gave rise for him to the racial idea as the self-evident foundation of National Socialist ideology."[71]

In 1946: "And now to the NS Teachers League. The *Volksschule* teacher Schemm undoubtedly had the intention of comprehending the meaning of biology; but he, too, lacked deeper positions of his own on biological questions, positions based on knowledge" (1946, p. 31).

In 1934: "To be sure, some positive things had already been accomplished earlier. We can think, for example, of selection by the SS, marriage permits for genetically healthy, superior, Aryan people" (1934a, p. 37).

In 1946: "However, today we are agreed that breeding goals, such as those which the SS formulated and substantially pursued, for example, in the 'Lebensborn' program, are not the goals of the German Volk" (1946, p. 82).

In 1937: "Only when the preservation of future generations is guaranteed through sufficient and well-educated children, are all preconditions for the successful waging of total war met . . . The army and the navy have their roots in the fatherland like the oak tree in the German soil."[72]

In 1946: "Based on this kind of thinking [a politically oriented biology], biologists agreeable to the party—and if possible militaristic ones to boot—had to replace other inconvenient ones right up to the end of the NS regime" (1946, p. 33).

In 1934: "Then the government started with the practical cleansing of the Volk from people of foreign races . . . This was the last chance to save Germany, which lies so close to the compact settlement areas of the Jews in the East, from becoming utterly Jewified; . . . We have freed our Volk from foreign races. And if we work on building up anew its racial composition, the next task, which arises inevitably from our basic biological knowledge, is to compensate within the ranks of our own Volk for the struggle for survival that operates freely in nature . . . and thus to take the eugenic measure of sterilization" (1934a, pp. 37–38,

In 1946: "Countless training lectures were given on racial science—'racial biology' . . . It was all based on the notion, implicit or indicated with a few words, that the matter of the genetic behavior of the human race was as simple as what happened say with the crossing of yellow and green peas. People thus believed that with the help of the simplest basic insights of genetics they could conquer the 'racial heavens.' However, this approach destroyed or masked what is in fact the true task of biology" (1946, p. 58).

On October 27, 1949, the State Commission on Political Purging (Staatskommissariat für die politische Säuberung), of Tübingen-Lustnau, pronounced the following decision in the matter of professor Ernst Lehmann: "The person in question is exonerated. The costs of the proceedings are borne by the government."[73] The explanation presented Lehmann as a victim of the SS, especially of Robert Wetzel, and with regard to Lehmann's position after 1933 it reached this astonishing conclusion: "This situation bears a certain similarity with that of someone accused in a political criminal trial; nobody will hold it against the latter if he seeks to defend himself by protestations of his political reliability and loyalty to the regime."[74]

4.5. Postscript, 1950–1955

The following account of the reactions to the granting of emeritus status to Lehmann is based on the files in the archives of the University of Tübingen.[75]

Lehmann was retired in 1949. In 1950 the cabinet of Württemberg-Hohenzollern, by an act of grace, granted him the legal status of a professor emeritus. Following protests from the university administration and the faculty of natural sciences, this decision was overturned. In 1952, however, it was implemented. Once again members of the faculty protested, among other reasons because Lehmann, as an emeritus, had the right to be listed in the university catalog. The small senate of the university then decided on February 11, 1954, that Lehmann should refrain from making use of this right. Lehmann filed a complaint with the Ministry of Culture of Baden-Württemberg. When the decision of the small senate was rescinded for technical legal reasons by decree on July 27, 1954, Lehmann had to be included in the catalog.

One of the professors who had protested against granting Lehmann the rights of a professor emeritus was Erich Kamke. As an associate professor of mathematics at the University of Tübingen, he had been dismissed in 1937 for unknown reasons (he was sent into retirement). After 1945 he was made full professor in Tübingen. Kamke made the following comments on the decision of the Ministry of Culture and its accompanying justification in a letter to the rector on November 15, 1954:

> The justification is purely formal in nature and proceeds from false assumptions in so far as it is based on customs and regulations that were passed at a time when nobody could have anticipated that a university teacher could behave like Mr. L. If one wanted to accept the formal justification, one would also have to say that the murder of millions of people during the NS period was legal (especially since even a minister of justice, on June 30, 1934, declared the murders to be "lawful"), and that those executed after July 20, 1944, are contemptible.

Kamke, along with Max Hartmann, Alfred Kühn, and Georg Melchers, requested that their names be deleted from the university catalog. The three biologists who joined him in this action were also directors at the Max Planck Institute for Biology. When the rector refused to persuade Lehmann to allow his name to be omitted, Kamke informed the rector that he would submit a request for retirement to the Ministry of Culture. Eventually a compromise, whereby Lehmann's name appeared in the catalog with the added comment "does not teach," was accepted by all parties concerned, except Lehmann, who tried (in vain) to sue for additional "rights."

We now move on to an examination of the influence of party and state on the promotion of scientific research in general. Which research projects and persons received support? And was this support dependent on political requirements? To answer these questions I will first introduce the most important organizations for the promotion of research, the German Research Society and the Reich Research Council.

5. The Notgemeinschaft (Emergency Association) of German Science, the German Research Association, and the Reich Research Council under National Socialism

The Emergency Association (Notgemeinschaft der Deutschen Wissenschaft) was founded on October 30, 1920, for the purpose of maintaining and promoting German scientific research, which was being seriously affected by the dire economic situation at the time. Five Academies of Science, the Association of German Universities, the Kaiser Wilhelm Society for the Advancement of Science, the German Association of Technical-Scientific Clubs, and the Society of German Natural Scientists and Physicians all participated in the founding. The Notgemeinschaft was set up as an incorporated society, that is, as a self-governing body. The former Prussian minister of culture, Friedrich Schmidt-Ott, headed the association from October 30, 1920, to July 23, 1934. It received financial support from the Reich, the states, and industry, and it handed out grants for individual research projects and scholarships for up-and-coming scientists.

The Notgemeinschaft was, together with the Helmholtz Society for the Advancement of Physical-Technical Research, the most important funding organization for German science. Its support and grants were of particular importance for experimental research at the universities because most university professors, unlike the directors of KWIs, got little basic support for research. Almost all of the money for research at universities came from funding agencies.

The Notgemeinschaft was "coordinated" *(gleichgeschaltet)* in 1934. On June 1, 1934, Bernhard Rust became Reich and Prussian minister for science and education, shortened to Reich education minister; on that day the Department of Culture in the Reich Interior Ministry passed into his jurisdiction. That same month Rust dismissed Schmidt-Ott from his position as president of the Notgemeinschaft and appointed as his successor Johannes Stark, the Nobel laureate and anti-Semitic cofounder of "German physics." A retroactive vote by mail among the members of the Notgemeinschaft—the relevant letters were sent out on July 17, 1934—gave this step at least the appearance of legality: all universities with the exception of the University of Munich voted yes (rectors acceptable to the Nazis had been appointed everywhere), four academies voted no, and the president of the Kaiser Wilhelm Society, Max Planck, abstained.

Stark introduced the führer principle that formed part of the *Gleichschaltung:* any arbitrary decision by the president and any interference by the REM was now permitted. In addition, he excluded Jewish scientists from being supported. In his opening speech in June of 1934, Stark emphasized economic self-sufficiency as the goal of scientific research: "The führer himself takes a lively interest in the organization of scientific-technical research. Given the current situation of the German Reich, he expects that this research will see its most important task in providing support for the economy, by helping to replace foreign raw materials as much as possible with domestic materials of equal quality, and by introducing products into the world market that are able to hold their own in competition by virtue of their quality or innovative nature" (Zierold 1968, pp. 174f.). In line with the führer principle, Stark appointed a new Central Committee at the beginning of 1935; it included, among others, the president of the Reich Health Office, Hans Reiter, and Professors Ferdinand Sauerbruch (surgery) and Alfred Kühn (zoology).

As a result of a quarrel between Stark and the REM, the income of the Emergency Association dropped from 4.4 million Reich marks (RM) in 1932 to RM 2 million in 1936 (Zierold 1968, p. 180). In Kurt Zierold's view Shark's dismissal in November of 1936 resulted from Stark's heavy preference for Alfred Rosenberg, who was already at that time relatively uninfluential, as well as Stark's underestimation of the importance of Himmler (1968, p. 212). On November 14, 1936, Rust appointed Rudolf Mentzel president of the Emergency Association.

Until 1937, SS Major Mentzel was consultant *(Referent)* for natural sciences in the REM, where he worked in an important capacity in the Office of Natural Sciences and became in 1939 *Amtschef* and ministerial director. In October of 1937 he changed the name of the Notgemeinschaft to the German Research Association (DFG). At the beginning of 1937, Rust established in addition to the DFG a Research council, which soon adopted the name Reich Research Council (RFR). It had the task of focusing certain branches of science on the goal set by the Four-Year Plan, to make Germany independent of foreign countries with respect to the most important raw materials, and to make the German economy ready for war in four years (Zierold 1968, p. 238). Hitler had announced the Four-Year Plan on September 9, 1936, and had appointed Hermann Göring as its plenipotentiary. General Karl Becker, holder of a doctorate in engineering, was appointed president of the RFR; he was given an important participatory role in drawing up and implementing that part of the DFG budget intended for the RFR and earmarked primarily for the natural sciences. The founding of the RFR involved the Kaiser Wilhelm Society as well, in so far as some of its influential members participated in decision making in the RFR. For example, since 1933 Becker had belonged to the senate of the KWG; his deputy in the RFR was State Minister Otto Wacker, who was shortly thereafter appointed first vice president of the KWG (Albrecht and Hermann 1990, p. 388).

The DFG and the RFR worked together very closely; the RFR remained organizationally dependent on the DFG and its operation was largely determined by Mentzel's decisions. Mentzel's friend Erich Schumann, head of the Science Section of the Reich War Ministry since 1938, was Becker's most influential adviser. The consultants *(Referenten)* of the RFR in the ministry also consulted to the DFG; the financial measures for implementing the appropriations for the RFR lay in the hands of the DFG. Mentzel, because he held the post of ministerial referent, later ministerial director, and because the advancement of science was incorporated into the Four-Year Plan, was much more successful than Stark in financial matters. He raised the Reich subsidies for the DFG from RM 2 million in 1936 to more than RM 7 million in 1937, and to RM 9 million in 1939. An important innovation introduced by the decree on the RFR was the appointment of scientists as heads of the specialized divisions (later sections) of the RFR. According to the führer principle, these heads decided all applications, could solicit expert reviews or choose not to, and were responsible for the promotion of research. Initially there were thirteen, later sixteen, of these specialized divisions. The vast majority of applications from biologists were decided by the following division heads: Professor Konrad Meyer from the Agricultural Science and General Biology Section, Professor Ferdinand Sauerbruch from the Medicine Section, and Professor Kurt Blome from the Hereditary Biology and Racial Care Section related to population biology. The grant files of biologists reveal that the Notgemeinschaft/DFG solicited expert reviews up until 1937, and that the reviewers were in most cases the same as those used prior to the *Gleichschaltung*. But after May 1937, with few exceptions, no more reviews are documented.

Konrad Meyer had an influential position within the RFR. After 1934 he was professor of agronomy and agrarian policy at the University of Berlin. He held a high rank in the SS and was also head of the Research Service (Forschungsdienst), which the agricultural scientists had combined to form in 1935 at the initiative of Richard W. Darré. Six Reich Working Groups (among them Crop Farming and Agricultural Chemistry) and four autonomous working groups (such as Fishery) were active within the Research Service. The business of the Agricultural Science and General Biology Section in the RFR was handled by the central office of the Research Service. This meant that behind Meyer in the RFR there stood a larger organization, which even had its own newsletter *(Forschungsdienst)*. After 1937, Meyer's section had at its disposal as much funding from the DFG or even more than all sections of the remaining natural sciences and technology taken together (Zierold 1968, pp. 233 and 234).

During the war, research had to be officially recognized as important to the state or the war effort in order to receive funding. (However, as will be shown below, this did not mean that only war-relevant or applied research still received funding.) One major problem was the shortage of materials; during the later years of the war almost every material was subject to controls. The Office of Wartime Economy of the RFR, housed within the DFG organization, assigned levels of priority and identifying numbers. There was priority level "S" and above that "SS." In July of 1942 Albert Speer, successor to Fritz Todt (who was killed in an accident) as armament minister, created the priority level "DE," reserving for himself the right to determine who would be admitted to this special designation. With each passing year the Office of Wartime Economy grew increasingly important for the scientists. It was, after all, easier to get funding from the RFR than it was to receive the highest priority level and an effective contract number from the Wehrmacht. Priority and contract, however, were necessary both for the procurement of materials and for having a scientist and his or her staff receive a "uk" (indispensable) classification, which exempted them from military service. As raw materials became increasingly scarce, it became more and more important for scientists to be admitted to priority level "DE," and consequently Speer acquired increasing power in the field of science.

To make the RFR more effective as an instrument for promoting scientific research, Speer ordered it reorganized in 1942; Hermann Göring was named the new president. A Planning Office for Scientific Research, set up in 1943 at Speer's suggestion, created, among other things, the precondition for finding a generous solution to the question of the "uk" classification of scientists (Zierold 1968, pp. 248f.). Professor Werner Osenberg, the head of the Planning Office (which was also placed under Göring's control), obtained "uk" classification for about fifteen thousand scientists (Zierold 1968, p. 250). Other than that, little changed in actual practice. The heads of the sections within the RFR remained largely the same, with three sections added in 1944, among them the Section for Biology headed by Professor Hermann Weber of Strasbourg. This section, however, had no more practical significance.

The Reich Office for Economic Development (Reichsamt für den Wirtschaftsausbau, RfW), was created within the framework of the Four-Year Plan as a section of the Ministry for the Economy and became an important agency for the promotion of research, especially in chemistry. A few biological projects were also supported by the RfW for a time during the war, but in 1944 most of them fell again under the jurisdiction of the RFR. The three branches of the military—the Army High Command (OKW), the Navy High Command (OKM), and the Air Force High Command (OKL)—financed major projects of applied research directly. The SS also carried out its own research projects. No figures are available for the money spent by industry to promote research. This support, especially from IG Farben, played in some areas a key role for biological research projects. For example, Alfred Kühn, in his report to the DFG in 1940, noted that he had received funds from industry (IG Farben) for his experiments with chemical mutagenesis and with animal viruses. With help from IG Farben, Adolf Butenandt, Kühn, and Wettstein began in 1940 to develop an institute for virus research in Dahlem (see Chapter 4, Section 2).

The Rockefeller Foundation supported a few larger projects by university institutes and the Kaiser Wilhelm Institutes in Germany. In

1932 it financed the rebuilding of the Institute of Zoology at the University of Munich. This support was not terminated across the board in 1933. Thus in 1934 the Rockefeller Foundation offered to support Kühn's work on genetically active substances in various animals. It is not known to me whether this support ever materialized.[76]

6. Funding for Biological Projects by the DFG and the RFR, 1933–1945, and the Significance of NSDAP Membership

How did support for biological projects by the Notgemeinschaft and the DFG develop after 1933, and what influence did the RFR have? To answer this question, I examined the amount of support given by the DFG and the RFR to biologists who submitted grant applications between 1930 and 1945. We should recall that experimental research at the universities was to a high degree dependent on outside support, while most of the Kaiser Wilhelm Institutes had their own research funds.

Prior to 1933, few individual grants from the Notgemeinschaft are documented. However, documents on provisional decisions concerning grants provide information about approvals prior to 1933.[77] According to these sources, biologists who form the basis of this study received RM 74,000 in 1930. The documents for grants after 1933 are also incomplete, but in most cases the grant approvals are continuously documented at least from 1937 on. The support for biologists rose from RM 300,000 in 1937 to RM 520,000 in 1944 (Figure 2.1). The number of approvals rose from 101 in 1937 to 135 in 1938, and declined to 80 in 1940 and 67 in 1944. As already mentioned, during the war all research had to be officially recognized as important to the war effort in order to have a chance of being supported. A process of concentration took place. Inflation did not change the value of the Reich mark between 1937 and 1944. To determine whether members of the party were given preference, I compared the amount of support for members and nonmembers for each year. A clear preferential treatment of party members emerges in 1934 (Figure 2.2). Beginning in 1935, however, the percentage of party members among the grant recipients corresponds roughly to the percentage of membership among all biologists who submitted an application to the DFG after

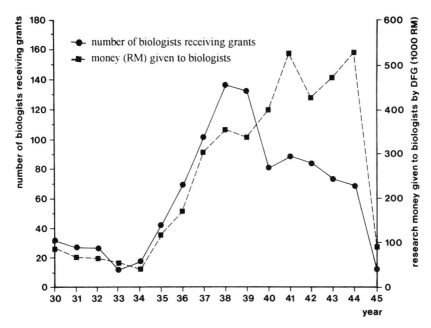

Figure 2.1 DFG/RFR funding of biologists between 1930 and 1945. Grants from January to May 1934, prior to the *Gleichschaltung* of the DFG (Notgemeinschaft), are listed under 1933. The data for the years 1930 to 1932 were taken from records about preliminary decisions on grants.

1933. After 1941 the percentage of party members among the grant recipients is even lower. Party membership was thus no prerequisite for financial support, and it did not represent a statistical advantage. Even if it can be assumed in some instances that scientists received funds because of their influential political position, on the whole it would appear that the research content of a proposed project and the scientific reputation of a scientist were of greater importance. Of course support was also given to people whose extraordinary scientific accomplishments went hand in hand with their political commitment to National Socialism. One example is Kurt Mothes. In 1935 he was full professor in Königsberg and the university teacher in botany who had the most research money at his disposal. Already at that time a good scientist in the field of the biochemistry of plants, Mothes, since 1933 a member of the NSDAP and later also consultant in the Office

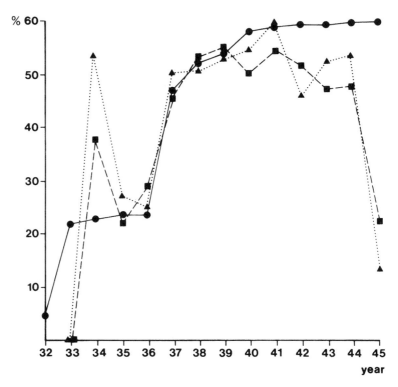

Figure 2.2   The relationship between membership in the NSDAP and the amount of DFG/RFR financial support. As in Figure 2.1, grants up to May 1934 are listed under 1933.

of Science in the NSDB in Königsberg, was considered an ardent Nazi.[78] He cultivated contacts with influential politicians, among them Erich Koch, the district leader of Eastern Prussia and Reich commissioner of the Ukraine, with whom he went hunting.[79] Through his policy of brutal Germanification, Koch was responsible for the extermination of hundreds of thousands of Poles and Jews.

7. Research Funding for Biologists at Universities and Kaiser Wilhelm Institutes

To determine whether universities and Kaiser Wilhelm Institutes benefited in equal measure from the rise in financial support, I totaled separately the DFG money and the RFR money (Figure 2.3). According to the documents on provisional decisions about grant proposals, between 1930 and 1932, biologists at the KWIs received in no year more than 11% of the sum that was given to biologists at universities. Biologists at the KWIs received 10% of the total financial support from the DFG and the RFR for biologists in 1932, 48% in 1941, and 49% in 1944. DFG funding for biologists at universities amounted in 1932 to about RM 55,000, in 1937 to RM 200,000, and in 1944 to RM 265,000: it thus rose by about a factor of five. KWI biologists received about RM 5,000 in 1932, RM 100,000 in 1937, and RM

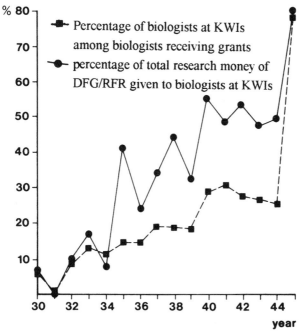

Figure 2.3 DFG/RFR funding for biological research at Kaiser Wilhelm Institutes as compared to universities. As in Figures 2.1 and 2.2, grants to May 1934 are listed under 1933. As in Figure 2.1, the data for the years 1930 to 1932 were taken from records about preliminary decisions on grants.

250,000 in 1944. This represents an increase by about a factor of fifty. In 1945 biologists at the Kaiser Wilhelm Institutes received 80% of the DFG support, but owing to the small number of grants (nine) in this year, a statistical evaluation is hardly possible. One curious fact should be mentioned, however. KWI biologists had more confidence in the continuing support of their research than did their university colleagues. The Kaiser Wilhelm Institutes received not only plenty of support from the DFG and the RFR but also a strong increase in financial support from the state. Owing to these subsidies, which came for the most part from special funds of the various ministries, especially the Ministry for Nutrition, the total budget of the Kaiser Wilhelm Society rose from RM 5.7 million in 1932 to RM 7.4 million in 1937 and to RM 14.4 million in 1944 (Albrecht and Hermann, pp. 377, 399).

After 1935, support for a biologist at a Kaiser Wilhelm Institute was, on average, more than twice that for a biologist at a university institute. In 1944, the two amounts differed by a factor of 2.9. The documents on applications that were turned down clearly reveal, even if they are not complete, that applications from biologists at Kaiser Wilhelm Institutes were less frequently turned down than those from biologists at universities. Between 1937 and 1939, 27 out of 326 applications (8.2%) originating in universities were turned down, as were 3 out of 71 (4.2%) from KWIs. From 1940 to 1941, there were 15 rejections out of 133 applications (11.2%) from universities, and 2 out of 52 (3.8%) from KWIs. After 1941, no further information about rejected applications is documented in the DFG files.

The Kaiser Wilhelm Institutes thus profited on the whole to a much greater degree from the rise in financial support for biological research than did the universities. Between 1932 and 1944, the share of the universities in the overall financial support for biologists declined drastically.

We might attribute the financial support of the Kaiser Wilhelm Institutes to the high regard, under National Socialism, for scientific research as a national task and the Kaiser Wilhelm Society as an eminent scientific institution best suited for fulfilling that task. First-rate biological research of economic and medical interest—such as work in genetics and radiobiology—was being carried out primarily at the Kaiser Wilhelm Institutes. Influential representatives of the RFR, such as Konrad Meyer, and of the Ministry for Agriculture, such as Herbert

Backe, supported the funding of this research, in some of which they had a personal interest. A strong increase in financial support can be noted not only for those directors of Kaiser Wilhelm Institutes who were members of the NSDAP (like Boris Rajewsky at the KWI for Biophysics and Wilhelm Rudorf at the KWI for Breeding Research), but also for those who were not (like Max Hartmann, Alfred Kühn, and Fritz von Wettstein, all at the KWI for Biology).

8. Research Funding and the Quality of Research

In Chapters 3 and 4, I analyze the content and results of biological research funded by the DFG and the RFR. Here I shall try to assess the quality and influence of this research by comparing the number of cited articles and citations in the Science Citation Index from 1945 to 1954 for a select group of biologists at universities and the Kaiser Wilhelm Institutes. Unless otherwise noted, the following analysis is limited to biologists who received grants from the DFG or the RFR between 1933 and 1945. Moreover, I have considered only scientists who, between 1937 and 1945, held chairs at universities (fifty-six individuals) or were directors or department heads at Kaiser Wilhelm Institutes (KWI for Biology with the Division for Virus Research, KWI for Breeding Research, KWI for Cultivated Plant Research, the genetics sections of the KWI for Brain Research, and the Hydrobiology Station in Plön)—seventeen individuals. The émigrés are thus not included. For each person I totaled the sums of support in the years 1937 to 1945, regardless of the year he was appointed director or full professor. I chose 1937 as the starting year because by this time the Jewish directors at the Kaiser Wilhelm Institutes had been replaced by non-Jewish successors from the universities.

8.1. A Comparison between Universities and Kaiser Wilhelm Institutes

The average number of citations per person varies little, from 96.5 at the universities to 110.8 at the Kaiser Wilhelm Institutes (Table 2.5). If we take the ratio of citations to articles as the criterion for the quality of research for the fifteen most frequently cited persons, three scientists stand out clearly above the rest (Table 2.6): Konrad Lorenz, Hans Bauer, and Gerhard Schramm. In 1973 Lorenz received

Table 2.5 DFG/RFR funding between 1937 and 1945 and the number of citations of all articles in the SCI 1945–1954 from biologists who were supported by the DFG/RFR between 1933 and 1945

	Univ.	KWG	Ratio of univ. to KWG
Biologists supported by the DFG/RFR			
Total	56	17	3.3
NSDAP members	27	9	—
Nonmembers	29	8	—
DFG 1937–1945 (RM)			
Total	818,139	899,191	0.9
NSDAP members	463,149	263,146	—
Nonmembers	354,980	636,045	—
Citations			
Total	5,402	1,866	2.8
NSDAP members	1,730	628	2.6
Nonmembers	3,672	1,258	2.9
Articles			
Total	2,642	836	3.2
NSDAP members	871	290	3.0
Nonmembers	1,771	546	3.2
Citations × 1,000 per RM			
Total	6.6	2.0	3.3
NSDAP members	3.7	2.3	1.6
Nonmembers	10.0	1.9	5.3
Ratio of citations to articles			
Total	2.0	2.2	0.9
NSDAP members	2.0	2.2	0.9
Nonmembers	2.1	2.3	0.9
Citations per scientist			
Total	96.5	109.8	0.9
NSDAP members	64.1	69.8	0.9
Nonmembers	126.6	157.2	0.8

Notes: KWG = directors and department heads of five Kaiser Wilhelm Institutes (see main text) between 1937 and 1945. Univ. = full professors who held a chair at a botanical, zoological, or genetic university institute between 1937 and 1945.

Table 2.6 The fifteen full university professors and directors or department heads at Kaiser Wilhelm Institutes with the highest number of citations in the SCI 1945–1954

Name	Field	Institution	Number of citations	Number of articles	Citations per article (avg.)
Schmidt, Wilhelm J.	zoology	Univ. of Gießen	480	241	2.0
Frisch, Karl von	zoology	Univ. of Munich	379	143	2.7
Timoféeff-R., N.	genetics	KWI for Brain Research	318	123	2.6
Schramm, Gerhard	virology	KWI for Biochemistry	287	68	4.2
Bünning, Erwin	botany	Univ. of Strasbourg	278	106	2.6
Lorenz, Konrad	zoology	Univ. of Königsberg	264	56	4.7
Höfler, Karl	botany	Univ. of Vienna	256	111	2.3
Bauer, Hans	zoology	KWI for Biology	252	56	4.5
Küster, Ernst	botany	Univ. of Gießen	238	155	1.5
Harder, Richard	botany	Univ. of Göttingen	237	93	2.5
Kühn, Alfred	zool./gen.	KWI for Biology	234	127	1.8
Oehlkers, Friedrich	bot./gen.	Univ. of Freiburg	233	70	3.3
Weber, Friedrich	botany	Univ. of Graz	226	95	2.4
Hartmann, Max	zool./gen.	KWI for Biology	212	78	2.7
Tischler, Georg	bot./gen.	Univ of Kiel	178	80	2.2

the Nobel prize in medicine together with Karl von Frisch (tied for fifth on this evaluation scale) and Nikolaas Tinbergen. Schramm got close to winning the Nobel prize through his work on the tobacco mosaic virus. The placement of the later Nobel prize winners shows that the evaluation measure on which this table is based is not inappropriate.

We can note in summary that the influence and quality of scientific research at universities and the Kaiser Wilhelm Institutes did not differ significantly. Since I have taken into consideration only biologists in certain positions who were also recipients of support from the DFG, the true costs of research on which the articles cited in the SCI were based are not known. For that reason we must exercise great caution in analyzing the relationship between citations and DFG money, a relationship that differs by a factor of 3.3 between universities and Kaiser Wilhelm Institutes (Table 2.5). Generalizing, we might say that biologists at the Kaiser Wilhelm Institutes spent 3.3 times more DFG research money than university professors and got the same number of citations. As already mentioned, this analysis considers only DFG grants and not research funding from the institutions of the Kaiser Wilhelm Society and the universities themselves.

With respect to the specific size of funding, non–party members at universities were cited 2.7 times more frequently than party members. Both at universities and Kaiser Wilhelm Institutes, party members received only about half the number of citations per person that non–party members did. By contrast, when it comes to the number of citations per article, there is no difference between them. Non–party members seem to have pursued research more intensively. However, three party members were among the fifteen most frequently cited scientists.

8.2.   DFG Funds and the Quality of Research at Universities

I have examined the possible influence of DFG support on the quality of research through the example of university professors (Table 2.7). A positive effect of DFG support on the number of citations is revealed in a comparison of professors who were funded and those who were not. The control group comprised all twenty-two biologists who held a chair between 1937 and 1945 and did not submit grant applications to the DFG between 1933 and 1945. The remaining fifty-four, more

Table 2.7   The influence of DFG funding on the quantity and quality of research at universities

Funding 1937 to 1945 (RM)	Number of professors	Citations per professor (avg.)	Articles per professor (avg.)	Citations per article (avg.)
0	22	28.5	18.4	1.5
1– 2,000	7	41.4	23.0	1.8
2,001– 4,000	10	137.2	65.0	2.0
4,001– 6,000	6	97.6	52.5	1.9
6,001–10,000	7	110.8	62.0	1.8
10,001–20,000	11	93.0	37.6	2.5
20,001–30,000	7	68.1	37.8	1.8
30,000 +	6	95.1	43.5	2.2

Note: The table lists the average number of citations per person for all articles in the SCI 1945–1954 in eight different groups, depending on the amount of funding received. All full professors who held chairs in botany, genetics, and zoology at German and Austrian universities between 1937 and 1945 are included.

than two-thirds of those who held chairs during this period, submitted at least one application to the DFG. If we take into consideration only those articles that were published between 1933 and 1945, DFG-funded professors, with 45.1 citations per person, were cited on average 3.8 times as often as nonfunded professors, with 11.8 citations per person. A rise in productivity and, to a lesser degree, in the quality of research shows itself with support above RM 2,000, that is, in individuals who submitted more than one application. According to these data, sufficient support was one of the preconditions for good research, for research resulting in a large number of citations.

However, this does not mean that generous support necessarily brought with it a large number of citations. The information in Table 2.7 shows that no rise in citations occurred with support over RM 4,000. The number of persons is too low for a statistical analysis; moreover, no information is available on the total costs of the research on which the study is based. Nevertheless, we can draw the conclusion that holders of chairs at universities were not supported in keeping with their scientific influence as measured by the number of citations. Reasons why the directors of corresponding Kaiser Wilhelm Institutes were supported on a much larger scale will be discussed at the end of the chapter.

9. Funding according to Individuals and Specialties

To see which fields of biological research were financially supported,
I assembled the projects supported by the DFG and the RFR according
to various research fields and added together the relevant individual
grants for the periods from June 1, 1934 (when Johannes Stark succeeded Schmidt-Ott as president of the DFG), to December 31, 1939,
and from January 1, 1940, to May 31, 1945 (research during the war)
(Tables 2.8 and 2.11).

9.1.   Botany

The strongest support was given to applied botany (Tables 2.8, 2.9,
and 2.10); 40% of the amount listed in the tables went to the KWI
for Breeding Research. According to a report of the RFR, of the work
supported in the second half of 1943 (including the projects in physics,
chemistry, and medicine), 25.5% was in the field of forestry and the
wood industry, and 44.8% in the field of agricultural science.[80] Here
we see the influence of the Four-Year Plan on decisions of the RFR,
an influence that found expression also in Konrad Meyer's powerful
position (described above) as head of the Agricultural Science and
General Biology Section. The policy of the "expansion of living space"
(Erweiterung des Lebensraums) (see Section 10 of this chapter), which
took effect in 1939, also had an impact on the support for research.

Against this background, the relatively strong support for basic research in genetics and the great increase in this support during the war
are remarkable. Of the traditional areas of genetic research in Germany, developmental genetics and evolutionary genetics received
larger sums. Research in cytoplasmic inheritance, which received meager support, was carried out mostly at the universities. Expeditions
undertaken by biologists to collect wild and primitive strains of cultivated plants, and genetic and breeding work with these specimens,
were supported beginning in 1940. Also in 1940, the DFG or the RFR
started to support the then modern research into viruses. The strongest
support within genetics went to experimental mutation research.

Morphological, ecological, and plant-physiological research was
carried out primarily at the universities. A statement from the chairman of the German Botanical Society, Friedrich Markgraf, indicated
how scientists, especially during the war, adopted economic and po-

litical goals. In 1942 he described the following topics in botany as up-to-date: "fat and fiber-yielding plants," "the botanical Balkan expedition of the RFR in 1941," "the rapid breeding of new species through chromosome doubling," and "afforestation in the German East."[81] Thus the kind of botanical research that was considered "important to the war" and modern was research to procure raw materials, research on the genetics of wild and primitive strains, mutation research, and ecological research in the service of German imperialism and the "policy of living space."

9.2. Zoology

With the exception of research in developmental physiology and morphology, support for zoological research increased in all areas after 1940 (Tables 2.11, 2.12, and 2.13). Cancer research was conducted in physiology, biochemistry, genetics, and virology. Cancer researchers were frequently exempted from military service. After 1940, physiologists justified their research in part by demonstrating the possibility of its medical application. The greatest financial support went to work in genetics and radiobiology. With the exception of some work on the genetics of developmental physiology and human genetics, the genetic and radiobiological research in zoology was carried out almost exclusively at Kaiser Wilhelm Institutes. I have taken into consideration only the human genetic research that was conducted by zoologists, leaving aside the work done by medical scientists and anthropologists at institutes of anthropology or racial hygiene. The fields of genetic research included, in addition to those already mentioned, classic genetics, cytogenetics, evolutionary genetics, and virology. Thus in zoology, too, both classical genetic areas of research—among them especially the genetics of developmental physiology—as well as modern areas of research were financially supported. As in botany, those areas included virus research and the field that received once again by far the strongest support, experimental mutation research.

10. The Political and Ideological Background to Research Funding

Before I discuss in detail, in Chapters 3 and 4, the content of research, I will evaluate the background factors of research funding in biology.

Table 2.8 DFG/RFR funding per person for botanical research in the indicated specialties in the periods June 1, 1934, to December 12, 1939, and January 1, 1940, to May 31, 1945 (research during the war)

Name	Morphology 34–39	Morphology 40–45	Physiology 34–39	Physiology 40–45	Genetics 34–39	Genetics 40–45	Mutation 34–39	Mutation 40–45	Ecology 34–39	Ecology 40–45	Medicinal plants 34–39	Medicinal plants 40–45	Applied botany 34–39	Applied botany 40–45	Total
Barthelmeß, A.	—	—	—	—	—	—	1,300	—	—	—	—	—	—	—	1,300
Bauch, R.	—	—	—	—	—	—	—	22,800	—	—	—	—	—	—	22,800
Biebl, R.	—	—	—	—	—	—	—	—	200	—	—	—	—	—	200
Boas, F.	—	—	—	—	—	—	—	—	4,000	5,000	—	—	5,460	—	14,460
Borriss, H.	—	—	2,875	—	—	—	—	—	—	—	—	—	—	—	2,875
Branscheidt, P.	—	—	—	—	—	—	—	—	—	—	—	—	16,540	1,900	18,440
Bredemann, G.	—	—	—	—	—	—	—	—	—	—	—	—	8,600	—	8,600
Buder, J.	—	—	1,800	—	—	—	—	—	—	—	—	—	—	—	1,800
Bünning, E.	—	—	5,360	—	—	—	—	—	—	—	—	—	—	—	5,360
Burgeff, H.	—	—	—	—	—	—	—	—	3,000	3,200	5,175	—	—	—	11,919
Czaja, A. T.	—	—	1,400	—	—	—	—	—	—	—	—	—	—	—	1,400
Denk, V.	—	—	—	—	—	—	—	—	—	—	—	—	—	11,000	11,000
Döpp, W.	—	—	2,000	—	800	1,200	—	—	—	—	—	—	—	—	4,000
Eckardt, T.	750	—	—	—	—	—	—	—	—	—	—	—	—	—	750
Egle, K.	—	—	300	—	—	—	—	—	—	—	—	—	—	—	300
Engel, H.	—	—	—	1,650	—	—	—	—	—	900	—	—	—	—	2,550
Faber, F.C.v.	—	—	700	—	—	—	—	400	—	1,000	—	—	—	—	2,100
Firbas, F.	2,000	—	—	—	—	—	—	—	300	400	—	—	—	2,000	4,700
Fitting, J.	—	—	9,660	—	—	—	—	—	—	—	—	—	—	—	9,660
Freisleben, R.	—	—	—	—	—	18,100	3,500	41,160	—	—	—	—	—	—	62,760
Funk, G.	—	—	—	—	—	—	—	—	—	—	—	—	3,400	—	3,400
Gaffron, H.	—	—	900	—	—	—	—	—	—	—	—	—	—	—	900
Gams, H.	—	—	—	—	—	—	—	—	300	—	—	—	—	—	300
Geitler, L.	—	—	—	—	550	—	—	—	—	—	—	—	—	—	550
Gruber, F.	—	—	—	—	—	—	—	—	—	—	—	—	6,500	1,500	8,000
Hackbarth, J.	—	—	—	—	—	—	—	—	—	—	—	—	30,000	10,700	40,700
Hannig, E.	—	—	—	—	—	—	—	—	—	—	—	—	750	—	750

Harder, R.	39,700	7,300	3,500	—	—	—	—	—	—	—	—	21,500	7,400	—
Heidt, K.	670	—	670	—	—	—	—	—	—	—	—	—	—	—
Hertzsch, W.	22,570	320	22,250	—	—	—	—	—	—	—	—	—	—	—
Höfler, K.	64,610	60,810	3,500	—	—	—	—	—	—	—	—	300	—	—
Hoffmann, C.	3,000	1,200	—	—	—	—	1,800	—	—	—	—	—	—	—
Hoffmann, W.	400	—	—	—	—	—	—	—	—	400	—	—	—	—
Hofmann, E.	3,150	—	—	—	—	—	—	—	—	—	—	—	—	3,150
Huber, B.	28,340	1,000	5,500	—	—	—	—	—	—	—	—	7,850	—	13,990
Husfeld, B.	85,857	21,700	56,557	4,400	—	—	—	—	—	—	—	3,200	—	—
Kappert, H.	8,200	—	1,200	—	—	—	—	7,000	—	—	—	—	—	—
Knapp, E.	51,400	—	—	—	—	—	—	37,500	13,900	—	—	—	—	—
Knoll, F.	5,300	—	—	—	—	—	—	—	—	—	—	—	—	5,300
Koernicke, M.	3,398	3,000	398	—	—	—	—	—	—	—	—	—	—	—
Krause, K.	1,900	—	—	—	—	—	1,900	—	—	—	—	—	—	—
Küster, E.	11,400	—	—	—	—	—	—	—	—	—	—	3,600	7,800	—
Laibach, F.	16,500	—	4,500	—	—	—	—	—	—	—	—	12,000	—	—
Lange, S.	250	—	250	—	—	—	—	—	—	—	—	—	—	—
Lehmann, E.	6,060	—	2,500	—	—	—	—	—	—	—	1,400	—	—	2,160
Leick, E.	1,300	—	800	500	—	—	—	—	—	—	—	—	—	—
Lorbeer, G.	2,440	—	—	—	—	—	—	—	—	500	1,940	—	—	—
Mägdefrau, K.	800	—	—	—	—	—	—	—	—	—	—	—	—	800
Markgraf, F.K.	2,400	—	—	—	—	—	1,500	—	—	900	—	—	—	—
Melchers, G.	33,643	—	—	—	—	—	—	—	—	29,143	4,500	—	—	—
Metzner, P.	24,065	11,165	—	—	—	—	—	—	—	—	—	2,900	10,000	—
Meusel, H.	2,350	—	—	—	—	—	2,350	—	—	—	—	—	—	—
Mevius, W.	37,460	10,300	11,900	—	—	—	—	—	—	—	—	9,760	5,500	—
Michaelis, P.	5,320	—	1,000	—	—	—	—	—	—	400	3,920	—	—	—
Möwus, F.	9,250	—	—	—	—	—	—	—	2,000	—	7,250	—	—	—
Moritz, O.	19,000	6,500	2,300	4,500	2,800	—	—	—	2,400	—	—	500	—	—
Mothes, K.	94,390	3,850	13,300	16,800	3,800	24,900	4,620	—	—	—	—	6,420	20,700	—
Noack, K.	49,500	19,800	3,000	—	—	—	—	—	—	—	—	16,150	10,550	—
Oehlkers, F.	18,600	—	—	—	—	—	—	7,000	—	8,400	3,200	—	—	—
Orth, R.	1,400	—	—	—	—	—	—	—	—	—	—	800	—	600

Table 2.8 (Continued)

Name	Morphology 34–39	Morphology 40–45	Physiology 34–39	Physiology 40–45	Genetics 34–39	Genetics 40–45	Mutation 34–39	Mutation 40–45	Ecology 34–39	Ecology 40–45	Medicinal plants 34–39	Medicinal plants 40–45	Applied botany 34–39	Applied botany 40–45	Total
Overbeck, F.	—	—	—	—	—	—	—	—	1,800	—	—	—	—	300	2,100
Pascher, A.	—	—	—	5,700	—	—	—	—	—	—	—	—	—	—	5,700
Pirschle, K.	—	—	—	—	—	—	—	11,600	—	—	—	—	—	—	11,600
Pirson, A.	—	—	—	2,500	—	—	—	—	—	—	—	—	—	—	2,500
Pisek, A.	—	—	300	—	—	—	—	—	—	—	—	—	—	—	300
Pohl, F.	—	—	—	—	—	—	—	—	—	—	—	—	—	900	900
Propach, H.	—	—	—	—	10,640	2,300	—	—	—	—	—	—	—	—	12,940
Rauh, W.	3,300	—	—	—	—	—	—	—	—	—	—	—	—	—	3,300
Reinmuth, E.	—	—	—	—	—	—	—	—	—	—	—	—	5,000	4,200	9,200
Renner, O.	—	—	—	—	7,380	5,200	—	—	—	—	—	—	—	—	12,580
Roberg, M.	—	—	2,400	—	—	—	—	—	—	—	—	—	800	—	3,200
Ruge, U.	—	—	2,100	2,000	—	—	—	—	—	—	—	—	—	—	4,100
Rudorf, W.	—	—	—	—	—	—	—	—	—	—	15,130	—	63,833	19,343	98,306
Ruhland, W.	—	—	5,200	11,600	—	—	—	—	—	—	—	—	9,060	—	25,860
Rutmer, F.	—	—	—	—	—	—	—	—	3,000	10,000	—	—	—	—	13,000
Scherz, W.	—	—	—	—	—	—	—	—	—	—	—	—	5,000	—	5,000
Schiemann, E.	—	—	—	—	6,475	27,250	—	—	—	—	—	—	—	—	33,725
Schiller, J.	—	—	—	—	—	—	—	—	—	—	—	—	750	—	750
Schmidt, M.	—	—	—	—	—	—	—	—	—	—	—	—	22,500	1,900	24,400
Schmitz, H.	—	—	—	—	—	—	—	—	—	—	—	—	9,500	4,800	14,300
Schmucker, T.	—	—	600	—	—	—	—	—	—	—	—	—	—	—	600
Schnarf, K.	—	—	700	—	—	—	—	—	—	—	—	—	—	—	700
Schratz, E.	—	—	—	—	—	—	—	—	—	—	—	13,900	2,950	—	16,850
Schulze, K.L.	—	—	1,200	—	—	—	—	—	—	—	—	—	—	3,600	4,800
Schumacher, W.	—	—	1,800	—	—	—	—	—	—	—	—	—	2,000	—	3,800
Schwartz, O.	—	—	—	—	—	—	—	—	—	—	—	—	1,300	—	1,300
Schwartz, W.	—	—	—	—	—	—	—	—	—	—	—	—	12,780	33,500	46,280
Schwarze, P.	—	—	—	—	—	—	—	—	—	—	—	—	10,000	11,500	21,500

															Total
Schwemmle, J.	—	—	—	—	6,600	—	—	—	—	—	—	—	—	—	6,600
Seybold, A.	—	3,100	—	—	—	—	—	—	—	—	—	—	—	—	3,100
Söding, H.	—	—	—	—	—	—	—	—	—	—	—	—	—	5,000	5,000
Stein, E.	—	—	—	—	—	—	4,400	—	—	—	—	—	—	—	4,400
Steiner, M.	—	—	—	—	—	—	—	—	5,500	—	—	—	—	6,700	12,200
Stocker, O.	—	—	—	—	—	—	—	—	950	—	—	—	11,100	15,500	27,550
Stosch, H.A.v.	—	—	200	3,100	2,150	—	—	—	—	—	—	—	—	—	5,450
Straub, J.	—	—	—	—	—	—	—	13,450	—	—	—	—	—	—	13,450
Strugger, S.	—	—	3,000	3,400	—	—	—	—	—	—	—	—	—	—	6,400
Stubbe, H.	—	—	—	—	—	26,400	27,724	41,612	—	—	—	—	1,500	—	97,236
Tischler, G.	—	—	—	—	2,840	—	—	—	—	—	—	—	—	—	2,840
Troll, W.	5,740	—	—	—	—	—	—	—	—	—	—	—	—	—	5,740
Ullrich, H.	—	—	2,000	—	—	—	—	—	—	—	—	—	—	24,460	26,460
Volk, O. H.	—	—	—	—	—	—	—	—	1,950	—	—	—	4,600	—	6,550
Walter, H.	—	—	—	—	—	—	—	—	—	—	—	—	12,850	14,200	27,050
Weber, F.	—	—	3,000	6,200	—	—	—	—	—	—	—	—	—	—	9,200
Weber, U.	—	—	—	—	—	—	—	—	—	1,000	—	—	3,500	2,500	7,000
Wettstein, F.v	2,600	—	—	—	40,000	61,500	—	52,000	—	—	—	—	—	—	156,100
Wettstein, W.	—	—	—	—	—	—	—	—	—	—	—	—	1,200	3,300	4,500
Wetzel, K.	—	—	8,000	—	—	—	—	—	—	—	—	—	4,100	2,500	14,600
Widder, F.	—	—	—	—	—	—	—	—	—	—	—	—	3,000	3,000	6,000
Winkler, H.	300	—	—	—	—	—	—	—	—	—	—	—	—	—	300
Zimmermann, W.	—	—	—	—	1,200	300	—	—	—	—	4,780	4,000	2,640	—	12,920
Total	20,950	19,740	143,575	98,100	100,845	181,993	55,224	234,522	33,170	45,400	36,585	40,870	393,668	331,248	1,740,434

Notes: Morphology = morphology, anatomy, systematics, paleontology, phylogeny, prehistorical and early historical botany. Physiology = metabolism physiology, developmental physiology, cell physiology. Genetics = classic genetics, cytoplasmic inheritance, collection and genetic work on wild forms of cultivated plants. Mutation = experimental mutation research, that is, the genetic effect of radiation and chemical mutagenesis, developmental physiological evaluation of mutation experiments, mutation breeding. Ecology = ecology, plant geography. Applied botany = research on pests and weeds, plant nutrition, plant breeding other than by mutation, colonial botany (research on useful tropical plants and tropical agriculture), botanical war research (on oil- and fiber-yielding plants).

Table 2.9   DFG/RFR funding for botanical research, 1934–1945

Field of research	1934–1939 amount (RM)	1940–1945 amount (RM)	1934–1945 % of total amount
Morphology, systematics	20,950	19,740	2.3
Physiology	143,575	98,100	13.9
Genetics	100,845	181,993	16.3
Mutation research	55,224	234,522	16.7
Ecology	33,170	45,400	4.6
Medicinal plants	36,585	40,870	4.5
Applied botany	393,668	331,248	41.7
Total	784,017	951,873	100.0

Note: This table summarizes the data in Table 2.8 with respect to the funding of individual specialties.

Table 2.10   The fifteen scientists in botany and botanical genetics who received the most funding between 1934 and 1945

Name	Institution	DFG (in RM)
Wettstein, F. von	KWI for Biology	156,100
Rudorf, W.	KWI for Breeding Research	98,306
Stubbe, H.	KWI for Cultivated Plant Research	97,236
Mothes, K.	University of Königsberg	94,390
Husfeld, B.	KWI for Breeding Research	85,857
Höfler, K.	University of Vienna	64,610
Freisleben, R.	University of Halle	62,760
Knapp, E.	University of Strasbourg	51,400
Noack, K.	University of Berlin	49,500
Schwartz, W.	TH Karlsruhe	46,280
Hackbarth, J.	KWI for Breeding Research	40,700
Harder, R.	University of Göttingen	39,700
Schiemann, E.	KWI for Cultivated Plant Research	37,725
Mevius, W.	University of Münster	37,460
Melchers, G.	KWI for Biology	33,643

Notes: Institution = place of last position up to 1945. DFG = DFG/RFR grants from June 1, 1934, to May 31, 1945.

I specifically look at what causes may have led to the previously described rise of support for biological research, from which especially
the Kaiser Wilhelm Institutes profited.

10.1. Political Reasons

Science and its organizations played a political role in Nazi Germany.
The Kaiser Wilhelm Society, in particular, was able not only to maintain its eminent position within German science but even to enlarge
it. The president of the Kaiser Wilhelm Society, Max Planck, explained
the KWG's goal with these words: "In the future, too, the Kaiser Wilhelm Society for the Advancement of Science will consider it among
its highest honors that it occupies this place (of scientific research) and
thereby contributes its part to the rise of the new Germany."[82] General
Director Friedrich Glum wrote on the occasion of the twenty-fifth
anniversary of the Kaiser Wilhelm Society: "No special justification is
required for the fact that the Kaiser Wilhelm Society has happily
placed itself at the disposal of Adolf Hitler's new Reich for his work
of reconstructing our German fatherland" (Glum 1936, p. 5). This
readiness by the Kaiser Wilhelm Society to make itself voluntarily
available to the new government was rewarded. Thus Glum was able
to write: "The Reich government, in particular the Reich minister of
science, has amply rewarded the readiness of the society to place itself
in the service of the great tasks of the Third Reich. It once again made
larger funds available to the society, which was suffering considerably
from the economic crisis, and it acknowledged its eminent place in
German scientific organization."

As mentioned above, owing to the increasing financial support from
the state (Albrecht and Hermann 1990, p. 377), the KWG budget rose
from RM 5.7 million in 1932 to RM 14.4 million in 1944. The establishment of new Kaiser Wilhelm Institutes for agricultural and biological research in the occupied countries indicates agreement with the
expansionist goals of the Nazi regime: the Bulgarian-German Institute
for Agriculture was founded in Sofia in 1941, and the German-Greek
Institute for Biology in Athens-Piraeus in 1940.

Despite the tendency of Hitler and segments of the National Socialist movement to be rather hostile to science, institutions of the
regime, in accord with the Kaiser Wilhelm Society, strove to appoint
eminent, internationally recognized scientists as directors of their in-

Table 2.11 DFG/RFR funding per person for zoological research in the indicated specialties in the periods June 1, 1934, to December 1939, and January 1940 to May 1945 (research during the war)

Name	Morphology 34-39	Physiology 34-39	Physiology 40-45	Experimental embryology 34-39	Experimental embryology 40-45	Genetics 34-39	Genetics 40-45	Mutation 34-39	Mutation 40-45	Ecology 34-39	Ecology 40-45	Cancer research 34-39	Cancer research 40-45	Applied zoology 34-39	Applied zoology 40-45	Total
Abel, O.	10,000	—	—	—	—	—	—	—	—	—	—	—	—	—	—	10,000
Ahrens, W.	—	—	—	—	—	400	—	—	—	—	—	—	—	—	—	400
Ankel, W. E.	—	—	4,200	—	—	—	—	—	—	—	—	—	—	—	—	4,200
Autrum, H.	—	2,000	—	—	—	—	—	—	—	—	—	—	—	—	—	2,000
Bauer, H.	—	—	—	—	—	—	—	2,400	10,400	—	—	—	—	—	—	12,800
Beleites, I.	—	—	—	—	—	—	—	800	11,100	—	—	—	—	—	—	11,900
Bergold, G.	—	—	—	—	—	—	—	—	—	—	—	—	—	3,850	1,400	5,250
Beutler, R.	—	—	—	—	—	—	—	—	—	—	—	—	—	2,800	—	2,800
Breider, H.	—	—	—	—	—	2,500	—	—	—	—	—	—	—	5,400	2,000	9,900
Brock, F.	—	800	6,900	—	—	—	—	—	—	—	—	—	—	—	—	7,700
Buchner, P.	1,000	—	—	—	—	—	—	—	—	—	—	—	—	—	—	1,000
Buddenbrock-H., W.	—	1,100	7,700	—	—	—	—	—	—	—	—	—	—	—	—	8,800
Danneel, R.	—	—	—	—	—	5,100	10,000	—	—	—	—	—	17,250	—	—	32,350
Demoll, R.	—	500	—	—	—	1,000	—	—	—	—	—	—	—	6,960	3,000	10,960
Diebschlag, E.	—	500	—	—	—	—	—	—	—	—	—	—	—	—	—	500
Döring, H.	—	—	—	—	—	—	—	6,760	3,000	—	—	—	—	—	—	9,760
Duspiva, F.	—	—	—	—	—	—	—	—	—	—	—	—	—	—	15,700	15,700
Eberhardt, K.	—	—	—	—	—	—	—	—	9,000	—	—	—	—	—	—	9,000
Erhardt, A.	—	—	—	—	—	—	—	—	—	—	—	—	—	4,238	4,200	8,438
Feuerborn, H.J.	2,400	—	—	—	—	—	—	—	—	—	—	—	—	—	—	2,400
Fischel, W.	—	7,720	—	—	—	—	—	—	—	—	—	—	—	—	—	7,720
Fischer, I.	—	—	—	3,600	—	—	8,170	—	—	—	—	—	—	—	10,000	21,770
Friederichs, K.	—	—	—	—	—	—	—	—	—	—	—	—	—	4,150	19,770	23,920
Friedrich-F., H.	—	—	—	5,150	—	—	—	—	—	—	—	3,350	9,900	—	—	18,400
Frisch, K.v.	—	4,100	—	—	—	—	—	—	—	—	—	—	—	500	—	4,600
Giersberg, H.	—	—	—	—	—	—	—	—	—	—	—	—	—	1,800	—	1,800

Name															
Goetsch, W.	—	—	—	—	—	—	—	—	4,200	—	—	—	1,800	7,200	13,200
Gottschewski, G.	—	170	—	—	2,000	26,600	—	—	—	—	—	7,200	—	—	35,200
Hämmerling, J.	—	1,500	—	—	—	—	—	—	—	—	—	—	—	—	170
Harms, J.W.	1,000	—	—	—	—	—	—	—	—	—	4,600	14,200	—	—	20,800
Harnisch, O.	—	—	—	—	—	—	—	—	2,500	4,500	—	—	—	—	7,000
Hartmann, M.	—	—	8,320	—	1,800	24,000	2,504	—	—	—	—	—	—	—	36,624
Heidermanns, C.	—	4,110	—	—	—	—	—	—	—	—	—	—	—	—	4,110
Hempelmann, F.	400	500	—	—	—	—	—	—	—	—	—	—	—	—	900
Henke, K.	—	—	—	—	13,935	18,460	—	—	—	—	—	—	—	—	32,395
Henseler, H.	—	—	—	—	—	—	—	—	—	—	—	—	5,600	—	5,600
Herre, W.	—	—	—	1,400	—	—	—	—	—	—	—	—	4,800	14,550	20,750
Höner, E.	—	—	—	—	—	—	3,600	—	—	—	—	—	—	—	3,600
Holst, E.v.	—	6,050	—	—	—	—	—	—	—	—	—	—	—	—	6,050
Holtfreter, J.	—	—	—	3,990	—	—	—	—	—	—	—	—	—	—	3,990
Just, G.	1,500	—	—	—	12,910	19,200	—	—	—	—	—	—	—	—	31,610
Koch, A.	—	—	—	—	—	—	—	—	—	—	—	—	1,500	5,000	5,000
Koegel, A.O.	—	—	—	—	—	—	—	—	—	—	—	—	—	1,800	3,300
Koehler, O.	—	750	—	—	3,500	500	—	—	—	—	—	—	—	—	4,750
Koehler, W.	—	—	—	—	4,000	—	—	—	—	—	—	—	—	—	4,000
Koller, G.	—	4,500	10,500	—	—	—	—	—	—	—	—	—	—	—	15,000
Kosswig, C.	—	—	—	—	500	—	—	—	—	—	—	—	—	—	500
Krieg, H.	—	—	—	—	—	—	—	—	—	—	—	—	9,200	—	9,200
Kröning, F.	—	—	—	—	—	—	—	—	—	—	64,456	102,800	—	11,000	178,256
Kronacher, C.	—	—	—	—	14,000	—	—	—	—	—	—	—	800	—	14,800
Krüger, F.	—	200	—	—	—	—	—	—	—	—	—	—	—	—	200
Krüger, P.	—	—	9,000	—	—	—	—	—	—	—	—	—	—	—	9,000
Kühn, A.	—	—	—	—	33,016	106,993	9,250	29,187	—	—	3,565	—	—	—	181,980
Kühnelt, W.	—	—	—	—	—	—	—	—	—	3,000	—	—	—	—	3,000
Kuhl, W.	—	5,920	—	—	—	—	—	—	—	—	—	—	—	—	5,920
Kuhn, O.	—	—	10,400	—	9,740	—	—	—	—	—	—	—	—	—	20,140
Lehmensick, R.	—	150	—	—	—	—	—	—	—	—	—	—	3,600	3,500	7,250
Leiner, M.	—	1,600	—	—	—	—	—	—	—	—	—	25,240	—	—	26,840
Lengerken, H.	—	1,750	—	—	—	—	—	—	—	—	—	—	1,200	16,650	19,450

Table 2.11 (Continued)

Name	Morphology 34-39	Physiology 34-39	Physiology 40-45	Experimental embryology 34-39	Experimental embryology 40-45	Genetics 34-39	Genetics 40-45	Mutation 34-39	Mutation 40-45	Ecology 34-39	Ecology 40-45	Cancer research 34-39	Cancer research 40-45	Applied zoology 34-39	Applied zoology 40-45	Total
Lorenz, K.	—	5,000	5,410	—	—	—	—	—	—	—	—	—	—	—	—	10,410
Ludwig, W.	100	—	—	—	—	4,610	7,600	—	—	—	—	—	—	—	—	12,310
Lüdicke, M.	—	3,225	—	—	—	—	—	—	—	—	—	—	—	—	—	3,225
Lüdtke, H.	—	—	8,300	—	—	—	—	—	—	—	—	—	—	—	—	8,300
Mangold, O.	—	—	—	7,320	—	—	—	—	—	—	—	—	—	—	—	7,320
Mattes, O.	—	—	—	—	—	—	—	2,000	—	—	—	—	—	—	—	2,000
Menner, E.	—	800	—	—	—	—	—	—	—	—	—	—	—	—	—	800
Mertens, R.	—	—	—	—	—	—	—	—	—	1,200	—	—	—	—	—	1,200
Meyer, A.	—	—	—	—	—	—	—	—	—	—	—	—	—	1,950	4,100	6,050
Nachtsheim, H.	—	—	—	—	—	6,700	5,000	—	—	—	—	—	—	—	—	11,700
Pätau, K.	—	—	—	—	—	—	—	—	3,500	—	—	—	—	—	—	3,500
Panschin, I.	—	—	—	—	—	—	—	—	7,200	—	—	—	—	—	—	7,200
Pax, F.A.	—	—	—	—	—	—	—	—	—	3,250	1,000	—	—	—	—	4,250
Radu, G.	—	—	—	—	—	—	—	—	10,350	—	—	—	—	—	—	10,350
Rajewsky, B.	—	—	—	—	—	—	—	44,940	448,451	—	—	—	—	—	—	493,391
Remane, A.	—	—	—	—	—	—	—	—	—	2,100	80,500	—	—	—	—	82,600
Ries, E.	—	—	—	1,000	2,000	—	—	—	—	—	—	—	—	—	—	3,000
Rietschel, P.	—	—	—	—	—	—	—	—	—	—	—	—	—	500	—	500
Scheuring, L.	—	—	—	—	—	—	—	—	—	—	—	—	—	4,000	600	4,600
Schlottke, E.	—	3,200	9,600	—	—	—	—	—	—	—	—	—	—	—	3,970	16,770
Schmidt, W.J.	—	2,300	—	—	—	—	—	—	—	—	—	—	—	—	2,500	4,800
Schramm, G.	—	—	—	—	—	—	14,250	—	—	—	—	—	—	—	—	14,250
Schubert, G.	—	—	—	—	—	—	—	—	11,000	—	—	—	—	—	—	11,000
Seidel, F.	—	—	—	12,350	24,640	—	—	—	—	—	—	—	—	—	—	36,990
Spek, J.	—	—	—	160	—	—	—	—	—	—	—	—	1,830	—	—	1,990
Spemann, H.	—	—	—	3,700	—	—	—	—	—	—	—	—	—	—	—	3,700
Stammer, H.-J.	—	—	—	—	—	—	—	—	—	1,600	—	—	—	—	—	1,600

Steinböck, O.	—	—	—	—	—	—	—	—	—	—	—	—	—	1,850	—	1,850
Steiniger, F.	400	—	—	—	—	—	—	—	—	—	—	—	—	—	—	400
Stresemann, E.	—	—	—	—	—	—	—	—	—	—	—	3,300	—	—	—	3,300
Stubbe, A.-E.	—	—	—	—	—	—	—	—	—	—	12,000	—	—	—	—	12,000
Studnitz, G.v.	500	10,200	—	—	—	—	—	—	—	—	—	—	—	—	—	10,700
Thienemann, A.	—	—	—	—	—	—	—	—	—	—	22,150	3,000	—	2,500	55,000	82,650
Timofeeff-R. N.	—	—	—	—	—	—	—	8,700	75,340	—	—	—	—	—	—	84,040
Uexküll, v.J.	—	3,700	425	—	—	—	—	—	—	—	—	—	—	—	—	4,125
Ulrich, W.	2,000	—	—	—	—	—	—	—	—	—	—	—	—	3,700	—	5,700
Wagler, E.	—	—	—	—	—	—	—	—	—	—	—	—	—	2,450	1,850	4,300
Weber, H.	—	—	—	—	—	—	—	—	—	—	—	6,940	—	—	—	6,940
Wettstein, O.v.	—	—	—	—	—	—	—	—	—	—	—	2,000	—	—	—	2,000
Wunder, W.	—	—	—	—	3,500	—	—	—	—	—	—	—	—	6,100	—	9,600
Zimmer, K. G.	—	—	—	—	—	—	—	12,900	25,200	—	—	—	—	—	—	38,100
Total	18,400	62,545	90,955	38,670	26,640	115,711	244,273	93,854	655,728	28,090	113,150	75,971	178,420	81,248	183,790	2,004,764

Notes: Morphology = morphology, anatomy, systematics, phylogeny. Physiology = animal physiology and animal psychology, neurophysiology, sensory physiology, muscular physiology, metabolism physiology, histology, medical physiology, animal psychology, ethology. Genetics = developmental genetics, cytogenetics, virology, genetic experiments for the purpose of breeding experimental animals with special constitutional characteristics for medical research, human genetics. Mutation = experimental mutation research, biological and genetic effect of radiation and radioactive isotopes on animals and humans, chemical mutagenesis. Ecology = ecology, animal geography. Cancer research = cancer research in physiology, genetics, virology. Applied zoology = plant pest control, parasitology, animal breeding, colonial zoology, (research on useful tropical animals and pests), zoological war research.

Table 2.12   DFG/RFR funding for zoological research, 1934–1945

Field of research	1934–1939 amount (RM)	1940–1945 amount (RM)	1934–1945 % of the total amount
Morphology, systematics	18,400	0	0.9
Physiology	62,545	90,555	7.7
Experimental embryology	38,670	26,640	3.3
Genetics, virology	115,711	244,273	17.9
Mutation and radiation research	93,854	655,728	37.3
Cancer research	75,971	178,420	12.7
Ecology	28,090	113,150	7.0
Applied zoology	81,240	183,790	13.2
Total	514,481	1,492,556	100.0

Note: This table summarizes the data in Table 2.11 with respect to the funding of individual specialties.

Table 2.13   The fifteen scientists in zoology, zoological genetics, and radiobiology who received the most funding between 1934 and 1945

Name	Institution	DFG (RM)
Rajewsky, B.	KWI for Biophysics	493,391
Kühn, A.	KWI for Biology	181,980
Kröning, F.	University of Göttingen	178,256
Timoféeff-Ressovsky, N.V.	KWI for Brain Research	84,040
Thienemann, A.	Hydrobiology Station of the KWG	82,650
Remane, A.	University of Kiel	82,600
Zimmer, K. G.	KWI for Brain Research	38,100
Seidel, F.	University of Berlin	36,900
Hartmann, M.	KWI for Biology	36,624
Gottschewski, G.	University of Vienna	35,800
Henke, K.	University of Göttingen	32,395
Danneel, R.	KWI for Biology	32,350
Just, G.	University of Würzburg	31,610
Leiner, M.	University of Berlin	26,840
Friederichs, K.	University of Posen	23,920

Notes: Institution = place of last position up to 1945. DFG = DFG/RFR grants between June 1, 1934, and May 31, 1945.

stitutes. For example, after the death of Carl Correns in 1933, Reich Education Minister Bernhard Rust urged Fritz von Wettstein, who was considered the best plant geneticist in Germany and never joined the party, to accept the position as "first director" of the KWI for Biology in Berlin. Wettstein, full professor of botany in Munich and head of the largest botanical institute in Germany, twice turned down the offer. In 1934 he wrote to the Bavarian minister of culture that shortly after he turned down the offer, "the Reich Minister for Science, Education, and People's Education informed me in a personal conversation . . . that he considered my move to Berlin absolutely essential."[83] After two additional meetings with Rust, Wettstein changed his mind: "I can now no longer turn a deaf ear to the views of the Reich authorities that my presence in Berlin is the greater obligation toward the entire Reich."[84] Together with the other directors of the KWI for Biology (Alfred Kühn and Max Hartmann), Wettstein succeeded in keeping the work at this KWI largely free from the direct influence of the party, in contrast to other institutes such as the KWI for Breeding Research (Melchers 1987).

In their grant applications to the DFG between 1937 and 1942, Kühn and Wettstein repeatedly emphasized the danger of being overtaken in the field of science by researchers in the United States. Kühn wrote in 1940: "There is thus the constant danger that we will be outpaced by efforts abroad, in particular in the United States, and that especially research into active substances, in which we have laid the foundation in Germany, in Göttingen and in Dahlem, will be taken out of our hands."[85] They received the funding they requested.

Nikolai Timoféeff-Ressovsky, director of the genetics division of the KWI for Brain Research, was also an internationally recognized geneticist. The authorities in Germany appreciated his reputation and took a special interest in his mutation research (see the next section). Although a Soviet citizen, he received from the Ministry of Education an increase in the budget for his research institute after the Carnegie Institution in Cold Spring Harbor approached him.[86] American colleagues, among them Milislav Demerec and Leslie C. Dunn, assumed that his position in Nazi Germany was unsure and that he was, for political reasons, looking for opportunities to leave Germany. To their great astonishment, Timoféeff-Ressovsky turned down the offer.[87] In the process he made clear that even as a Soviet citizen he felt politically completely unmolested in Germany, and that he considered the conditions in Germany normal. In his view only "great catastrophes, such as war or Communist revolution," would be able to endanger his position.[88] It is remarkable that he could completely ignore the just completed expulsion of Jewish scientists and the dismissal of the non-Jewish director of the KWI for Brain Research, Oskar Vogt.

An assessment of Timoféeff-Ressovsky by the NSDAP in 1938 emphasized that he was "without a doubt one of the best geneticists," that the enjoyed "world renown," and that foreign colleagues who came to Berlin usually made a point of looking him up. With respect to politics the report stressed that he, as a Russian, was "a decided opponent of Communism."[89]

After the first German military successes, National Socialist representatives of science policy ascribed an important political function to German science in connection with the goal of German hegemony in Europe, a goal that had now moved closer. This was clearly demonstrated at a conference summoned by the REM on November 12, 1940, in Berlin, where scientists and representatives of the ministry, under the leadership of Oberregierungsrat Dr. Scurla, were supposed to clarify the "future position of official Germany and of German science"[90] on the question of international cooperation. Rudolf Mentzel explained the expectations placed on science: "On the other hand, we must emphasize that scientific work means political work, above all also foreign policy work."[91] This task was to be fulfilled through scientific achievement: "(We) will only have success if we can come up with the necessary scientific successes as part of this mighty cultural propaganda. The German scientist will be the best scientist only if he has the best accomplishments to show in comparison with neighboring peoples."[92]

The majority of scientists participating at this conference were directors of Kaiser Wilhelm Institutes: Viktor Bruns (KWI for Foreign Public Law and International Law), Eugen Fischer (KWI for Anthropology), Ernst Heymann (KWI for Foreign and International Private Law), Richard Kuhn (KWI for Medical Research), Peter A. Thiessen (KWI for Physical Chemistry), and Fritz von Wettstein (KWI for Biology). An illness prevented Alfred Kühn (KWI for Biology) from participating. Konrad Meyer attended as another representative of biology. In each field the scientists and scholars offered up for discussion their ideas about the future approach with regard to the international scientific associations that had existed since 1918, the transfer of their

permanent seats from Brussels and Paris to Germany, and the useful-
ness of these associations for German science. In addition, they were
asked to talk about how they intended to respond to the "competitive
efforts in America" in the field of science. The question of whether "a
German firm could succeed against an American one" was accorded
great foreign political importance for the future.[93] Scurla voiced his
conviction that "in the restructuring of Europe," the Germans "would
simply take in hand the claim to leadership in intellectual matters and
implement it, since we are convinced that we have the most powerful
science at our disposal."[94] As the following statement shows, even
participants who were not party members supported this view. Wettstein declared: "I think the most correct way would be that of ordering
the European sphere under German leadership, as the European section [of the Union of Biology]. America can be left to itself. The battle
with America will have to be fought after the war in any case, and
achievement will decide whether the right of leadership will lie in
America or Europe."[95]

The goal of cultural hegemony in Europe had been a permanent
part of German politics since the previous century. The Kaiser Wilhelm Society, too, had striven since its founding to preserve Germany's
international predominance in the field of science or to regain it after
World War I. The implementation of this goal, though not National
Socialist in its origins, in the year 1940 presupposed the German offensive war and the occupation of Eastern Europe, and it meant a
German hegemony under Hitler. The acceptance and support these
and the further expansionist plans of the Nazis received from influential scientists amounted to a strengthening of this policy. Presumably this widespread support further encouraged the generous funding
given to basic research at the Kaiser Wilhelm Institutes, at least in
biology.

10.2. Ideological Reasons

Support for genetic research, which accounted for a large part of biological research at the Kaiser Wilhelm Institutes, also had an ideological basis. Most of the leading members of the party and the SS,
and many of those in lower positions, saw basic research in genetics
as important for National Socialist racial ideology. It did not matter
whether a scientist supported this ideology and the racial policy. Lead-

ing geneticists like Alfred Kühn, Hans Stubbe, Nikolai Timoféeff-Ressovsky, and Fritz von Wettstein were among the scientists who received the most funding, even though they were not members of the
NSDAP. In part they were even known for their disapproving attitude
toward National Socialist ideology. For example, rumors circulated
that Alfred Kühn had been a member of the Communist Party. In
response the regional personnel office of the NSDAP made inquiries
about his political past. The NSDAP leadership in South Hannover-
Braunschweig made clear that these rumors were false, and it also
emphasized Kühn's important role as a geneticist in spite of his indifference to National Socialist politics: "His conduct toward the
[NSDAP] is today what it has always been, if not hostile then entirely
uninterested. His work as a scientist is without a doubt exceptional.
Professor K. has also done good work in the field of hereditary research. Last fall I had an opportunity to attend a course of his on
hereditary research, which I quite liked. From this perspective I would
like to say that he is doing his work along our lines without thereby
feeling any attachment to the NSDAP."[96]

The SS Captain Walter Greite was of the opinion that the Vienna
botanist Fritz Knoll should not become a member of the SS, since he
had denied the necessity of basic research in genetics.[97]

Nowhere did geneticists voice public disagreement with the racial
politicians. On the contrary, some used the ideological interest in their
field to make their work appear socially relevant. For instance, Timoféeff-Ressovsky and Stubbe probably emphasized the importance
of their modern basic research for racial hygiene to get financial backing. In *Der Erbarzt* (The genetic doctor), the leading journal for racial
hygiene, Stubbe presented genetic diseases of plants and animals as
models for the "genetic doctor" and went on to say: "Quite simply
the breeder, once he has recognized the nature of hereditary diseases,
has available to him entirely different possibilities for a truly positive
racial selection than the physician and human hereditary researcher,
who in most cases must simply select without being able consciously
to create the particularly healthy and resistant combination" (Stubbe
1935, p. 69).

In the same journal, Timoféeff-Ressovsky also called for measures
against the genetic burden in human populations, which he believed
was increasing owing to reduced natural selection.[98] He was invited
to give a lecture on "experimental mutation research" in a training

course on "racial science" and genetics, which the NSDAP's Office of Race Policy organized in October of 1938.[99] Afterward, the participants in the course visited his institute in Berlin-Buch.

These examples are not intended to give the impression that the discussion about the growth of a genetic burden on society took place only in Germany at the time. In many countries geneticists were pointing out the increase in harmful mutations within human populations and were calling for eugenic measures (see Adams 1990). However, warnings about the genetic burden meant something quite different in a democratic country like the United States than did demands for racial-hygienic measures in 1938 in Germany, where compulsory sterilization had been introduced in 1933 as a eugenic measure and where the Nuremberg Laws had been passed in 1935.

There are other examples for the cooperation between geneticists and racial hygienists: in 1938, Timoféeff-Ressovsky reported on his collaboration in the field of population genetics with Hermann Boehm of the Institute for Genetics in Alt-Rehse.[100] This institute formed part of the SS Leadership School of German Physicians (SS Führerschule der deutschen Ärzteschaft), which had been set up in 1935 by the National Socialist League of Physicians. At this institute, participants in the physician's courses in Alt-Rehse were to be trained in the basics of genetics and racial care. Timoféeff-Ressovsky and Stubbe supported Boehm in planing and setting up the institute (Boehm 1937; see also Proctor 1988, pp. 83–86). They also placed at his disposal a lot of material on mutations of corn, *Antirrhinum,* and *Drosophila* for the courses. Boehm, an old party member who described Timoféeff-Ressovsky as his teacher in genetic science, in turn helped "his teacher" when he refused to adopt German citizenship, despite political pressure to do so.[101] In this context it is worth mentioning that Timoféeff-Ressovsky did not even get into any significant trouble when statements by him, in which he expressed doubt about the German victory over the Soviet Union, reached the REM in September of 1943.[102] He was merely admonished by the Kaiser Wilhelm Society to express himself more cautiously on political matters, seeing as he was a foreigner.[103] The zoologist Walther Arndt, professor and curator at the Zoological Museum in Berlin, fared very differently in a similar situation. He was denounced for having stated on July 28, 1943: "Now it is all over with the Third Reich; the only thing left is punishment of the guilty."[104] In response the People's Court sentenced him to death for demoralizing the armed forces, and he was executed on June 26, 1944; the petition from Hanns von Lengerken, director of the Zoological Museum, and other professors in Berlin to the REM, as well as the plea for mercy from his lawyer, had no effect.[105] Incidentally, this is the only case known to me of a biologist kept on or newly hired in 1933 or 1938 who was sentenced by a German court for political reasons during the Nazi years.

10.3. Economic and Military Reasons

Representatives of science policy considered research in botany and radiobiology important and worthy of support for economic, medical, and military reasons as well. In 1937 the director of the KWI for Breeding Research, Wilhelm Rudorf, assigned political tasks above all to genetic and plant breeding research. He justified his approach with the statement that the "German people, compared to the size of its population, is settled on much too narrow a living space. This living space encompasses large tracts of nature comprised of poor and very poor soils with a climate which does not permit the growth of all cultivated plants that are necessary to provide foodstuffs and raw materials." Rudorf demanded the "breeding of strains of cultivated plant species that would ensure on German soils and in German climatic zones the food supply and the provisioning with important raw materials, fiber materials, oils, cellulose, and so on" (Rudorf 1937a, p. 4). With these goals he saw himself in agreement with National Socialist policy: "National Socialist agrarian policy, with its foundation in 'blood and soil,' has clearly brought out the political tasks of plant breeding: safeguarding the conditions of life of the German Volk on German soil" (1937a, p. 15).

As these statements indicate, above all the goals of economic self-sufficiency and the expansion of German living space became important for the support of botanical research, an emphasis that became noticeably stronger during the war. The "Germanification" of the annexed part of Poland, and of other areas whose annexation was planned, began officially in October of 1939 under the leadership of Himmler. Konrad Meyer emphasized the fundamental importance of the concept of living space for all future political decisions in a speech entitled "Planning and Reconstruction in the Occupied Eastern Territories," which he delivered at an event put on by the Kaiser Wilhelm Society at the Harnack-House (Berlin) on January 28, 1942:

> One must therefore be able to think ahead using the criteria of the future order, which rise above the previous, spatially narrow thinking. One must thus have the faith, the imagination, and the courage to participate in the task of reorganization in the East. We must therefore also try to break free from the half-solutions of earlier times, and instead of treating the symptoms have the courage to undertake surgical intervention, that is to say, to venture a reordering of the entire structure of the Volk from the perspective of national politics and national economy. The Eastern task [*Ostaufgabe*] is the unique opportunity to realize the National Socialist will and make it a reality without compromises.[106]

Meyer held a high rank in the SS. Head of the Planning Division of the Reich Security Main Office, the headquarters of the National Socialist police and security forces, he was given the task by Himmler to work out settlement and area planning for Poland and the Soviet Union (Heiber 1958). The result was his memorandum "General Plan East [*Generalplan Ost*]—Legal, Economic, and Spatial Foundations of the Reorganization of the East."[107] This plan contained the entire concept for the legal, economic, and spatial reordering of the occupied and still to be occupied territories between the Oder River and the Ural Mountains. As Meyer envisaged it, during the period of reconstruction, new settlement areas for German farmers would be placed under Himmler's direct control, and the new settlers would be selected according to the racial criteria of the SS. Meyer emphasized before the Kaiser Wilhelm Society: "The highest goal of settlement planning must be to utterly Germanify the entire area" (Meyer 1942, p. 256).

The concept of the General Plan East was part of the SS's policy of Germanification in the East. It presupposed the successful "final solution of the Jewish question" and provided for the dispossessing of millions of citizens from the Soviet Union, Poland, and other Eastern European countries and their expulsion from these regions.[108] Alternatives such as the complete deprivation of rights with the obligation to perform slave labor (for Slovenes, for example, "not capable of being Germanified"), deportations to Siberia, and the "scrapping through work" of people who were "racially" undesirable, were discussed in various committees, which included at least one member of the Kaiser Wilhelm Society in the person of Eugen Fischer.[109]

The botanical expeditions of the RFR after 1941, the research at some botanical university institutes, and the research at the KWI for Cultivated Plant Research and the KWI for Breeding Research were supported with reference to the political goal of *Lebensraum* (expanding the living space). To appear particularly worthy of support, botanists, in their grant applications, emphasized the importance of their research for the *Lebensraum* policy the way the geneticists mentioned above highlighted the importance of their work for racial hygiene. For example, botanists emphasized the need for mutation research in botany, in particular polyploidy research, for the fast breeding of new strains of plants suitable for growing in the kind of extreme climate that prevailed in the East. In his application to the RFR for funds for a camera for a botanical excursion, Hans Stubbe of the KWI for Biology wrote in 1942: "The systematic collection and preservation of such [wild and primitive strains of cultivated plants] is seen by German breeding research as an urgent task, because the extraordinary variety of these plants in the still unexplored mountains of the Balkans and their adaptation to extreme habitats offer the guarantee that we will find among them strains that are resistant to frost, drought, and disease. With their valuable characteristics, these strains play a decisive role in breeding hardy strains for the German East."[110]

These botanical expeditions, carried out in 1941 and 1942, received, in addition to support from the RFR, funding from the REM and the Reich Ministry for Nutrition. Since their target areas were Greece and Albania, areas that were occupied in part by German and in part by Italian troops, they could not take place without military protection. The high command of the Wehrmacht outfitted both expeditions correspondingly.[111]

One man who played an important role in the support for botanical and agricultural research at the Kaiser Wilhelm Society was the state secretary in the Reich Ministry for Nutrition and Agriculture, Herbert Backe, who was also a member of the senate and from 1941 to 1945 vice-president of the Kaiser Wilhelm Society. His role explains the establishment of the large KWI for Cultivated Plant Research near Vienna during the war, and the support it received from Backe.[112] The examples that will be given later make clear the central function that Backe had for the Kaiser Wilhelm Society, for which in turn his growing influence in the Nazi power apparatus was one precondition. Backe, a member of the NSDAP since 1923, was given responsibility in 1936 for coordinating the potential of agricultural and industry within the framework of the Four-Year Plan, and later he organized the food supply for the war against the Soviet Union. In 1942 he took Richard Darré's place as head of the Ministry for Agriculture, and in 1944 Hitler appointed him Reich nutrition minister.[113] Convinced of the genetically based inferiority of the Russians, Backe tried to secure the increasingly difficult task of feeding Central and Western Europe by carrying out genocide in the Soviet Union.[114] After Robert Kempner, the prosecutor of the military court in Nuremberg, confronted him with his relevant appeals and plans, Backe committed suicide in the Nuremberg prison in 1947 before he could be charged (Kempner 1987, pp. 282ff.). The following excerpt from a statutory declaration by Ernst Telschow in 1949 shows that members of the Max Planck Society, which had just emerged from the Kaiser Wilhelm Society, saw no reason even after the war to reevaluate its cooperation at the highest level with one of the originators and perpetrators of the policy of extermination:

> [Backe's] election to the senate [of the Kaiser Wilhelm Society] was done by the free will of the members of our senate, namely in consideration of Mr. Backe's known friendly attitude toward scientific research and the agricultural research institutes run by the Kaiser Wilhelm Society (thus not under political pressure). Mr. Backe was among those who were especially supportive of German scientific research. It was clear to him that it had to be kept free of all political influences as well as of the influence of state authorities and the bureaucracy. I personally know that he frequently defended the independence of the Kaiser Wilhelm Society and its institutes against such tendencies, even in his own ministry. In keeping with this attitude, he never let himself be guided by political considerations in senate deliberations that dealt with the appointment of scientists.[115]

In 1942, after the Germans had occupied parts of the Soviet Union, many Soviet institutes of plant breeding, among them those of the Vavilov organization with world-famous collections of wild and primitive strains of cultivated plants, lay within the German sphere of power. After what has been said above, it is understandable that these collections were of great importance to the KWI for Cultivated Plant Research, the KWI for Breeding Research, and the SS (see Chapter 5, Section 3). The Kaiser Wilhelm Society was ready to cooperate with Alfred Rosenberg's ministry in order to obtain control of these institutes.[116] It was Fritz von Wettstein who, in a conference with Backe, Konrad Meyer, and Ernst Telschow on February 9, 1942, recommended that "the stations be placed under the care of the Kaiser Wilhelm Society in order to save the valuable material and to make further work there possible."[117] He presented a plan in which he compiled "the most important points for the development of the network of biological stations of the Kaiser Wilhelm Society on the basis of the Institute for Cultivated Plant Research."[118] In this plan he discussed in detail the possible importance of the nine Soviet institutes from Murmansk to Baku, that is from the Kola peninsula to the Caspian Sea. We read his reference to

> Djetskoje-Selo with branch station in Novgorod. Until now a Russian institute of the Vavilov organization. For us an important base in the northern Russian forest and cultivation region for the sea and moor areas of the Valdai Hills as well as for marine-biological questions concerning the northeastern Baltic. I would consider it valuable to withdraw as much as possible from the difficult ground of the big Russian cities and to develop the branch at Novgorod into a base. A move to Novgorod is also valuable with respect to the question of the later political affiliation of Leningrad, whether it will be handed over to the Fins. This must be clarified on the spot.[119]

Five of these institutes, among them those in Murmansk and Baku, lay outside the territories occupied by the Germans at this time. Wettstein penciled them into the accompanying map as "to be added," since "they are for the time being not yet in our possession."[120] Backe supported Wettstein's suggestion completely. Since the stations in question were under the jurisdiction of the Ministry for the East (Ostministerium), Teltschow had to obtain a decree from Rosenberg. He was authorized to use Backe's name in the planned talks.[121]

It is not without a certain irony that Wettstein's plan to expand the Kaiser Wilhelm Society's network of biological stations through the incorporation of the Soviet institutes on the basis of the KWI for Cultivated Plant Research seemed to founder at the very outset: in a handwritten addition to his memo of February 10, 1942, Telschow noted on March 31, 1942, with regard to Wettstein's plan and the planned talk with the Rosenberg ministry: "Moot, since Professor Rudorf took up the matter with Minister Rosenberg at the East conference without our knowledge."[122]

The "reordering of Europe," whose precondition—the expulsion of millions of Soviet citizens as well as the "final solution of the Jewish question"—was taking shape in many points as early as 1942, was already a reality. People as different as Wettstein and Rudorf, as well as other scientists, were concerned to make the quickest possible use of the new opportunities.[123] German military defeats thwarted the realization of the plans of Wettstein and Rudorf. In 1943 the botanist and SS Second Lieutenant Heinz Brücher, however, succeeded in stealing part of the Vavilov collection with the help of a special detachment of the SS in order to use it for his research in one of the institutes set up by the SS (see Chapter 5).

According to official statements by the Kaiser Wilhelm Society, "important tasks in the agricultural development of the occupied eastern territories" fell to the KWI for Breeding Research.[124] Plant breeders at this KWI conducted research on fat and oil-yielding plants and on plants that were suitable for cultivation in the East. After Joachim Hackbarth had become head of the section on oil plants and grain-yielding leguminosae at the branch of the KWI for Breeding Research in East Prussia, all his projects that were recognized as "important to the war" centered on cultivation in the East.[125] This branch of the KWI was enlarged during the war with the help of "Eastern laborers."[126]

Scientists of this KWI and biologists of other institutions did not shrink from cooperation with Himmler and those in charge of the Auschwitz concentration camp. After Himmler had received from Hitler the task of pressing ahead by every means with the cultivation and breeding in Germany of the rubber plant "Kok-saghyz," the KWI for Breeding Research placed itself under Himmler's leadership.[127] An SS lieutenant colonel was charged with searching for the Kok-saghyz plant in the Soviet Union, under the code name "plant 4711," and the first plants were located in 1941.[128] The KWI took over the scientific breeding work.[129] Rudorf used fifty Slovenes from a nearby resettlement camp for work in the plantations.[130]

On June 25, 1943, a working conference of persons interested in the Kok-saghyz rubber project took place in the SS Main Office for Economy and Administration (SS Wirtschafts- und Verwaltungshauptamt). The following scientists participated in the working groups on cultivation, breeding, and basic research: Professors Christiansen-

Weniger and Ries von Pulawy from Lublin (Research Institute for Agriculture), Professor Walter (with the scientific office of the army), Professor Rudorf and Dr. R. W. Böhme (KWI for Breeding Research), Professor Kappert (Institute of Genetics, University of Berlin), Dr. Amlong (represented by March and K. Lauche, District Research Institute (Gauforschungsanstalt) for Plant Physiology, Posen), Dr. Zimmermann (Forestry College, Eberswalde), Professor Tobler (TH, Dresden), and Professor Krause (Botanical Garden, Berlin). Four participants came from the commander's office of the Auschwitz concentration camp: the chemist Ruth Weinmann, SS Captain Kudriawtzow, SS First Lieutenant Dr. Schattenberg, and SS Major Dr. Joachim Caesar.[131]

As the representative of the SS Main Office of Economy and Administration, Caesar presided; Rudorf presented the working program.[132] Caesar was head of the Agricultural Section of the Auschwitz concentration camp. Since 1942, its newly established plant breeding station had focused its research on the Kok-saghyz plant.[133] In accordance with the agreed-upon division of labor, the KWI sent samples from all seed batches to the Auschwitz concentration camp, where tests of strains and purity were carried out in the breeding fields, the existing strains were tested for their rubber content, and breeding selection was to take place.[134] Later, Rudorf never mentioned his close contacts with Auschwitz.[135]

Beginning in December of 1939, all Kaiser Wilhelm Institutes in Berlin and Brandenburg were declared to be institutes important to the war *(kriegswichtige Institutionen)*. In this way they could be supplied with raw materials for their research.[136] In all of the Kaiser Wilhelm Institutes mentioned, at least some of the research was assigned urgency classifications by the Office of Wartime Economy in the RFR; as described in Section 5 above, during the war such a classification became to an increasing degree the prerequisite for grant approvals and the allocation of research materials and equipment. However, this research did not necessarily have anything to do with the war; at least in the area of biology, only a few of the research projects with a high-urgency classification dealt with war-related matters. For instance, during the war, materials for building an ultracentrifuge were authorized to the Division (Arbeitsstätte) for Virus Research, and the "investigations in the field of virus research" were given the high-urgency classification "SS" in 1942, even though this was pure basic research.[137]

In the Genetics Section of the KWI for Brain Research and in the KWI for Biophysics, with their well-equipped biophysics laboratories, research in the fields of radiobiology and radiation protection was recognized as important to the war. The high-urgency classification given to research on the biological effects of neutron radiation and the artificial production and use of radioactive isotopes also had something to do with the attribution of a great many capabilities to radioactive radiation in the forties, especially to neutron radiation. Otto Hahn and Fritz Straßmann had bombarded uranium and thorium atoms with neutron radiation to produce "transuranic elements" with even higher atomic numbers, and in the process they had discovered nuclear fission. The Americans J. Aebersold and Ernest Lawrence had shown that the cells of malignant tumors were damaged more effectively by neutron radiation than by X-ray radiation. Many different applications were expected from radioactive isotopes (artificially produced with the help of neutron radiation) in the field of drug research (distribution in the body) and in the military sphere, for example in measuring the permeability of filters for poisonous gases.[138] Thus one project in fact connected to a war application was the development— by Hans Joachim Born and Karl Günter Zimmer in the Genetics Section under Timoféeff-Ressovsky—of a new testing procedure for the transmission strength of gas masks against suspended matter using radioactive phosphorus.[139]

Zimmer was given several military contracts from the Army High Command and the Air Force High Command, some of which were carried out in the biophysics laboratory of the Auer Society, some in the laboratory of the Genetics Section.[140] Nothing is known about the content of this work. At the KWI for Biophysics, questions relating to the tissue damage due to radiation were studied at the request of the NSDAP's Central Office for Public Health and the German Labor Front (Deutsche Arbeitsfront). In 1940 Boris Rajewsky emphasized the special importance of the study of the biological effects of radiation in the war by pointing out that the use of radioactive substances (phosphorescent paint) as well as the technical use of X rays (testing of materials and tools) had risen sharply.

From 1943 to 1945 some biophysical research received support from the plenipotentiary of the Reich marshal for nuclear physics in the RFR (Professor Abraham Esau and, from the end of 1943, Professor Walther Gerlach). The plenipotentiary was responsible for the entire uranium project, including questions about the biological effects of radiation and radiation protection. According to Gerlach, neutron dosimetry had a special importance to nuclear physics, since it had been discovered that the effect of neutron and gamma radiation on the human organism was far greater than had been initially assumed.[141] The Genetics Section in Buch was given research contracts to study the biological and genetic effects of fast neutrons on small mammals, especially rats, and to establish a neutron dosimetry.[142]

In response to the emergency program "Energy Production from Nuclear Processes," introduced in January of 1945 by a decree from Hitler, the plenipotentiary for nuclear physics concentrated research in this area in six working groups. Only these groups were still to receive energy, material, and personnel. The Radiation Protection and Dosimetry Group included the Genetics Section in Buch and the Physical-Technical Reich Institute.[143] The plenipotentiary for nuclear physics supported the research of the KWI for Biophysics only until 1944. According to Boris Rajewsky, the director of the institute, the goal of this work was to clarify "the biological effects of corpuscular radiation, including neutrons, *taking into consideration their possible use as a weapon,* but above all to clarify the biological basis of radiation protection."[144] Despite the emphasis on the possible military use, the application was turned down. Gerlach was not interested in Rajewsky's research.[145] However, the possibility of nuclear explosives was something he envisaged himself. At a meeting in October of 1944, Walther Gerlach, Rudolf Mentzel (RFR), and Erich Schumann (plenipotentiary for the physics of explosives), decided that the three million volt facility ordered by Rajewsky and approved by the RFR but not yet built, should not be given to Rajewsky but should be used for the production of neutrons in physics and for "experiments in the physics of explosives."[146] This high-voltage facility was important to physicists because the other neutron generators had been lost or completely destroyed in the war.

The discussion so far has shown that the willingness on the part of influential scientists to cooperate with the party and the state was an essential precondition for the substantial financial assistance given to biological research. This willingness manifested itself in the support for the expansionist plans of the regime and in the absence of any objections to anti-Semitic, racial-hygienic, and racist measures or support for them in the name of biology. Here I should point out once more the particularly striking example of cooperation between biologists under the leadership of Wilhelm Rudorf and those in charge of the Agricultural Section of the Auschwitz concentration camp.

The centralization of research support through the abolition of democratic control within the DFG and the establishment of the RFR led to stronger direct cooperation of individual scientists with National Socialist politicians. Thus the interest of Herbert Backe and Konrad Meyer, for example, in advances in basic research in biology or in specific application possibilities, especially during the war, led to the described support for the plans of various institute directors of the Kaiser Wilhelm Institutes.

Did the centralization of research funding and the influence of politics change the content of biological research? This question will be examined in the following chapters.

3

The Content and Result of Research at Universities

The FIAT-Reviews of German Science give a comprehensive account of the content of biological research between 1939 and 1945. The *FIAT-Reviews* ("FIAT" stands for Field Intelligence Agency, Technical) were compiled by German scientists shortly after the war on orders from the Western occupying authorities: the scientists were called upon to give a complete and exact account of the research that had been carried out in their field during the war. Biophysics is discussed in volumes 21 and 22 (1949, edited by Boris Rajewsky and Michael Schön), biology in volumes 52–55 (1948, edited by Erwin Bünning and Alfred Kühn), biochemistry in volumes 39–42 (1947–1953, edited by Richard Kuhn). Special mention should be made of the essays by Hans Friedrich-Freksa, "Forces in the Building of Biological Structural Units" (vol. 21, pp. 44–49), Georg Melchers, "Phytopathogenic Viruses" (vol. 52, pp. 111–129); and G. Piekarski, "The Electron Microscope in Biology and Medicine" (vol. 22, pp. 173–209).

The *FIAT-Reviews* do not contain any contributions on research in racial hygiene. Volume 22 (p. 350) cites experimental results of the director of the Hygiene Institute of the Waffen-SS, Joachim Mrugowsky, who was sentenced to death in the Nuremberg physicians trial, but does not provide the relevant references. The same volume (p. 219) assembles August Hirt's publications on fluorescent microscopy but makes no mention of his crimes (murder of concentration camp inmates to build up a "Jewish skull collection," mustard gas experiments with inmates, and more).

In this and the following chapter I limit my discussion of the results of the biological research funded by the DFG and the RFR to a small selection of scientific publications. The most detailed section concerns genetic research, since it was a relatively young field of research whose methods and approaches, moreover, underwent a strong development with the rise of molecular biology in the forties. In all areas my investigation has focused on the research of scientifically influential biologists (as measured by the number of citations in the SCI 1945–1954 and, wherever available, the verdict of professional colleagues) and on research that manifests a connection to National Socialist ideology and politics. Research at the universities is described in the present chapter; research at the Kaiser Wilhelm Institutes in Chapter 4. In four cases I have not adhered to this division: I treat all research in the field of chemical mutagenesis and of cytoplasmic inheritance in Section 1.2 of this chapter, in the field of ecology in Section 2.2 of this chapter, in cancer research in Section 2.6 of this chapter, and in the field of wild and primitive plants in Section 3 of Chapter 4. All information comes from the DFG files on the relevant scientists, unless attested in other sources.

The year 1933 was not a turning point in biological research. In purely scientific terms, I have picked a somewhat arbitrary starting date for my analysis of scientific research.

1. Botany

1.1. Applied Botany and Research on Medicinal Plants

To illustrate the special support for agricultural botany mentioned earlier, I shall point out that a single institute, the Institute for Plant Cultivation and Plant Breeding in Halle, received RM 306,962 for plant-breeding research under Theodor Römer in the period 1935 to 1944. No botanical or zoological university institute was supported on such a scale. Römer, a student of Erich von Tschermak-Seyssenegg (Vienna), founded his own school of plant breeding in Germany; he and his collaborators bred many strains of barley, wheat, rye, and legumes.

However, questions relating to plant nutrition, plant diseases, colonial botany, the drought resistance of crops, and research into oil- and fiber-yielding plants were also studied at some (nonapplied) botanical university institutes. It is difficult and often impossible to distinguish applied research from ecological and physiological research, as the following examples show. The director of the Botanical Institute in Münster, Walter Mevius, had been given the assignment by Konrad Meyer to set up the working group in agricultural botany, which, beginning in February of 1937, handled all DFG applications in the area of plant physiology in the Forschungsdienst (Research Service, see Chapter 2 Section 5).[1] Mevius himself worked in the field of nutritional physiology. He studied the effect of trace elements on plants and the physiology and drought resistance of nitrogen-fixing soil bacteria. According to Karl Höfler (Vienna), research on drought resistance stood at the time in the forefront of interest in pure and applied plant physiology. His own experiments supplied comparable data on the drying-up ability of various open land and cultivated plants; in contrast to the general term "drought resistance," he defined the concept of "drying-out resistance" purely in terms of cell physiology (Höfler, Migsch, and Rottenburg 1941).

Otto Stocker (TH Darmstadt) had been the first ecological physiologist to, at the end of the twenties, study the water balance of Egyptian desert and salt plants, and he had published quantitative data on the water metabolism of plants under conditions of extreme aridity (Lange 1989). Beginning in 1937, the research he conducted on the drought resistance of grain plants, which was important to agricultural crop farming only by virtue of his experimental objects, also related to the analysis of the general problem of the adaptive manifestations of xerophytic plants. His studies of various strains of oats, barley, and wheat showed that the general basis of drought resistance is a physiological mechanism—different from variety to variety and species to species—which works toward a relative increase of the assimilation excess concurrent with a relative curtailment of transpiration (Stocker, Rehm, and Schmidt 1943).

Heinrich Walter (TH Stuttgart, after 1941 Posen) began receiving support in 1937 for colonial-biological research, as part of which he studied botanical questions of farming in "German Southwest Africa." Kurt Mothes (Königsberg), who among all botanists at universities received the highest level of DFG funding and who worked in various areas of physiology and ecology, was also supported for experiments on drought resistance as well as for plant-breeding experiments involving wild plant material from the 1935 botanical expedition to the Hindu Kush. However, I know of no publications relating to his applied research.

The studies of Friedrich Boas (TH Munich) on blueberry plants formed part of the scientific program of the German Forestry Association for Nutrition from the Forest; the practical work was to be carried out by the Reich Working Group on Nutrition from the Forest. In an another project, Boas wanted to inventory the "national biological wealth" of Germany in a "physiological and dynamic flora of Germany." In DFG applications he highlighted the national economic importance of the research for the "German Reich of farmers" *(das deutsche Bauernreich)* and spoke in this context of "the highest biological defensive will." His claim to be creating a new science with his "dynamic botany" was rejected by Walter Mevius, Kurt Mothes, and Wilhelm Ruhland in their expert reviews. They criticized his inexact, physiologically unscientific approach (since metabolic processes are dependent on external conditions, they are inconstant and cannot be inventoried), the way he dissipated his energies on uncoordinated efforts, and his "showy kind of self-promotion." The project did not receive any funding from the DFG.

Apart from breeding research, work aimed at "closing the protein and fat gap," extracting other raw materials from plants, and preventing the exhaustion of raw materials was also recognized as important to the war.[2] The following botanists received larger amounts of support in this connection: Kurt Mothes for studying the biological value of fodder yeasts, Paul Metzner (Greifswald) for research into potato diseases and the grafting of plant fibers, and Kurt Noack (Berlin) for research on active substances and cell physiological studies on cold-resistant strains being bred by Wilhelm Rudorf at the KWI for Breeding Research. Noack also indicated that he was conducting physiological experiments on fodder preservation. Richard Harder (Göttingen) received grants to study the synthesis of fat through microorganisms. Together with Hans von Witsch he devised a method for a particularly favorable mass development of diatoms, whereby they reached a fat content of 40–50% of the dry substance (Harder and Witsch 1942). Otto Moritz in Kiel examined the production of food and fodder from mushrooms. Viktor Denk (German University in Prague) worked with algae cultures, Bruno Haber (Forestry College Tharandt near Dresden) won the urgency classification "SS" for the mass production and selective breeding of staghorn sumac as a tanning agent. The morphologist Wilhelm Troll (Halle) got a "war-important" research contract to study the catabolism of cellulose. Publications and DFG documents do not indicate any results of this "war research" worth mentioning.

To what extent this research in fact related to the war can no longer be determined. It can be assumed that normal basic physiological research was concealed behind many applications for applied research. Werner Rauh has spoken of the efforts to formulate DFG applications with as much orientation toward application as possible, for otherwise there was no chance of having an application approved.[3] Occasionally reviewers recognized this strategy. In an expert opinion on Friedrich Laibach, Mevius noted that the title "Practical Investigations along the Lines of the Four-Year Plan" was an exaggeration.[4]

During the war, a number of botanists did contractual research for the Wehrmacht. In 1943 Franz Bukatsch (Munich) was asked to study, in free collaboration with the Wehrmacht Institute for Food Research and by using the tools of plant physiology, the "causes for the degradation of the quality of vegetables during the drying process."[5] The assignment derived from the importance of dried vegetables in feeding the troops. From 1942 on, Ernst Hentschel (Hamburg) and Curt Hoffmann (Kiel) were involved in a research contract of the OKM to develop underwater paint that prevented marine growth.[6]

At the Institutes of Hereditary Science and Botany in Strasbourg, Edgar Knapp and Erwin Bünning carried out an assignment from the OKL with the urgency classification "SS" on the topic "Cultivation, Selection, and Breeding Experiments with the Fiber Plant Asclepias and the Oil Plant Perilla."[7] Its importance to the war lay in Perilla oil's newly discovered value in technical uses, since crude oil could no longer be imported. Bünning had found that the yield of the Perilla plant was good in the Alsace and that cultivation there could be recommended. He had brought back the seed from the Ukraine and East Asia. Heinz Brücher at the SS Institute for Plant Genetics in Lannach (see Chapter 5, Section 3) was also working on a research project for extracting oil from plants, using the Lafi plant.[8]

Research into medicinal plants during the Nazi period, supported with a total of RM 76,000, was integrated into the ideological and practical measures of National Socialist medicine. Subsumed under the designation "New German Art of Healing" *(Neue Deutsche Heilkunde)*, these measures took a "holistic" approach to medicine by combining natural and folk methods of healing with modern scientific methods. After 1937, botanists at nearly all universities offered courses in medicinal plants and field trips. In 1935, the leadership of the Reich Physicians League established the Reich Working Group for the Art of Healing with Medicinal Plants and the Procurement of Medicinal Plants, whose advisory board included Ernst Lehmann (Tübingen), chairman of the German Association of Biologists (see Chapter 2, Section 4).

The Reich Working Group was able to engage schools, groups affiliated with the party, and other organizations in the collection of medicinal plants.[9] Research into medicinal plants was to make possible a stronger use of native drugs and thus, as in other areas of applied biology, serve the goal of economic self-sufficiency. Cheap labor was needed to make the cultivation of these plants economical. Even concentration camp inmates were put to work collecting plants, for example in the herb garden in Dachau (Wuttke-Groneberg 1983). This garden had been set up around 1937 on Himmler's orders. Plants from the German and homeopathic pharmacopoeia as well as folk-healing plants were cultivated on forty acres of land, along with wild vegetables and wild lettuce. This plot of medicinal plants by the SS formed the basis for Himmler's planned Institute of Medicinal Plant Research in the Dachau concentration camp.

Walter Zimmermann (Tübingen) carried out breeding research with the castor-oil plant. The 1936 Four-Year Plan called for research to alter this tropical plant through breeding to the point where it could be economically grown in Germany. The castor-oil plant was to supply oil for medicinal and technical purposes, for example as a lubricant for airplane motor, needs which had hitherto been met primarily by imported oil.[10]

From 1939 to 1944, Kurt Mothes received DFG funds for research on the physiology and cultivation of medicinal plants. He conducted studies on the physiology and biochemistry of the synthesis of nicotine and other alkaloids. He proved that nicotine is formed above all in the root (Mothes and Hieke 1943), that the root of lupines requires only carbohydrates and no preliminary stage for the synthesis of alkaloids. He also questioned the theory that alkaloids were formed during the breakdown of protein (Mothes and Kretschmer 1946). Eduard Schratz of Münster studied the possibility of cultivating subtropical medicinal plants in the occupied eastern territories. Wilhelm Rudorf of the KWI for Breeding Research, who was working on the morphines of poppy, coordinated collaborative work in the field of medicinal plant research. Walter Zimmermann participated in this work, as did Otto Moritz (Kiel), who studied whether medicinal and aromatic plants were suitable for cultivation in Schleswig-Holstein. Hans Burgeff (Würzburg) conducted seed and cultivation experiments with soil-growing orchids for the extraction of the salep drug used to treat diarrhea in children. Karl Heidt (Gießen) studied the knowledge and use of native medicinal plants in the Hessen-Nassau district. With the help of teachers and students, nearly every family in the villages was questioned.

The collection and breeding of medicinal plants also played an important role during the war. In 1943 Schratz became acting head of the Reich Working Group for the Art of Healing with Medicinal Plants and the Procurement of Medicinal Plants and head of its scientific division. The Reich Office for Economic Development charged him with setting up and running a collection program for medicinal plants in the entire Reich. "The impossibility of importing into Germany important medicinal plants from hostile Europe or non-European countries has forced us to collect German wild-growing medicinal plants with the help of schools on a scale never seen before."[11] Over and above this, Eduard Schratz was given research contracts for breeding and chemically analyzing medicinal plants.

1.2.   Genetics and Mutation research, Chemical Mutagenesis

The majority of genetic research in Germany was carried out at the Kaiser Wilhelm Institute for Biology, where the work in all departments was to a stronger or lesser degree oriented toward genetics, and in the Genetics Department of the Kaiser Wilhelm Institute for Brain Research (see Chapter 4). Genetic research at the universities centered on the fields of cytoplasmic inheritance and the experimental induction of mutations.

Cytoplasmic Inheritance. The study of cytoplasmic inheritance concerned itself with the genetic substance outside the cell nucleus. Mendelian heredity is based on the genes in the chromosomes of the cell nucleus. The genes of the ovum and the sperm are equally involved in this hereditary transmission. By contrast, cytoplasmic hereditary factors are passed on almost exclusively by the ovum, since it contains much more cell cytoplasm than the sperm. Thus, when cytoplasmic inheritance is involved, reciprocal crossbreeding produces differing results. In the twenties, research on cytoplasmic inheritance formed the main area of genetic research in Germany (Barthelmeß 1952; see also Harwood 1985 and Sapp 1987). To facilitate an understanding of this research, I will briefly summarize its main results before describing its development after 1933. As mentioned above, the research at the KWIs in this area will be included here and will not be discussed separately. DFG support for cytoplasmic inheritance dwindled after 1933 and stopped almost completely during the war (Table 2.8).

Carl Correns, like many botanists, worked on the causes of chlorophyll variegation—a dark green to yellow pattern of spots on the leaves—which is found in a number of plants. In 1909 he published for the first time an explanation according to which non-Mendelian inheritance, based on plasmatic factors, caused the variegation in the four-o'clock *Mirabilis*. Erwin Baur attributed the same phenomenon to genetic changes in the plastids, correctly so according to what we know today, a notion that Correns rejected. At this time, only a few scientists outside Germany were doing intensive research on cytoplasmic inheritance. Most of these geneticists, including until about 1915 also Thomas Hunt Morgan, shared Baur's view.

Most of the questions and controversies in research concerned with cytoplasmic inheritance in Germany developed out of the approaches of Correns and Baur. In addition, the hypothesis formulated by Jacques Loeb around the turn of the century—according to which the cytoplasm was responsible for the characteristics of higher taxonomic groups while the genes of the nucleus supposedly determined exclusively differences of species and varieties—was still influential.

Correns's only student, and after the death of Baur in 1933 the most influential plant geneticist in Germany, was Fritz von Wettstein. In

1926 he introduced the term "plasmon" for the entire genetic material of the cytoplasm, analogous to the term "genome," which Hans Winkler had formulated in 1924 for the totality of the genes in the nucleus. The work of Otto Renner, professor of Botany in Jena, and Peter Michaelis, until 1927 Renner's assistant, then assistant at the TH Stuttgart, and after 1933 department head at the KWI for Breeding Research, confirmed the existence of a plasmon. Along the way Renner supported Baur's supposition that changes in plastids were responsible for non-Mendelian inheritance. On the basis of a large number of experiments with the evening primrose *Oenothera,* he formulated the hypothesis that plastids, as self-reproducing bodies, showed genetic differences and that these changes were comparable to gene mutations. The totality of genetic factors in the plastids he called "plastidom." Lehmann rejected the notion of a "plasmon" and tried to explain the differences in reciprocal crossbreedings by positing certain "inhibiting genes" *(Hemmungsgene)* in the cell nucleus.

In the twenties, Friedrich Oehlkers in Freiburg had pursued research into cytoplasmic inheritance in *Epilobium.* Owing to the rarity of tangible plasmonic differences, he subsequently changed the object of his experiments and studied the plasmon of *Streptocarpus.* In the thirties and forties, however, he worked mostly in the areas of cytogenetics and experimental mutation research. Otto Renner continued his studies of *Oenothera,* also treating the question of species formation. He discovered a high degree of heterozygosis in the final stages of species formation, which he explained through preceding interspecific hybridization and subsequent inbreeding. In his view the ring formation of chromosomes in the meiosis resulting from reciprocal translocations brought about, as a structural heterozygosis, a stabilization of the genetic heterozygosis. In *Oenothera* he saw confirmation of the view of evolution according to which "nature, in the creation of new forms, does not plan, it plays" (Renner 1946). Julius Schwemmle (Erlangen), also working with *Oenothera,* supported Renner's hypothesis of the genetic properties of plastids, and he was able, through plastid mutations, to ascertain the influence of plastid genes on the formation of certain traits. These genes caused an activation of chromosomal genes; other plasma factors played no role (Schwemmle 1944).

Fritz von Wettstein, in his summary of the genetic and developmental-physiological significance of cytoplasm (1937), concentrated on the carriers of hereditary transmission in cytoplasm; he did not say much about hereditary transmission through plastids, which he considered to have been proved. In his view genes did not exist in the cytoplasm as individual elements as they did in the nucleus. The plasmon was homogenous and was passed on to the offspring without division. He could not find any evidence for the thesis of Michaelis that the cytoplasmic carriers of inheritance could be modified by genetic factors in the nucleus over the course of generations. According to Wettstein, genome and plasmon were parts of the idioplasm (the totality of hereditary factors) that were equal in importance but not in kind ("gleichwertig aber nicht gleichartig"). In the mid-thirties Wettstein began to concentrate on the fields of developmental genetics, mutation research, and evolutionary genetics, which I will discuss in Chapter 4, Section 1.

As the head of the Cytoplasmic Inheritance Section in the KWI for Breeding Research, Peter Michaelis received DFG funding for research on cytoplasmic inheritance in *Epilobium hirsutum* until 1939. After extensive studies on the diffusion of plasmon differences, he found phenomena that ran counter to Wettstein's notion of the plasmon as a homogenous genetic entity, but which were compatible with an interpretation of plasmon as the sum of numerous entities (Michaelis 1942). He believed that he had discovered in some varieties of *Epilobium* individual "plasma-sensitive" alleles of the cell nucleus, which could work unimpeded only in conjunction with a specific plasmon.

Almost all botanical geneticists in Germany considered it proven that plasmon effects were the cause of reciprocal differences. Ernst Lehmann was the only one who, in interpreting the differences (which he, too, had observed) in reciprocal crossbreedings, clung to the "monopoly of the nucleus" in inheritance.[12] He worked, among other things, with *Epilobium* varieties that, when crossed with a test variety, differed to the same extent to which their cytoplasm impeded the growth of hybrids. Lehmann attributed the reciprocal differences in the crossbreeds to growth or developmental hormones whose formation was conditioned by the genome, more precisely, by "unifactorially dividing inhibiting genes with a pleiotropic effect"[13] of the cell nucleus (Lehmann et al. 1936). He assumed that these active substances were for some time dragged along by the cytoplasm and thus produced the differences in the growth retardation of the hybrids; here he suspected interferences in the synthesis of auxin (Lehmann 1941).

Peter Michaelis and Heinz Brücher failed to confirm Lehmann's data. Michaelis showed that the cytoplasmic effects of growth retardation were preserved even after backcrossings over fourteen generations, which argued against a control through nuclear genes (Michaelis 1938). Brücher, Lehmann's doctoral student, disproved Lehmann's supposition of "unifactorially dividing inhibiting genes" as the cause of reciprocal differences. Starting with the observation that reciprocal differences appeared only when certain strains of *Epilobium* were crossbred, Brücher introduced the genome of a development-retarding *Epilobium* strain into the plasm of a nonretarding strain and vice versa by crossbreeding and backcrossing with the male. In these experiments, the result of reciprocal crossbreedings with the newly created forms was determined solely by the cytoplasm and not the nucleus of the specific strain (Brücher 1938). Although his theory had been clearly disproved, Lehmann clung to it, even later in a modified form.[14]

The work of Otto Renner and Peter Michaelis in the field of cytoplasmic inheritance received international recognition. If we include research after 1945, the work of Michaelis is among the most comprehensive in this field anywhere.

With the clarification of the structure and function of DNA, the questions and methods of research into cytoplasmic inheritance changed fundamentally. All organelles and other structures of cytoplasm that contain DNA play a role in the formation of traits. In addition, it has so far not been disproved that certain, not genetically determined, submicroscopic structures of the cell cortex of the old cell have an influence in the formation of the structures of the daughter cell. One example was the phenomenon of cytotaxis in paramecium discovered by Tracy S. Sonneborn (see Sapp 1987, Chapter 7). The phenomenon of variegation is today attributed to various mechanisms, which include the dissociation of genetically different plastids, instability of genes, positional effects of genes, and chromosome mutations.

According to Ernst Mayr, the early specialization of German botanical geneticists in "unorthodox genetic mechanisms" such as cytoplasmic inheritance contributed to their weak interest in other genetic problems (Mayr and Provine 1980, pp. 280f.). Germany in the twenties did not see the creation of a school of "transmission genetics" (the study of the structure and doubling of genetic material), such as were established by Thomas Hunt Morgan in the United States and

William Bateson in England. However, for the Nazi period we can note that research into cytoplasmic inheritance received comparatively small sums. The total support for this research from the DFG for the group of scientists under examination was between RM 20,000 and RM 30,000; during the war meaningful sums went only to Otto Renner. Instead, large amounts went to what was, in the wake of Hermann Muller's discovery in 1927 of the mutagenic effect of X-ray radiation, the most modern branch of genetic research—experimental mutation research. However, little of this research was carried out at the universities.

Mutation Research. Since 1919, Emmy Stein had been working at the Institute for Research in Heredity and Breeding at the University of Berlin (until 1935 the Agricultural College) on the generation of hereditary carcinomas in *Antirrhinum* by means of radium. Erwin Baur, her teacher, had founded the institute in 1914, and it remained the only genetics institute at German universities until 1935. Stein was one of the first scientists to, as early as 1921, at the first meeting of the German Society for Research into Heredity, report on the effects of radium radiation on living organisms *(Antirrhinum).* Her experiments, in which she radiated vegetative points of *Antirrhinum,* concerned the induction of mutations through artificial influences (Stein 1922). It is likely that when it came to interpreting her results she was influenced by the then ongoing discussion about the permanent modifications in paramecia, which Viktor Jollos at the Kaiser Wilhelm Institute for Biology had discovered, and about the cytoplasmic inheritance factors in botany. In any case, she theorized (1926) that the radiomorphogeneses she had discovered, as well as other radiation-related changes, were not based on mutations but tended to correspond to cytoplasmically conditioned permanent modifications. A year later Hermann J. Muller presented his discovery of the induction of mutations through X-ray radiation at the International Congress of Genetics in Berlin (1927). A few years later, Stein replaced her question about the nature of the radiomorphogeneses with the question about the nature of the somatically induced hereditary phytocarcinomas she had discovered in 1930. As the cause for this she assumed changes in the cell nucleus, now ruling out a cytoplasmic basis (plasmon mutation) for these hereditary tissue degenerations (Stein 1935). Beginning in 1938, Rudolf Freisleben in Halle was supported in his work to produce polyploid strains of cultivated plants and in his experiments to use X-ray-induced gene mutations for breeding purposes. In 1942 he was able, by radiating air-dried kernels, to produce a mildew-resistant barley mutation; the resistance was polygenetically conditioned, and the mutations were fully fertile (Freisleben and Lein 1942).

Following Hermann Muller's discovery of the artificial induction of mutation through radiation, the question of a chemical triggering became the focus of intensive research at many institutes. Some geneticists combined the search for chemical mutagens with the hope of being able to induce a specific change in genetic material, in contrast to the nonspecific effect of radiation.[15]

The experiments that Nikolai V. Timoféeff-Ressovsky and his collaborators at the Genetics Section of the KWI for Brain Research carried out to give proof of a chemical mutagenesis had no results. They examined possible mutagenic effects of iodine, DNA, and carcinogenic substances (not further specified) on *Drosophila*. In 1938 Hans Stubbe at the KWI for Biology began testing various inorganic and organic substances on *Antirrhinum*. He found no mutagenic effect using strychnine, nitrite, arsenic compounds, caffeine, nicotine, ethanol, and carcinogens.[16] A clear effect could be noted with chloral hydrate, phenol, and potassium thiocyanate, with a rise to $5.92 +/- 0.94$ mutations per 100 F_1-plants as compared to $1.72 +/- 0.51$ in the control groups (Stubbe 1940). Stubbe and Helmut Döring were able to increase the frequency of mutation by a factor of two by using nitrogen, phosphorus, and sulfur deficiencies (Stubbe and Döring 1938).

Robert Bauch of the Botanical Institute in Rostock was working on the experimental induction of mutations in yeasts and other technically important fungi. With the help of mitosis-retarding substances, such as camphor, acenaphtine, and colchicine, he induced polyploidization, which led to gigantism. In a series of experiments with corn and onion roots, he discovered that certain sulfonamides had an antagonistic effect toward colchicine.[17] Beginning in 1942 he also used carcinogenic substances to induce mutations. On yeast he tested the mutagenic effect of 3,4 benzopyrenes, methylcholanthrene, and 1,2,5,6 dibenzoanthracene, substances he obtained from Adolf Butenandt (at the KWI for Biochemistry) He was able to induce gross mutants with benzopyrenes and methylcholanthrene. After these experiments, Bauch believed it was justified to speak of a mutation-inducing effect of carcinogenic substances (Bauch 1942). He recommended certain mitosis poisons (he did not mention any names but only emphasized that they were less toxic than colchicine) for cancer therapy.[18] The first tests in Rostock clinics using a mitosis poison derivative were said to have gotten under way successfully.[19]

Gerhard Schramm and Hans Müller discovered that nitrous acid led to a loss of biological activity of the tobacco mosaic virus (Schramm and Müller 1942). However, at this point they had not yet realized that a mutation had been induced by changes in the RNA (see Chapter 4 Section 2).

In 1940 Friedrich Oehlkers (Freiburg) began studying the effect of certain chemicals on meiosis. These experiments led to his discovery of chemical mutagenesis. He was the only scientist in Germany to use chemicals to obtain a mutation rate comparable to X-ray radiation (Oehlkers 1943). In his work he examined the induction of translocation through inorganic salts and various other organic substances. With aluminum nitrate he obtained an increase of 8.2 in the rate of translocation, with a certain combination of ethyl urethane and potassium chloride an increase of 38.0. However, he did not discover in individual substance a connection between concentration and effect. Charlotte Auerbach in Edinburgh furnished the first proof for the existence of chemical mutagens that was quantifiable with regard to concentration and effect on the phenotype (Auerbach and Robson 1946). With the help of the CLB-method developed by Hermann Muller, she demonstrated in *Drosophila* a mutagenetic effect of mustard gas on the creation of lethal mutations; this effect was on the same order of magnitude as that of X-ray radiation. She sent her first report on the work concerning the mutation induction of mustard gas, which had been carried on since 1941, to the Ministry of Supply in March of 1942. Since she was not allowed to make her results known during the war, they were not published in *Nature* until 1946.

1.3. Physiology

Most of the nongenetic basic research funded by the DFG and the RFR was physiological research. However, in 1939 many DFG grants in the field of physiology were stopped. In response some scientists shifted their research to applied botany, or at least they emphasized its orientation toward practical application in their DFG applications.

Metabolism Physiology. Kurt Noack of the Institute of Plant Physiology in Berlin worked in the area of photosynthesis. He developed model experiments for the catabolism of chlorophyll, which suggested a longer reaction chain *in vivo* (Noack 1943). As the first step he indicated the proteolysis of plastid stroma assisted by the hydrogen peroxide of respiration, which already at the beginning of yellowing was no longer catalytically broken down. He attributed the occasional self-ignition of hay to the reduction of the separated chloroplastic iron to pyrophoric iron. Comparative biochemical analyses of chloroplasts and cytoplasm in spinach leaves revealed that chloroplasts contain 82% of the total leaf iron. With regard to proteins, however, he and E. Timm found hardly any difference between cytoplasm and chloroplasts (1942). In connection with findings that suggested a function of the chloroplasts as carriers of heredity, the authors believed it was possible that the nucleic acid discovered in chloroplasts in 1939 played an important role.

August Seybold (Heidelberg) conducted a large number of chromatographic and spectrophotometric analyses of assimilation pigments in various plant groups. According to his findings, there were frequent deviations from what was in older publications regarded as the normal 3:1 distribution of chlorophyll a and b. Certain classes of algae had no chlorophyll b but were capable of photosynthesis (Seybold, Egle, and Hülsbruch 1941). Seybold discovered a species-specific pigment transformation as the leaves turn in the fall, whereby in many cases secondary carotenoids are formed in conjunction with the metabolism of proteins and lipids; these carotenoids are also said to cause the red coloring of certain petals and fruits. He interpreted the creation of chromoplasts out of chloroplasts as a symptom of the metabolism of proteins and lipids (Seybold 1943).[20]

Kurt Mothes, already mentioned several times, was also financially supported for studying the nitrogen fixation in legumes and the protein metabolism in plants. With respect to the biosynthesis of the acid amides asparagine and glutamine he discovered that they were formed directly from the corresponding amino acids and hardly at all from ammonia and urea. To him the importance of these amides was above all in their function as storage bodies of the corresponding amino acids and not in the detoxication of ammonia (Mothes 1940). Mothes was a student of Wilhelm Ruhland. His papers in the field of nitrogen metabolism and the secondary plant substances are among the most important in Germany. After the war he became one of the most influential plant physiologists in the German Democratic Republic.[21] Paul Metzner (Greifswald) and Walter Mevius (Münster) received DFG funding for research on nitrogen bonding and nitrogen metabolism.

Among the internationally best known German plant physiologists around 1930 was Wilhelm Ruhland (Leipzig). He received DFG grants for various research projects concerning metabolism physiology, for example on aerobic fermentation, the metabolism of vitamins C and B, and the salt resorption by root cells. His most important scientific papers were published before 1933. Ruhland was the editor of the eighteen-volume *Handbuch der Pflanzenphysiologie* (Handbook of plant physiology, 1955–1967). On the importance that Ruhland and the school of chemical physiology of metabolism he founded had on modern plant physiology, see Höxtermann 1991.

Cell Physiology and Developmental Physiology. Erwin Küster (Gießen) examined the physical characteristics of protoplasm, for example changes in the viscosity of plasm through external factors. In 1939 he discovered vacuole contraction, and after 1940 he was supported for his work on "the effect of alkaloids on the living plant cell." Siegfried Strugger (Veterinary College of Hannover) analyzed the absorption and storage of substance by the plant cell. He developed (in 1940) a staining method for yeast, fungal, and bacterial cells, with the help of which one could quickly determine whether a cell had died off or was alive.[22] This method promised to have applications in medicine and was important for the development of disinfectants.

The DFG and the RFR gave the strongest support for work in developmental physiology to the areas of growth factors, photoperiodism, and vernalization. Hermann von Guttenberg (Rostock) proved that heteroauxin, whose effect is largely identical to that of auxin, is itself not a growth factor but stimulates the production of auxin (Guttenberg 1942).[23] Friedrich Laibach (Frankfurt) carried out growth-factor studies with heteroauxin on slips. He suspected that auxin normally formed in the plastids in the dark from monosaccharides (Laibach 1941). The studies for the practical application of heteroauxin that were carried out by Laibach and others did not produce results of any significance with regard to increasing the yield of cultivated plants.[24]

The support for work in photoperiodism and vernalization must also be seen in the context of the thirties, when these were considered the fields of physiology with the greatest importance for plant breeding (Kuckuck 1939, pp. 112f.). "Vernalization" refers to the influence of temperature on physiological processes that have a clear impact on the course of development. What we are dealing with here are non-hereditary phenomena. In 1918 Gustav Gassner made the first systematic study of temperature as a factor in the developmental physiology of plants. In the twenties vernalization research was carried out on a smaller scale in several places. In the thirties such research became a main focus not only of plant breeding but also of plant physiology. Richard Harder, Georg Melchers, and Anton Lang (KWI for Biology), Wilhelm Rudorf (KWI for Breeding Research), and Wilhelm Ruhland worked in this area with larger DFG grants. In 1937 Seybold described vernalization as trendy research *(Modeforschung)*.[25] Rudorf, who received the most DFG support for vernalization, also cited the results of the Soviet agriculturalist Trofim Lysenko (Rudorf and Schröck 1941, p. 129). Lysenko had developed a practical method for vernalization on a large scale, and since the beginning of the thirties it had been used in agricultural practice in the Soviet Union, though later it proved to be unworkable.[26] (For the influence of Lysenko on Soviet biology see Chapter 4, Section 7).

Many biologists examined the influence of light and questions of photoperiodism. Johannes Fitting (Bonn) worked on the induction of dorsiventral symmetry and the influence of light on development. Harder received substantial financial support for his research, which he carried out with many collaborators, on pattern formation in blossoms, reactions to long and short days, and the occurrence and effect of animal sexual hormones in plants. He carried out extensive studies on the influence of photoperiodic phenomena on plant forms (Harder 1946). Harder and his collaborators were not able to induce photoperiodically conditioned changes of the vegetative organs separate from changes in the reproductive organs (Harder and Bode 1943). The theory that favorable photoperiodic conditions exert their effect via the optical influencing of blossoming hormones was repeatedly substantiated by Harder and Georg Melchers (KWI for Biology), though they were not yet able to explain how this influencing came about (Harder 1946; Melchers in *FIAT-Reviews,* vol. 53, p. 142).

Erwin Bünning (Königsberg, after 1941 Strasbourg) had begun

work in 1934 on the physiological mechanism of the endogenous circadian rhythm in plants. In 1938, at the request of the RFR and the Research Service, he led a year-long trip to Indonesia to clarify the circadian periodicity in plants. Bünning first analyzed the mechanism of the endogenous circadian rhythm that was based on physiological self-regulation and occurred also when the external conditions were constant. He discovered a circadian succession of synthetic and catabolic processes in which the endogenous variations in carbon dioxide production and hence of assimilation played a key role. These circadian variations in acidity also caused circadian movements of growth cells and closing cells (Bünning 1942). He developed a theory that was supposed to explain all known photoperiodic phenomena. According to this theory, photoperiodic reactions are created through the interaction of external and endogenous (autonomous) rhythms, in that the light has a qualitatively different effect on the two phases of the endogenous rhythm (*FIAT-Reviews,* vol. 53, p. 144).

1.4. Morphology

After the death of Karl Göbel (Munich) in 1931, Wilhelm Troll in Halle became the most influential morphologist in Germany. He received grants for the morphology and systematics of higher plants and for histological studies of the development of flower organs. His former assistant Werner Rauh (Heidelberg) studied growth forms of alpine and domestic woody plants. On the whole the support for morphological research was low, and one must take into account that genetic and physiological research costs a good deal more.

In contrast to Göbel, Troll advocated idealistic morphology, which had its roots in the natural philosophy of Goethe and contained elements of Plato's theory of ideas.[27] According to Troll, morphologists faced the general problem of trying to comprehend the diversity of organic forms from an archetype common to them all (Troll 1935/ 1936). Goethe, so Troll, had coined for this archetypal unity the concept of the type *(Typus).* Troll rejected a causal-analytical approach for morphology: "For while the type is, within the framework of mechanism, a foreign body which some believe can be understood by analyzing the phenomena on which it is based and thus find a causal explanation for them, in the process of which the real problem is merely pushed aside, in morphology the unity of the type itself is the principle that brings clarity into what is a rather obscure and confusing diversity of organic forms. The task of morphology is therefore not to analyze the type but to bring it, self-contained as it is, to light" (Troll 1935/1936, p. 355).

Troll rejected Darwinism as "crass externality of the English perception of things" and hoped that by publishing his *Gestalt und Urbild* (Form and archetype) he could contribute to the "rebirth of morphology from the spirit of German science" (Troll 1941, preface).

Fritz Knoll (Vienna) received funding for his studies of the morphology of fruit and for the rebuilding of the botanical garden in Vienna only after Fritz von Wettstein interceded on his behalf. Bruno Huber's application for studying historical and prehistoric wood artifacts, including his work on a Central European chronology of annual rings, was initially turned down as having no importance to the state or the war. This verdict changed once the DFG found out about his collaboration with Professor Hans Reinerth, the leader of the Reich League for German Prehistory, who had been charged by Hitler himself with working on prehistorical issues.

Elise Hofmann in Vienna received a research contract in 1944 for paleobotanical studies; it was to be used for studying coal deposits and yield clues to the way of life of prehistoric and early historical people.

2. Zoology

2.1. Applied Zoology

In 1934 alone, the DFG granted just under RM 100,000 for joint work in the field of animal breeding, which was carried out at various institutes of applied and agricultural zoology.[28] However, since the present study is limited primarily to institutes of nonapplied zoology, only a small portion of all the applied research after 1933 will be described (as was also the case in botany). Wolfgang Herre (Halle) received grants for work in the field of hormone physiology and research into domestication. He analyzed the effect of additional implanted pituitary glands on salamander larvae and studied the pituitary glands and gonads of pigs with respect to hormone production and its effect on the constitution and skeletal peculiarities of the animals. As result of his studies on domestication, he noted (1943) that domestication influences in various domesticated animals had developed along parallel lines even in histological details.[29] One main area of research of Carl Kronacher (Institute for Animal Breeding and Genetics of Domesticated Animals of the Agricultural College, since 1936 of the University of Berlin) from the end of the twenties was twin research involving cattle. His successor Jonas Schmidt compared the growth of identical cattle twins fed different diets. The director of the Institute for Hereditary Biology and Racial Hygiene of the University of Frankfurt, Otmar von Verschuer, described research on cattle twins under varying environmental influences as very valuable for research on human twins.[30] During the Nazi period, twin research itself was carried out above all to prove the heredity of all essential traits of a person, including intellectual and emotional characteristics; it thus served to provide scientific legitimation for National Socialist racial doctrine.

Werner Ulrich (Berlin,), Bruno Geinitz (Freiburg), Ruth Beutler (University of Munich), Peter Rietschel, and Hermann Giersberg (Frankfurt) received grants for research on apiculture. Friedrich Kröning (Göttingen) received grants in 1943 and 1944 for work on the science of hunting. Erich Wagler (University of Munich) and Wilhelm Wunder (Breslau) worked in fish farming. Karl Friederichs (Rostock, after 1941 Posen) was given financial backing to study forest pests, including the habits of the pine looper and possible ways of combating it.

Institutes of animal breeding, too, could run into difficulties with materials. The head of the Institute for Animal Breeding and Breeding Biology at the TH in Munich, Heinz Henseler, controlled a yearly budget of RM 565 in 1935. In 1936 he received RM 1,800 from the DFG but had to return it since he was not able to use it, owing to a lack of a research farm for animal breeding. Arnold Scheibe from the Institute for Plant Breeding at the TH in Munich had similar problems: the Ministries of Finance and Education had approved the acquisition of land for experiments, but the Bavarian State Ministry (Landesministerium), which had jurisdiction in this question, did not see itself in a position to furnish proof of such land let alone turn it over.[31] At the beginning of 1942, Himmler, who had studied agriculture with Henseler in Munich, therefore pleaded the case of his former teacher with the Bavarian minister-president, though without success, at least until 1944. He was able, nevertheless, to assure Henseler that General Governor Hans Frank was willing to make a large experimental estate in Galicia available to the TH in Munich.[32]

Heinz Henseler used his connection to the SS for more than getting help from Himmler. He apparently also took his students on study trips to the agricultural installation of the Dachau concentration camp. SS Lieutenant Colonel Heinrich Vogel wrote to Himmler in 1942 that Henseler "has expressed the wish to undertake again this year a study trip to our installations in Dachau. I have told him that he can certainly visit the installations with his students with prior appointment."[33]

Since health, food, and house pests became an increasingly serious problem during the war, many of the zoological projects officially recognized as important to the war after 1940 addressed issues of pest control. The SS and Himmler, too, took a special interest in research on biological pest control (see Chapter 5). What follows are outlines of some of the projects. Hanns von Lengerken (Berlin) studied, as "war-important research," the increase in the effectiveness of gaseous insecticides through the addition of carbon dioxide. The economic value of this work related to the war in that carbon dioxide increased the effect of chlorinated hydrocarbons, which made it possible to save important raw materials. In November of 1944 a second project was approved, for which Lengerken also received money from industry: "The use of substances containing hydrogen thiocyanate to combat rats and mice while taking into consideration at the same time the effect of scent and color substances." In justifying his work, Lengerken mentioned the war-related loss of the most common rat poison, the sea onion, and the declining production of other substances such as thallium sulfate. Lengerkern thus wanted to introduce new, previously unused substances into rat control, for example certain highly toxic compounds that occurred as waste products in large-scale chemical industry. These substances were to produce hydrogen thiocyanate in the intestinal tract of rats. More detailed information or reports about the results of this work are not available.

Research projects dealing with the destruction of raw materials by pests were supervised by Franz Duspiva (Heidelberg) and Egon Schlottke (Rostock, after 1942 Strasbourg). In this capacity Duspiva worked first as research commissioner for the Reich Office for Economic Development, later on with financing from the RFR. In 1943 he was exempted from military service, that is, given "uk" (indispensable) status, in order to continue his war-important research with the urgency classification "SS." The topics of his research contract were as follows: (1) the biological and chemical processes that occur when pests destroy raw materials (digestion of keratin, wax, collagen, and so on); (2) the formation and catabolism of animal spinning fibers; (3) the occurrence and effect of enzymes during cell growth. Duspiva demonstrated that not only the larvae of the clothes moth but other insect larvae as well digest keratin by making the keratin molecule susceptible to an enzymatic breakdown through the reduction of the sulphur bridges. The study of the enzyme-containing secretion of the silkworm moth led to the precise characterization of the enzyme that splits silk sericin.[34] Egon Schlottke worked on a research contract concerning the digestive enzymes of wood-eating beetle larvae.

In 1940 Wilhelm Jürgen Schmidt (Gießen) submitted a proposal to the RFR to use the work of the institute in the field of polarization optics for war-important research to study the minute structure of certain fibers. He sought to determine important characteristics such as tensile strength and to supervise the production process of artificial fibers. The RFR approved his request. Adolf Remane (Kiel) received RM 80,000 in 1942 for the purchase of a small research ship. The OKM gave its approval in view of "the cooperation which . . . has already occurred on several occasions between the navy and the Institute for Marine Science in Kiel."[35] Remane focused on marine ecology. The DFG files provide no insight into the nature of the cooperation with the OKM. Ilse Fischer (Münster) in 1944 studied the effect of industrial types of dust in tissue cultures; the DFG files do not contain a research report.

In 1943 Wilhelm Goetsch (Breslau) received a war-important research contract on the following topic: "Studies of Termites and Other Tropical and Subtropical Pests." The significance of these studies, with the urgency classification "SS," emerges from the information in his 1944 research report to the DFG: he aimed not to control pests but to find a way to make usable certain active substances of termites and of fungi used by termites.[36] He carried out testing of termite-fungi extracts as growth accelerators in animals and important nutritive yeasts. At the same time the discovery of the inhibitory effect of some termite fungi on tissue and bacteria led to the beginning of research on antibiotics. Goetsch, like all scientists working on fungal substances, belonged to the working group "Penicillin Circle," headed by

Paul Rostock, the commissioner of the führer for medical science and research. According to Goetsch, American antibiotic researchers had focused on practical and chemical aspects of the active substances of fungi. In contrast, he wanted to work out proposals in which the possibilities inherent in the antagonism of microorganisms could also be used to fight pathogenic bacteria in humans in a biological way. We have no knowledge of any results of this work.

Research into antibiotics accelerated during the war. The Nobel prize winner and biochemist Richard Kuhn from the Kaiser Wilhelm Institute for Medical Research in Heidelberg received about RM 25,000 in 1943, alone, specifically for antibiotic research.

The conquest of colonies in Africa or Latin America was, if at all, only a secondary long-term goal of National Socialist policy, which from the beginning focused on the expansion eastward. Nevertheless, in the late thirties many scientists proceeded on the assumption that colonies would be acquired in the near future. In DFG applications the speedy recovery or new acquisition of colonies was considered necessary and justified. Some colonial institutes were established, one in 1938 in Hamburg. The Institute for Colonial Agriculture in Göttingen (1940) was founded despite the opposition of the central authorities in Berlin. Behind these moves stood the hope that in the wake of victory over England, trained personnel would be needed once again for the German colonial service (Becker 1987). Zoologists also carried out research projects in colonial zoology; until 1942, these projects were supported primarily by the DFG and the RFR. Projects were approved by the RFR in consultation with the NSDAP's Office of Colonial Policy.

The following examples illustrate the hopes for a great German colonial empire that were shared by many scientists, as well as the racism of these scientists. Adolf Meyer (after 1940 Meyer-Abich), associate professor of philosophy and the history of science at the University of Hamburg, was charged in 1936 with setting up and heading the German-Dominican Tropical Research Institute in Ciudad Trujillo (Santo Domingo, Dominican Republic). In his DFG application in 1938, Meyer indicated that the institute would train scientists in tropical work, so that "we shall have many young scientists when the time comes for the return of our colonies, which we hope will happen soon."[37] Via Meyer, the DFG supported the scientific work of this institute between 1938 and 1942.

At the request of the NSDAP's Office of Colonial Policy, Adolf Meyer wrote a memorandum in 1940 entitled "Thoughts on the Organization of Scientific Research in the German Colonies."[38] In it he emphasized that in contrast to the first German colonial empire, in which scientists hardly took an interest in the colonies, "in the coming greater German colonial empire, scientific development will be the pacemaker and pioneer of economic work in the colonies." With regard to his own experience that many young workers in the medical colonial service in Santo Domingo had proved failures, he noted: "Part of the failure of the men was attributable . . . to the failure of the women. Years ago, competent people in Germany argued that only married civil servants should be sent to the colonies, and surely for good reason, namely to prevent the dissipation of valuable German blood." Meyer thought differently: in his view, women, out of a longing for the larger cities, tended to impede their men, "because the vast majority of women have very little understanding for the greatness of nature that surrounds them everywhere in the tropics." Thus it would be better to send out unmarried men, "after all, our colonies are no longer a wild land inhabited only by Negroes. Fortunately there are still, and once again, many German farmers in the colonies, whose pure-blooded [*reinblütige*] daughters have no longing for Berlin . . . (and) will make the best conceivable wives."[39]

The institute got into trouble because German researchers behaved disrespectfully and in some cases unscrupulously toward native school directors and young girls. Despite their protest, the girls had to undress completely so that anthropologists could take their measurements.[40]

The DFG-supported research in colonial zoology included studies of diseases of colonial domestic animals by Anton Koegel (TH Munich), research on coffee pests in "German East Africa" by Gernot Bergold (Entomological Branch of the Kaiser Wilhelm Society in Oppau), and Rudolf Lehmensick's zoological, parasitological-medical work in East Africa on the epidemiology of trichinosis.

Like many other scientists, Lehmensick firmly believed in Germany's legitimate claim to colonies: "By carrying out their often pathbreaking studies in the field of tropical research, unswayed by the bitter fact that they do not possess their own colonial territories and serving only the cause itself, German scientists have demonstrated before all the world the legitimacy of our claim to colonial possessions" (Lehmensick 1940). During his research trip to Tanganyika, he also appeared as a speaker at party events of the local NSDAP. In his 1938 report to the DFG he said of that part of his trip: "Thus on the same evening I gave a lecture to the local group of the NSDAP in Dar es Salaam . . . on the scientific basis of the eugenic laws of the Third Reich. (I had already given a similar lecture in Mbeya to the German colony under the chairmanship of party comrade Dr. Eckardt.)"[41]

Other biologists also worked with the NSDAP organizations abroad on their trips to the tropics. Thus Hans Krieg (University of Munich), in his DFG application in 1937 for funding for his South America expedition, emphasized that he considered cooperation with NSDAP organizations abroad very valuable. In his view it was "necessary that the people who go abroad are uncompromising and honest supporters of the Third Reich without appearing from the outset as propagandists. We scientists are best suited for this task . . . Our positive work has more effect in the long run than all journalistic noise."[42]

Krieg's research centered on systematic and ecological-geographical work on birds. Albert Ehrhardt (Rostock, since 1941 Posen) was funded for parasitological studies of worms (including hookworm) and ways of combating them pharmacologically. In his DFG reports he indicated that he had demonstrated in animal tests that a preparation from a certain company was a universal vermicide, and that he now wanted to try it on humans in the colonies. The DFG documents do not reveal whether such tests were carried out.

In the early forties, the policy of expansion of *Lebensraum* in the East, pursued especially by the SS, came to the fore as the primary goal of National Socialist politics. The following examples illustrate the efforts, beginning around 1940, to restrict colonial research in favor of research on the "German East." Heinz Henseler, mentioned above, was head of the working group on colonial science at the Institute of Technology (TH) in Munich. In this capacity he visited, in 1941 at the request of the RFR, the installations of colonial science in the occupied parts of France, Belgium, and Holland. When he wanted to present the colonial work on the occasion of a university celebration in Munich in December of 1941, he was given "a hint from above . . . to tread softly."[43] After that time public colonial events could not be held, though colonial research and teaching were continued. Henseler, who continued to have hopes for colonies (such as Cameroon), sent Himmler his plans on the future collaboration of the TH in the tasks of colonial science. In Himmler's response in March of 1942, we read the following: "As I see the overall development, the colonies in Africa will be a sidelight; for the greatest colony the German Reich has ever had lies outside its door: Russia. And this colony will one day no longer be a colony but will become our settlement region . . . It would be most agreeable if the Technische Hochschule in Munich, specifically the Agriculture Department, but possibly also the entire college as such, were to orient its research toward providing and training capable manpower for the East, in the first place for the General Government, Poland, and Galicia."[44] The files of the DFG and the BDC provide no information about Henseler's subsequent research. He retired with emeritus status in 1948.

2.2. Ecology

Questions about the interrelationship between organisms and their environment have exercised minds since antiquity. Ernst Haeckel suggested the term "ecology" as the teaching of the "household of nature" in 1866. The naturalist Alexander von Humboldt (1769–1859) is considered the founder of plant geography. Many ecological studies of the nineteenth and early twentieth centuries dealt with the habitat conditions of individual species and the influence of ecological factors on their diffusion. They included Karl August Möbius's 1877 book on an oyster bank, in which he introduced the term "biocenosis"; Karl Semper's *Die natürlichen Existenzbedingungen der Thiere* (The natural conditions of existence of animals, 1880); and later Richard Hesse's *Tiergeographie auf ökologischer Grundlage* (Animal geography on an ecological basis, 1924). Ecology in Germany did not exist as a separate field at universities until the 1960s. Until then, zoologists and botanists addressed ecological questions. During the Nazi period, comprehensive ecological treatises establishing a theoretical foundation for ecological laws were published by August Thienemann, on the basis of studies in hydrobiology, and Karl Friederichs, in applied entomology (that is, pest research). I will discuss these two works before mentioning some other ecological research and its significance at the time.

August Thienemann was head of the hydrobiology station of the Kaiser Wilhelm Society in Plön from 1917 on. He developed fundamental concepts for the ecology of waters and clarified interrelationships between the members of limnological biocenoses, that is, biocenoses of lakes. In 1921 he introduced the terms "oligotrophic," "eutrophic," and "dystrophic" for the condition of lakes, and in 1931 he published the essay "Der Begriff der Produktion in der Biologie" (The concept of production in biology). In it he laid the basis for the quantitative analysis of the life processes of a biotope (see Jahn, Löther, and Senglaub 1982, p. 614). Like Karl Friederichs, Thienemann emphasized the dynamic character of a biocenosis, whose capacity for self-regulation was aimed not at favoring individual members but at the preservation of the system as a whole (1939). On the basis of his limnological research he described ecology as a steplike structure (1941). In it "autoecology" as the study of single lives, "biocenosis" as the study of life within a community, and "general ecology" as the study of the household *(Haushalt)* of nature (adopting Ernst Haeckel's old term) build on each other. The DFG gave Thienemann funding above all for applied research. In 1940 he received support for "a) studies of pond management on a ground of primary rock (influence of fertilization, draining, etc.), b) studies of the growth of economically important fish, c) studies of the physical, thermal, and chemical conditions of wet soils and their animal colonization, and d) studies of the limnological nature of a lake that was transformed into a marsh through blasting."[45] In addition, Thienemann was working on a taxonomy of Chironomidae and water mites.

The entomologist Karl Friederichs (Rostock, after 1940 professor in Posen) received funding from the DFG and the RFR from 1938 to 1944 for research on the habits and possibilities of combating various forest pests, in particular the pine looper and spruce sawfly. In his publications between 1927 and 1950, Friederichs worked on clarifying basic concepts and theories of general ecology, starting from the analysis of active ecological factors among animals. As early as 1927 he described the close connection between biocenosis and biotope and emphasized the self-regulatory achievement of a biocenosis. His view of ecology, which approached holism, will be briefly discussed, since it represents what was for the twenties and thirties a typical attempt to transcend so-called materialism in science, which was often equated with "Jewish science."

Unless otherwise noted, the following quotes are taken from Friederichs's 1934 essay "Vom Wesen der Ökologie" (The nature of ecology). "Within the natural sciences, ecology is the highest synthesis of . . . all the natural sciences into the science of nature." Ecology is a path "to the total world view, a view of the world in which everything is related to everything else, everything directly or indirectly affects everything else, and in which everything is simultaneously in motion and in a process of transformation." The ecological method common to all sciences was, in his view, "that one does not—as used to be commonly done and as is readily the habit of the specialist—detach every phenomenon from its larger context." This "ecological" conception was spreading to all spheres of life: "the fostering of local traditions and nature conservation, city planning, the Volk as community, the economy as organism, etc."

Friederichs thought he was justified in believing that goal-oriented, teleological processes led to the preservation of a norm in nature. In his 1927 essay he had already spoken of the self-regulation of an ecological system with the goal of a biocenotic balance. In 1934 he wrote: "In natural science, purposefulness, teleological happening is nothing other than the direction of occurrences that is inherent in a natural phenomenon and strives for the preservation of a whole in an existing state, which can also be a development, an augmentation, a becoming toward the norm." This norm is determined by what Friederichs assumed to be the "harmony of the life-form with its environment"; he also spoke of the "regional determination of every life-form" and of the "dependence of the race on soil and atmosphere." From such a perspective the world can no "longer be thought of as a sum of beings and things which, created by chance, stand in random relationships to each other and are governed by chance." Rather, "people have remembered once again that [the world] is a cosmos, an orderly structure, a whole. With the repressing of materialism, this notion, once so self-evident, is regaining recognition as a basic foundation."

As for the practical tasks of ecology, Friederichs listed solutions to the problems of the "enormous increase in damage from insects (dying forests), the polluting of rivers, the decline of fish stocks in rivers and in the sea, and everywhere the negative side effects that are linked with the unrestrained advance of technology."

Friederichs's conception of nature, in which teleological processes create a harmonious whole with a common norm, in which there is no room for chance, and in which the regional determination of each life-form is of central importance, shows many similarities to the racial ideology of the National Socialists. And Friederichs certainly wanted his efforts toward spreading ecological thinking to be seen as a contribution to the politics of National Socialism. "Such an effort on the part of the ecologist would be useless if it were not entirely in agreement with the current of the time, especially the political one" (1937, p. 86). He quoted Hitler (without a source reference)[46] and described ecology as the "doctrine of blood and soil" (1937, p. 91).[47]

In the fifties, the definitions of the terms "biotope" and "biocenosis," and their theoretical justifications as formulated by August Thienemann and Karl Friederichs, among others, were criticized and in part rejected as "anthropocentric" (see Jahn, Löther, and Senglaub 1982, pp. 615–617).

Aside from Thienemann and Friederichs, the following zoologists worked on ecological issues: Adolf Remane (Kiel) studied biocenoses on beaches and sandy tidelands. Max Auerbach's ecological studies of Lake Constance, which he had been carrying out since the twenties, did not receive any further DFG funding after 1934. Hermann Weber (Münster) was funded until 1939 for his work on the ecology of animal and book lice and for studies of the parasitical hymenoptera and *Drosophila*. Wilhelm Goetsch (Breslau) carried out research on ecological and sociological questions in South American ants. He and Rose Stoppel (Hamburg) succeeded in determining the kind of fungi cultivated by leaf-cutting ants and in cultivating these fungi in the institutes in Breslau and Hamburg (Goetsch and Stoppel 1940).

The applications for much of the DFG funding for ecological research in botany referred to its importance for the new "German eastern territories." From 1939 on, Kurt Mothes received funding for the topics "Causes of the Dying Off of Pine Trees" and "Forest History and Ecological-Physiological Studies in East Prussia." After 1943, he continued this research "with special consideration for the new German eastern territories." Mothes suspected that a virus was the responsible for the dying of the pine, which was regarded as a serious economic problem in some areas of East Prussia. However, this hypothesis was not proven until 1945. At the Congress of Botanists in 1942, Hans Burgeff, who became known through his studies of mosses, took on the task of establishing a school to train experts on the mycorrhizae of forest trees. In 1944 he received a DFG grant to study mycorrhizae in marsh and forest. In his application he stated that these studies, "though not directly important to the war, will be of importance for rebuilding the forest in the East; on this you might want to consult with Professor Mothes, Königsberg." He did not explain in greater detail what he meant by the "East." In this context it should be mentioned that since 1940 there had existed SS plans to settle Germans in the occupied countries of Eastern Europe; to that end a part of the population living there was to be driven out and exterminated. According to the opinion of one of the anthropological experts for settlement questions, Professor Eugen Fischer, in some areas the settling of Germans was, for climatic reasons, only possible if rich forests were established and in this way a change in climate brought about (Heiber 1958). Ecology thus had a permanent, valued place within the imperialistic policy of National Socialism.

2.3. Collaborative Work in the Field of Genetic Damage from Radiation

Some of the mutation research carried out during the Nazi period, for example the work of Alfred Kühn and Nikolai Timoféeff-Ressovsky, was the continuation of a research program of the old Emergency Association (Notgemeinschaft, prior to the presidency of Stark) as later administrated by the DFG and RFR divisions on medicine and "racial care and hereditary biology relating to population policy" but not the division on general biology. In September of 1933, a number of geneticists and doctors gathered in Berlin at the invitation of the Emergency Association to outline a working plan for tackling further experimental studies on the question of germ damage through X-ray and radium radiation (Martius and Kröning 1936). Joint efforts involving experiments on *Drosophila* were to provide further clarification in particular of the effect of radiation on germ plasm, the relationship between mutation rate and the dose of radiation, the effect of various radiation qualities, and the significance of the length of exposure for the genetically damaging effect of the radiation dose. Experiments with mammals (especially guinea pigs and mice) on the question of genetic damage through X-ray radiation were to follow.[48]

Alfred Kühn, at the time professor of zoology in Göttingen, became chairman of this Notgemeinschaft commission. The commission members included his assistant Friedrich Kröning, Heinrich Martius (head of the Women's Clinic in Göttingen), Lothar Löffler (since 1934 professor for racial biology in Königsberg), Paula Hertwig (associate professor at the Institute for Hereditary Research of the University of Berlin), and Nikolai Timoféeff-Ressovsky (head of the Genetics Department of the KWI for Brain Research in Berlin-Buch). As the discussion at the meeting of the commission in 1934 reveals, the scientists emphasized—at least publicly—the possible practical use of their research as it related to genetic damage in humans.[49] This explains the interest that National Socialist racial hygienists took in this zoological mutation research. At the meeting, Timoféeff-Ressovsky reported on the induction of mutations in *Drosophila* through X-ray radiation. Vitality-altering mutations, that is mutations without any morphological changes, were induced two and half times more frequently than lethal or sublethal factors, most causing a reduction in vitality. Thus he concluded: "Such mutations, applied to humans, must be considered as particularly undesirable from the perspective of racial hygiene, for they cause a hereditary constitutional weakness that is not serious enough to eliminate itself from further propagation through a quick death, and they show no crass and clear pathological characteristics by which one could easily recognize them."[50] He also mentioned the significance of the frequent, small vitality mutations for issues relating to population genetics and evolution. According to Kühn, who reported on the effect of mutated genes in the flour moth *Ephestia,* most vitality-reducing mutations were dominant, unlike visible mutations. Hertwig presented preliminary results of her radiation experiments with mice, which were intended to find the best radiation dose for experiments concerning genetic damage. Relevant preliminary tests with guinea pigs were carried out by the Zoological Institute in Göttingen (Kühn and Kröning) in collaboration with the Women's Clinic (Martius).

At this meeting, the commission decided to expand mutation experiments involving mammals, to carry out studies on the threshold values of the effects of radiation, and to conduct preliminary tests with invertebrates on chemical mutagenesis (Kühn and Timoféeff-Ressovsky). Since many scientists in biological and medical disciplines were interested in genetically pure breeds of mammals, it was urged that the breeding installations of the Notgemeinschaft, located in Göttingen and on the government estate Plauerhof and headed by Kühn and Kröning, be expanded. In November of 1933, the Notgemeinschaft approved grants in the amount of 30,000 to RM 40,000 for these joint efforts.[51] The new president, Johannes Stark, completely agreed with the commission's suggestion for the direction of research, and scientists continued to decide on their own the content of their research.

Until 1945, Alfred Kühn, Friedrich Kröning, and Nikolai Timoféeff-Ressovsky were among those who received the largest DFG grants in the field of genetic research.

The following information and quotes come from the DFG files on Kühn. During his time in Göttingen, Kühn was, until 1936, funded with about RM 40,000 from the Notgemeinschaft for work on "hereditary experiments and collaborative work with the goal of breeding laboratory animals with special constitutional characteristics for use in various special fields of medicine including that of genetic damage through X-ray radiation." Together with his assistant Friedrich Kröning, he set up a genetically based breeding program for laboratory animals, in which specific traits were to exist homozygously through continuous inbreeding. According to Kühn, the breeds of laboratory animals were intended for use in carrying out "model experiments on questions concerning the cause of human genetic diseases and for experimental therapy." Zoologists, Chemists, and Pathologists collaborated on questions of tuberculosis resistance, the testing of serums for diphtheria, the genetic predisposition for cancer, and genetic damage from radiation. For example, Kröning, in collaboration with the Institute for Experimental Therapy in Frankfurt, used eighteen inbred strains to carry out experiments on susceptibility to tuberculosis. Mice and guinea pigs from the Göttingen program were bred for a number of users, among them the Reich Health Office (RGA), the Hygienic Institute, and the Pharmacological Institute in Marburg. The racial hygienist Ernst Rodenwaldt (Heidelberg) studied genetically conditioned resistance to diphtheria in animal breeds from Göttingen. Mice with hereditary cancer were supplied to the Cancer Section of the State Institute for Experimental Therapy in Frankfurt for research purposes and to the Pathology Institute of IG Farben for testing cancer drugs. In 1935 the president of the RGA, Hans Reiter, expressed his strong support to the DFG for Kühn's plans concerning the breeding of laboratory animals.

Kühn directed another large-scale installation of the Notgemeinschaft and the DFG for breeding laboratory animals on the prison estate Plauerhof. "The collaboration with the administration of the prison estate . . . continues to be positive."[52] After Kühn went to Berlin, the breeding was continued under the direction of Kröning. I describe Kühn's other research in Chapter 4, Section 1.

2.4. Genetics

As with botanical research, most genetic research was carried out at Kaiser Wilhelm Institutes. Very little experimental mutation research and no virus research at all was done at universities. Research at universities concerned primarily developmental genetics. Work in this field was done by Alfred Kühn and his school. Like the research of Richard Goldschmidt, who was forced to emigrate in 1935, it stood in the tradition of "phenogenetic" research in Germany. Valentin Haecker (1864–1927), a student of August Weismann, coined the term "phenogenetics" in 1912. He understood it to mean an area of research that studies "in terms of morphogenetics and developmental physiology the creation of a finished organism's external characteristics" and seeks "to trace their roots back into the earliest possible developmental stage by going back step by step to the intermediate processes that are effective during development, and to the temporary characteristics" (Haecker 1918, p. 4).

According to Haecker, the two main elements of Mendelian genetics, the "visible, mature external characteristics of the finished organism" and the "invisible, hypothetical hereditary traits embedded in the germ cells," which so far had been only logically connected, were to be "related to each other through a chain of actual observations." Richard Goldschmidt, too, rejected a purely formal theory of inheritance. From the genetic and developmental-physiological analysis he sought to deduce a theory of heredity that was supposed to shed new light also on evolutionary problems (Goldschmidt 1927b). According to Georg Melchers, genetics oriented toward developmental physiology also formed the basis on which the beginnings of a general biology in Germany took shape through the collaboration of Alfred Kühn and Fritz von Wettstein when they were, respectively, professors of zoology and botany in Göttingen in the late twenties (Melchers 1987).

Of Kühn's students, Karl Henke (successor to Kühn as professor of zoology in Göttingen after 1937), Otto Kuhn (professor in Cologne since 1935), and Wilhelm Koehler (since 1936 assistant in Munich), were supported by the DFG for phenogenetic research. From 1933 to

1935, Henke was assistant to Goldschmidt at the KWI for Biology, and in Göttingen he continued Goldschmidt's developmental genetic analysis of the wing pattern of butterflies. His work, carried out especially with *Ephestia* and after 1937 also with *Drosophila,* contributed to the appreciation of the insect wing as the best-known complex animal organ in terms of developmental genetics (Kühn 1957). Through temperature stimulation of the eggs, larvae, and pupae of *Drosophila,* he and his collaborators induced various modifications in the fully developed fly (Henke, Finck, and Ma 1941). Numerous modifications bore a strong resemblance to the phenotypes of certain mutations and were called "phenocopies" of these mutations, a term Richard Goldschmidt had created in 1935. Henke and his collaborators discovered that in some modifications of the size, form and tension, and structure of the interior of the wing not only the final state but the course of the development, as well, corresponded to those of certain mutants. Henke called these modifications "true phenocopies"; their sensitive periods come before the phases in which, through mutation, the change in the development of the relevant trait becomes visible, or they overlap with them. The appearance of polar fluctuations of genetic manifestations, in which the same mutation can cause contrary changes of the phenotype, was also found with temperature modifications.

Rolf Danneel was funded for phenogenetic research involving primarily vertebrates. Through his work on genetically conditioned pigment formation in rabbits, he clarified two successive gene-dependent cell-physiological processes and demonstrated the cold modification of one enzyme involved (Danneel and Paul 1940). Danneel began his research as an assistant to Otto Koehler in Königsberg and continued it after 1939 at the KWI for Biology in Kühn's department. An agreement between Wettstein (then director of the Kaiser Wilhelm Institute) and Koehler had led to a trade of Danneel and Georg Gottschewski (until 1939 at the Kaiser Wilhelm Institute).[53]

Gottschewski (from 1939 to 1942 in Königsberg, thereafter head of the Genetics Section at the Institute for Racial Biology in Vienna) was one of the few university biologists working on *Drosophila* genetics. He analyzed processes during subspecies and species differentiation in twenty wild populations of *Drosophila.* He discovered two factors on the X chromosome that were responsible for the genetically conditioned isolation mechanism of two populations (Gottschewski

1939). In 1944 he began drawing up a cytogenetic chromosome chart of *Drosophila obscura*.

Paula Hertwig described six new recessive mutations in the house mouse (four after exposure to X-ray radiation), nearly all of which had a pleiotropic effect. Five of the mutations were pathological, three could be safely integrated into existing linkage groups (Hertwig 1942).[54]

During the National Socialist period, research in human genetics *(humangenetische Forschung)*—this term established itself in Germany only after 1945—served primarily the goals of racial hygiene. It was carried out mainly by anthropologists and medical scientists in institutes of anthropology or racial hygiene. I will not discuss this research. Among the handful of zoologists working in the field of human genetics were Hans Nachtsheim, whose career and research I describe in Chapter 4, Section 7, and Günther Just. In December of 1933, Associate Professor Just was made director of the Institute for Human Genetics and Eugenics at the University of Greifswald (after April 1936 it was called the Institute for Hereditary Science). He also became head of the Reich Health Office's Research Institute of Hereditary Science in 1937 and head of the Institute of Racial Biology in Würzburg in 1942. He was one of the few geneticists to study genetics in humans not suffering from hereditary diseases, for example questions relating to the genetic basis of a person's constitution.[55] For that reason I will discuss his approaches in somewhat greater detail. Until 1937, the main areas of his research were "studies in the border area between the science of heredity and pedagogy," later "basic questions relating to the genetic and constitutional biology of the entire person," among them the genetic basis of achievement. In addition, until 1940 he received smaller grants for experimental genetic studies of *Drosophila* and Orthoptera and for work on the phenogenetics of the cat.

Three examples illustrate Just's working method and the direction of his research. First, he studied the genetic basis of constitutional types in the manner of Ernst Kretschmer. He put forth the hypothesis that, just as gender was genetically determined by a quantitative-effect-relationship between male and female dispositions, so the direction the psychophysical building plan would take was determined by the ratio of leptosomatic-schizothymic to pyknic-cyclothymic genetic potential (Just 1939). Second, he studied the connection between performance in school and aptitude type by grouping former gymnasium graduates into various school achievement groups—based on the percentage share of the various professions—and comparing them. He discovered that teachers at universities and higher schools were clearly more strongly represented in the higher school achievement groups, while the opposite distribution was found among doctors. Primary school teachers *(Volksschullehrer)* were in between. From this he inferred the existence of different aptitude types; theoretical tendencies were supposedly more strongly present in teachers, while practical, life-oriented aptitudes corresponded to a basic disposition that focused more on practical medical work. He found the same distribution also among the children of the various professionals: children of doctors were clearly behind children of primary school teachers when it came to academic performance. He explained this in part by noting that children for the most part shared their parents' predisposition for greater or lesser aptitude for schoolwork (Just 1939). Third, in his DFG report of February 3, 1944, Just related that he had ascertained, through a series of tests, positive relationships among constitutional type, average school achievement, and IQ (Binet-Bobertag method).

He planned other series of tests to determine the relationship between intellectual norm or hereditary mental illnesses and the distribution of papillary patterns and constitutional types. For this purpose Just determined the frequency of papillary patterns and the genes on which they were based for the population of Würzburg. In addition, he performed a larger test series involving juveniles to find out how this pattern correlated to the hereditary morphology of the nose. We have no information about the results.

2.5. Experimental Embryology

Experimental Embryology of Vertebrates. Wilhelm Roux was the first biologist (around 1885) to use the physiological investigation of causes in the study of the development of individual organisms (ontogenesis). His approach stood in contrast to the embryology of Ernst Haeckel and Carl Gegenbaur, which was exclusively descriptive or phylogenetic in orientation. Roux introduced the concept "developmental mechanism of organisms" *(Entwicklungsmechanik der Organismen),*[56] which was in Germany soon replaced by "developmental physiology" *(Entwicklungsphysiologie),* a concept coined by Hans Driesch. (The English term "developmental biology" was not introduced until the fifties and refers to the next great transformation of this discipline.)

Until about 1910, questions relating to experimental embryology dominated zoology. As examples I shall mention the debates between Hans Driesch and Wilhelm Roux about the vitalistic or mechanistic interpretation of blastomere separation or blastomere elimination experiments, and the discovery of mosaic and regulation eggs by Hans Spemann.

As a result of the difficulties that arose in the experimental analysis of complex physiological processes, many developmental physiologists in the United States turned to genetics and general physiology in the mid-twenties (Harrison 1937). Thomas Hunt Morgan, who had collaborated with Driesch for some time in the field of developmental physiology, began, as early as 1910, to work on heredity—specifically, the transmission of genetic information from generation to generation—separate from embryonal development. Morgan wanted to place biology on an equal footing with physics and chemistry as an experimental science. He therefore thought it necessary to solve genetic questions before turning to more complex questions of developmental physiology in order to free biology of its speculative and metaphysical thinking (Allen 1985).[57]

In Germany the most important contributions to experimental embryology in the twenties came from Hans Spemann, the only experimental embryologist ever to win the Nobel prize (in 1935). After 1918, experimental embryologists focused on questions relating to the timing and course of determination (the determination of the subsequent developmental direction of an embryonal part) in early amphibian development. In 1924 Spemann and Hilde Mangold published the results of their transplantation experiment, in which a certain part of the amphibian embryo was able to induce the development of new embryos even when transplanted into embryos of other species (Spemann and Mangold 1924). This experiment is widely regarded as the most outstanding achievement of classical experimental embryology. It led to many experiments on the nature of the organizer (see Chapter 1, Section 5).

After 1933, experimental embryology that was not explicitly genetic did not receive much funding from the DFG. As full professor in Freiburg, and since 1937 as emeritus, Spemann, until his death in 1941, carried out studies on embryonal induction effects in amphibian embryos with little DFG funding. In 1937 his application for a research assistant was turned down, and in 1940 the remainder of his grant was canceled since his research was not recognized as important to the war. Spemann was not able to verify whether determination fields still existed in the mature organism—for example, whether an inducing liver field still existed in the salamander—by transplanting presumptive ectoderm between liver and peritoneum. However, his experiments gave cause for assuming that the implant forms into a tumorlike growth (Spemann 1942). Johannes Holtfreter, who became one of the internationally leading experimental embryologists (see, among others, Hamburger 1988; Oppenheimer 1970), received, until his dismissal in 1938, little DFG funding for experiments on induction and differentiation processes in the embryonal development of vertebrates. His research in Germany, and in Canada and the United States after he was forced to emigrate in 1939, are covered in Chapter 1, Section 5.

Otto Mangold, from 1923 to 1933 department head at the KWI for Biology, was made full professor of zoology in Erlangen on September 1, 1933, and in 1937 he became Spemann's successor in Freiburg. Until 1939 he received grants for studying the determination and the organizing factors in chimeras of primary germ layers in amphibians. He did not publish any important papers after 1933. Erich Ries, in 1934 lecturer in Leipzig and in 1941 professor in Münster, carried out histochemical studies relating to developmental physiology.

Developmental Physiology of Invertebrates. Ilse Fischer, Max Hartmann's collaborator until 1938, received a research fellowship of the DFG at the Kaiser Wilhelm Institute for Biology. In 1942 she became assistant professor at the Zoological Institute in Münster. In addition to pancreatic cells, she was able to culture insect tissues that could be transferred in several sequences. Together with Georg Gottschewski, she demonstrated in explantation experiments that the eye, leg, and wing primordia of *Drosophila* differentiated themselves *in vitro* up to a certain developmental stage in an organ-specific way (Fischer and Gottschewski 1939).

The only developmental physiologist whose work was recognized as war-important after 1939 was Friedrich Seidel (Berlin). His experiments on the organization of the insect egg were therefore funded until 1944, even though they were of no significance to the war. It can be assumed that Seidel's active work in the party and the SA was decisive in getting his work recognized as important to the war and thus worth funding. Seidel was one of the few developmental physiologists using insects as experimental material as early as the twenties. He studied how the "center of differentiation" in insects took effect; the term refers to an embryonal area from which the formation processes in insect development originated, and whose primordium could frequently be seen already in the structure of the egg.[58] No papers of any importance exist for the period after 1933. During this time Seidel worked on developing a medium for *in vitro* culturing of embryonal parts of invertebrates. Seidel and his students, including Gerhard Krause, were among the most influential developmental physiologists in Germany after the war.

2.6.   Cancer Research

With one exception, medical and biochemical cancer research are not included in this brief review of cancer research. The DFG supported biologists such as the following conducting projects in cancer research: Rolf Danneel (Kaiser Wilhelm Institute) received grants from 1941 to 1943 to study "virus-caused tumors"; the metabolic physiologist Michael Leiner (Berlin) collaborated after 1941 with the Tumor Institute of the Charité to study the effect of the enzyme carbohydrate anhydrase on human tumors. The geneticist Georg Gottschewski was given a research contract in 1944 for "genephysiological studies of carcinoma strains in mice."

After 1939 grants for cancer research increased sharply; like war-important research, cancer research sometimes provided an exemption from military service and a justification for soliciting DFG funds. Many scientists evidently used the claim of conducting cancer research to secure exemption from military service, a maneuver criticized by the director of the KWI for Medical Research in Heidelberg, Richard Kuhn, who was also the head of the Organic Chemistry Section of the Reich Research Council. In 1942 he wrote to the president of the RFR (and Reich Education Ministry), Bernhard Rust, that "work in the field of cancer . . . must not always provide a pretext for 'uk' [exempt] status."[59]

The cancer research of the "half-Jewish" cell physiologist Otto Warburg supposedly allowed him to work at the KWI for Cell Physiology in Berlin largely unmolested during the entire period of the Third Reich.[60] Hitler, despite his hostile attitude toward science, supported cancer research directly: funds from a "führer donation" paid for the buildings of the "tumor farm" at the General Institute for Tumor Research at the Rudolf Virchow Hospital in Berlin;[61] "an ongoing endowment of the führer and reich chancellor" also supported work at this institute.[62]

In 1937 the DFG made an attempt to set up a central organizing office for the work in cancer research that was being carried out in various places.[63] This cancer research program covered the areas of tumor genesis, diagnosis, therapy, and others (for example statistics). As compared with RM 110,000 for all other areas, RM 120,000 was earmarked for research on tumor genesis.[64] The following scientists, among others, received grants for research on tumor genesis in 1937: Friedrich Kröning (Göttingen): breeding of tumor mice; E. Haagen (Berlin): virus research; Max Borst and Wilhelm Stepp (Munich): vitamins and cancer; Heinrich Cramer (Berlin): cultures of human cancer tissue; Boris Rajewsky (Frankfurt): radioactive substances as carcinogenic agents; Robert Rössle and Else Knake (Berlin): interrelationships between normal and malignant tissues.

The director of the KWI for Biochemistry, Adolf Butenandt, who received the Nobel prize for chemistry in 1939 for his work of isolating and determining the structure of sexual hormones,[65] conducted after 1937 research on "cancer and hormones." Three recipients of DFG fellowships, all of whom were given "uk" status, participated in the research.[66] Among them was the zoologist Hans Friedrich-Freksa, who was given his fellowship for studies on the "physiology of normal and pathogenic growth." In research carried out over several years without any conclusive results, he and Heinz Dannenberg attempted to determine whether carcinogenic hydrocarbons occurred in the liver of people suffering from cancer.[67] The question of whether hydrocarbons played a role also in the appearance of spontaneous tumors had arisen after English researchers (such as James W. Cook) had shown dibenzoanthracene and benzopyrenes to be the first carcinogenic hydrocarbons in animal tests in 1929 and 1933.[68]

Together with Hans Aurel Müller and Carl Kaufmann, director of the Women's Clinic at the Charité in Berlin, Friedrich-Freksa began in 1940 an experimental investigation of whether the follicular hormone itself could cause tumors. In thirty-three test series, 5,770 mice of genetically different strains—obtained from, among others, Paula Hertwig and Friedrich Kröning—were treated with varying concentrations of the natural estrogen hormones estrone and equilin, the synthetic substance stilbestrol, and esterified hormone derivatives (Kaufmann 1949). The researchers found no qualitative difference between these active substances; the effect depended solely on their estrogen potency. Only with the application of unphysiologically high doses of these hormones over longer periods of time was it possible to alter the appearance of mammoth tumors beyond the spontaneous occurrence. However, it became clear that in animals with a genetic predisposition to cancer, the follicular hormone could support the formation of malignant tumors. The authors therefore spoke of a "conditionally cancer-causing" rather than a carcinogenic effect. Depending on genetic predisposition, these substances could cause an increased production of various tumors; their effect was thus less specific than that of carcinogenic substances. At almost the same time (1947), Isaac Berenblum, in his croton oil experiment, furnished the first proof for the existence of noncarcinogenic substances that could trigger tumors in conjunction with a subthreshold dose of a carcinogen. He called them co-carcinogens (Süss, Kinzel, and Scribner 1970, pp. 40f.). On the molecular genetic research of Hans Friedrich-Freksa, see Chapter 4, Section 2.

Despite the costs involved, the DFG had plans for the creation of a central institute devoted to breeding and supplying tumor-diseased animal material for the working group on tumor genesis: "This laboratory animal farm should be capable of supplying at any time to all German research institutes all kinds of tumors—tumors caused by immunizations, spontaneous tumors, tumors caused by irritations, etc.— at all necessary stages from flawless large-scale breeding with breeding records."[69] Until 1937 the DFG had two breeding installations for cancer-diseased animals: the breeding in Göttingen (headed by Alfred Kühn and later Friedrich Kröning) of genetically pure breeds of mice with mammoth tumors and of cancer-free breeds of mice and guinea pigs, and the breeding carried out by Friedrich Holtz in the Rudolf Virchow Hospital with genetically pure animal breeds for tumors caused by immunization. Kröning received about RM 160,000 for breeding carcinomatoid breeds of mice and rabbits.

In 1937 Kröning and the organic chemist Hans Lettré took over the Biology Section of the Institute for Tumor Research at the Rudolf

Virchow Hospital as successors to Friedrich Holtz, who had to leave his post on June 1, 1937, because of internal problems.[70] Kröning headed the "tumor farm" until his appointment as associate professor at the Institute of Hunting Science at the University of Göttingen in 1942. In November of 1942, Lettré was also given a professorship at the University of Göttingen. Subsequently Richard Kuhn saw to it that Lettré kept his "uk" status.[71] Kuhn's justification deserves quoting: "I therefore take the liberty to suggest that you give Professor D. H. Lettré, Chemical Institute of the University of Göttingen . . . certification that the work on the cancer problem involving chemically tagged antigens is war-important. This problem is being actively worked on in the United States, and there a number of suggestions have been made for the military use of carcinogenic substances."[72]

The "tumor farm" was taken over by the Kaiser Wilhelm Institute for Biochemistry in 1942;[73] it was later to become a part of the planned Central Institute for Cancer Research. At the meeting of the senate of the Kaiser Wilhelm Society on April 24, 1942, Mentzel, who at this time was still president of the RFR, stated: "From the führer has come the suggestion to set up a central research institute [for cancer research] at a suitable location."[74] He went on to say that the deputy Reich medical führer, Kurt Blome, was already working on a cancer index file housed in Dahlem. The institute was to be set up at the University of Posen and on a nearby estate. The existing working group for cancer research was to collaborate with the Central Institute. In 1942 this working group was made up of Adolf Butenandt, Boris Rajewsky, Friedrich Kröning, Nikolai Timoféeff-Ressovsky, and four other professors.[75] The senate of the Kaiser Wilhelm Society consented to having its general administration take over the administration of the Central Institute; the scientific responsibility for the institute was not to lie with the KWG.[76] Friedrich Holtz was made director of the institute. On the significance of this institute for research on biological warfare under the direction of Kurt Blome, see Chapter 6.

2.7. Animal Physiology and Ethology

Most of the physiological research funded by the DFG lay in the fields of the sensory, nerve, and hormone physiology of mammals, birds, and amphibians. Between 1939 and 1945, this work involving vertebrates received about 75% of the grants for physiological research.

The comparative physiology of invertebrates was not widely present in zoological research and teaching in Germany.[77] It received little financial support.

Hormone and Metabolism Physiology. Gottfried Koller (in 1934 at Tung-Chi-University in Shanghai, in 1940 at Kiel, and in 1941 at Prague) and Otto Kuhn (Cologne) worked in the field of the hormone physiology of vertebrates. In 1944 Koller carried out two research assignments from the Wehrmacht: the influence on wound healing of hormonal growth factors (urgency classification "SS"), and damage to the functioning of the pituitary gland through loss of eyesight (classification "S"). In reports to the DFG in 1944 he indicated the following results: starting from the observation that wound healing in animals accelerated during pregnancy, he showed through experiments of his own that testosterone, the follicular hormone, and the corpus luteum hormone had a positive effect on healing in older animals. He therefore requested that such preparations be given only to wounded who were over the age of thirty. Histamine did not result in any clear acceleration of healing. In experiments with frogs he demonstrated that the effectiveness of the water-balance hormone of the posthypophysis (HHL) is light-dependent. For that reason he believed it would make sense to treat a person who is totally blind with HHL extracts if there were problems with the water balance.[78]

Kuhn obtained funds for studying the interrelationship between hormones and vitamins in the formation of specific characteristics especially in salamanders. In addition, approval was given to applications for hormone-physiological studies concerning the differentiation of the thyroid gland and the timing of its functioning.[79] Work in the field of the hormonal and metabolic physiology of invertebrates was done by Wolfgang Buddenbrock-Hettersdorf in Halle (internal secretion of invertebrates), Curt Heidermanns in Bonn (excretory metabolism of invertebrates), Wulf-Emmo Ankel in Gießen (cytology of the silk gland cells of the silkworm), and Franz Duspiva in Heidelberg (histophysiology of the intestinal tract of insects).

Sensory and Nerve Physiology. Gotthilft von Studnitz (Halle) studied basic aspects of the process of vision in vertebrates, considered war-important research. Studnitz and his collaborators discovered that the spectral absorption of color substances in frog, chick, and guinea pig retina cones was largely identical. They demonstrated the existence of three cone color substances, finding the maximum of the yellow substance at a location that was considered by electrophysiologists as the maximal site of the red substance (Studnitz, Loevenich, and Neumann 1943). Hansjochem Autrum (Berlin) criticized Studnitz's work on the absorption values as methodologically inadequate and riddled with great uncertainties (*FIAT-Reviews*, vol. 52, p. 176). In opposition to Studnitz, Autrum maintained that any significant changes in Hermann L. Helmholtz's theory of three components would be premature (vol. 52, p. 178).

From 1942 to 1944, Studnitz carried out two war assignments for the OKW with the urgency classification "SS" on the problem of night blindness. He reached the conclusion that high doses of vitamin A, administered in an oil emulsion over a period of sixteen days, should improve human night vision ten- to fortyfold.[80] Autrum checked these findings against his own experiments with people and animals and came to a different conclusion: in a normal diet, vitamin A had no effect on dark adaptation.[81] In 1942 Autrum had received a research contract to study dark adaptation of the eye. He carried out experiments at the Research Institute for Aeronautical Medicine of the Reich Ministry of Aviation. Heinz Lüdtke (Königsberg) received grants between 1941 and 1944 for studies of the development of the eye in insects, fish, mammals, and humans. Paul Krüger (Heidelberg) carried out research on the topic "Tetanus and Tonus in Skeletal Muscles"; it was recognized as important to the war.

Autrum was the only physiologist working in the field of the comparative physics and biology of hearing. In 1936 and 1937 his applications to the DFG were turned down; thereafter he received a minor grant to study physiological acoustics especially in invertebrates. He analyzed the characteristics of the organ of hearing of locusts and developed a quantitative method with the help of which it was possible to distinguish the performance of the subgenualorgan as an organ for the perception of vibration from that of the organ for the perception of sound transmitted through the air (Autrum 1941).

Between 1936 and 1938 Erich von Holst received DFG grants for experiments on the comparative physiology of the locomotive reflexes in fish. Several of his grant applications were, however, turned down.

Experiments on triggering fin rhythm after severing the spinal cord in fish had the following results with regard to the nervous coordination of movement: rhythms produced in the central nervous system (spinal cord) lead to the movement of a fin with a specific beat. A visible "wave" that runs over a fin is caused by the concatenation of several centrally produced rhythms. Concatenation occurs also between the central elements of distant rhythms (of different fins). Holst assembled the various periodic forms that arise from the overlapping of rhythmic oscillations and lead to various frequency relationships of the fin movement. When there are more than two rhythms involved, new "system characteristics" (periods) are created, which he called "forms" *(Gestalten)* in the sense of Gestalt psychology. These *Gestalten* operate relatively independently of outside influences, including influences in the central nervous system with the exception of the automatic-rhythmic elements that are involved. Only when the boundary conditions for such an equilibrium are crossed through the continuous change of an external factor does the system jump into a different condition of equilibrium (Holst 1939).

Holst's explanation of fin movements as automatisms of the central nervous system that are fundamentally independent of the environment refuted the then widely held interpretation of these movements as reflex chains. According to Holst, reflexes could at most modify these movements. He also transferred the model of automatism to more complex movements. When Konrad Lorenz gave a lecture in Berlin in 1936 at the invitation of the Kaiser Wilhelm Society, Holst criticized his notion of an instinctive act *(Triebhandlung)* as a "complex chain reflex." Instead, he suggested that instinctive acts be seen as "automatisms," that is, as the product of physiological systems that, like the pacemaker in the heart, work independently of the environment. After lengthy discussions Lorenz accepted this suggestion and integrated it into the theory of instinctive acts. Beginning in 1939, Holst conducted experiments at the Reich Aviation Ministry's Institute for Aerodynamics with the goal of clarifying the mechanics of the flight of animals.

Until 1938, Karl von Frisch, professor in Munich, received a small DFG grant for sensory physiological research with bees and fish. He showed that Cyprinidae (carp) emit a defensive substance in the presence of enemies, and that certain species of fish, for example minnows, possess a sense of smell comparable to that of dogs. In the process he proved that the fright reaction of minnows triggered by dead fish is caused by a perception of smell and not taste (Frisch 1938).

Frisch's DFG grants were in part severely cut back or turned down. Beginning in 1936, the NSDAP tried to have Frisch sent into retirement as a "quarter-Jew." In 1941 his dismissal was prevented by a research contract from the Reich Nutrition Ministry to combat the nosema disease in bees. As mentioned in Chapter 1, this disease had been spreading since 1940, and in 1941 alone it destroyed hundreds of thousands of bee colonies in Germany (Frisch 1973, p. 116). Frisch's studies largely failed to generate results. But the research contract was soon broadened. He was asked to investigate and expand upon experiments, reported in Russian beekeeping journals, in which bees had been induced to visit specific types of flowers. The scent-guiding experiments led him in 1944 to return to his work on the language of bees, a topic he had been investigating experimentally since the beginning of the twenties (Frisch 1973, p. 125ff.). In 1946 he was able to demonstrate that the round dance and waggle dance of the bees do in fact contain information about the distance and the direction of a food location and do not merely indicate the existence and nature of the food source (Frisch 1946). He was awarded the Nobel prize in medicine in 1973 (jointly with Konrad Lorenz and Nikolaas Tinbergen) for his work on bee communication. A detailed account of his scientific accomplishments and of the political intrigues against him can be found in Chapter 1, Section 7.

Animal Psychology and Ethology. Behavioral research appeared as a separate discipline in Germany only in 1936. On January 10, 1936, biologists and psychologists working in the field of behavioral research founded the German Society for Animal Psychology. As the purpose of their association they mentioned "the study of the psyche of animals and the practical use of insights into animal psychology. These insights shall help to clarify the relationship of man to the animals, support animal protection, and, through the appropriate use of animals, enhance the variety of services animals render to mankind."[82] The society published the *Zeitschrift für Tierpsychologie* (Journal of animal psychology).

DFG- and RFR-funded zoologists in the field of animal psychology or in the field of comparative behaviorism or ethology, which emerged from it, included the following: Friedrich Brock, assistant professor at the Institute of Environmental Research in Hamburg, worked on the smelling capacity of the dog with special consideration of urine excretion as a means of communication in the canine world. After 1940 he indicated as the goal of his work improvements in the training of seeing-eye dogs. Hanns von Lengerken (Berlin) studied the brood care instinct in various insects. After 1939 he received DFG support only for work on improving insecticides. Otto Koehler (Königsberg) analyzed the counting ability of the pigeon with the help of a DFG grant. He showed that while pigeons do not count in the human sense, they can comprehend sequences of events up to six (Koehler 1937). One raven was able to differentiate sequences of seven. This animal was able to distinguish groups of points, regardless of their location, that differed only by a single point, the highest group being seven; if the groups differed by more, it could distinguish higher values (Koehler 1943).

The thirties saw a controversy between Otto Koehler and Friedrich Alverdes (Marburg) about Alverdes's claim (1937) that paramecia had the capacity to learn and distinguish simple forms.[83] Arguing from a holistic perspective, Alverdes maintained (1938) that even animals of simple organization represent a unity of body and psyche and are thus capable of higher accomplishments, such as those necessary for spatial perception. Ulrich Grabowski, who repeated Alverdes's experiments at Koehler's suggestion, came to the conclusion that paramecia were not capable either of learning or of recognizing forms (1938). Werner Fischel, as the head of the Research Institute for Animal Psychology in Münster, received a DFG grant from 1935 to 1939 for the study of "emotion and memory in vertebrates."

The best-known, and to this day most influential, behavioral scientist in Germany and Austria is Konrad Lorenz. At the end of the thirties he helped the science of comparative behavior (ethology), a field founded by Oskar Heinroth, to achieve a breakthrough as a separate branch of biological research. The ethological approach constituted a change of course in contemporary behavioral research through the radical attempt to make animal and human behavior comparable and capable of investigation through causal analysis on the basis of the theory of evolution.

What follows is a detailed discussion of the career and research up to 1945 of Lorenz, an eminent scientist who was very supportive of the goals of National Socialism.

3. Konrad Lorenz, Ethology, and National Socialist Racial Doctrine

3.1. The Development of Ethology to 1945

Konrad Lorenz, who was born in Altenberg near Vienna on November 7, 1903, and died there on February 27, 1989, is known as the co-founder of the science of comparative behavior (ethology). His scientific accomplishments earned him the Nobel prize in medicine in 1973. His 1973 book *Die Rückseite des Spiegels* (The back of the mirror) is considered one of the earliest works on "evolutionary epistemology." He became popularly known as the "father" of the greylag geese and as the author of many books on human and animal behavior. In his last years he was in the news as a conservationist, an exhorter against ecological catastrophe in the wake of increasing technologizing, and an opponent of nuclear power plants. He was hailed as one of the few "committed scientists" who "came out of the ivory tower . . . and mounted the barricades with us nature conservationists and ecologists" (Weinzierl and Lötsch 1988). The theologian and biologist Günter Altner (1988) appreciated that Lorenz "in the desperate situation of the crisis of survival . . . gave forceful voice to the melody of life."

Konrad Lorenz maintained that he always kept away from politics. In an interview in 1988 on the occasion of his eighty-fifth birthday, he stated: "In fact I always avoided all politics because I was absorbed with my concerns. I also shirked a confrontation with the Nazis in a very disgraceful way, I simply did not have time for it . . . I reproach myself for it. On the other hand: if I had remembered my political duties early on, I would have never accomplished many of the things for which I was awarded the Nobel prize" (Brügge 1988).

Lorenz conducted a large part of the research that led to the establishment of ethology as a separate branch of research in biology during the thirties in Austria and Germany. In the following discussion I seek to shed some light on the beginnings of Lorenz's research in ethology and on his career during that time.

The development of ethology as a scientific discipline was based fundamentally on the scientific accomplishments of Tinbergen, Lorenz, and Holst. They combined precursor concepts and a wealth of individual hypotheses on animal behavior into a theoretical structure.

As early as 1898, Charles O. Whitman had used patterns of behavior as taxonomic characteristics; that is, he had drawn on patterns of behavior for classifying animals into a biological system and had called for the phylogenic-comparative study of animal behavior. Independently of him, Oskar Heinroth (1871–1945), in his study of duck courtship display, had found that some patterns of behavior differ from species to species and genus to genus in a way that was comparable to the differentiation of physical characteristics (Lorenz 1965). Lorenz was strongly influenced by Heinroth. The work of Jakob von Uexküll (*Umwelt und Innenwelt der Tiere* [The outer and inner world of animals], 1909) also played a large role in the formulation of ethological theories. Uexküll's work stimulated Lorenz to make observations and to adopt some of the concepts Uexküll had introduced (such as "functional circle" and "companion") (Lorenz 1935).[84] In contrast to the other precursors of ethology, Uexküll rejected Darwin's theory of evolution.

Ethology differed fundamentally from contemporary currents in behavioral research. One was behaviorism, founded in the United States by John B. Watson (1878–1958) and later considerably expanded by B. F. Skinner (1904–1990). Behaviorists attempted to trace all forms of behavior back to unconditioned and learned reflexes and thus to see them as conditioned by the environment. Behaviorism generalized the theory of Ivan P. Pavlov (1849–1936) of unconditioned and conditioned reflexes for all behavior. By contrast, ethological studies focused on the spontaneous activities of an organism that occurred independent of external stimuli. According to Lorenz, instinctive movements could not be influenced by learning.

Ethology also set itself apart from vitalistic currents in animal psychology.[85] Vitalists regarded the instinct as a directional factor giving a goal to the inborn behavior of the animal. Inborn patterns of behavior could accordingly be changed in pursuit of this goal. By contrast, Lorenzian ethology started from the assumption that instinctive movements were absolutely fixed and rejected their orientation toward a goal. Beyond that, Lorenz also opposed the psycho-Lamarckian ideas that could be found in the work of some animal psychologists: thus he rejected the widespread belief that the capacity of higher animals to make free decisions constituted an important factor in species change in evolution. One of these scientists was Werner Fischel, who wanted to make the soul and the psyche of animals the topic of biological study and who proceeded from the assumption that higher animals, "by free choice," could expose themselves to influences that, like geographic factors, played a phylogenetic role: "The aspiration of a living being decides racial [*rassische*] changes that can occur in it" (Fischel 1934).

To this speculative, psycho-Lamarckian approach of the animal psychologists, which Fischel combined with insights from the theory of evolution through the importance given to geographic factors on the formation of race and species (according to Bernhard Rensch), Lorenz counterposed strict adherence to the scientific formulation and testing of hypotheses. He considered it "fundamentally necessary to keep the still far too philosophical animal psychologists to pure inductive science, to praise them if they come up with something through causal analysis, and where possible to discourage them when they show a tendency to the luxurious growth of hypotheses which, like a malignant tumor, permeate what is in and of itself nice experimental work."[86] Lorenz defined ethology as "the discipline which applies to the behavior of animals and humans all those questions asked and those methodologies used as a matter of course in all the other branches of biology since Charles Darwin's time" (Lorenz 1981, "Introductory History").

Lorenz's essay "Über den Begriff der Instinkthandlung" (The concept of the instinctive action, 1937b) is regarded as the beginning of the comparative study of behavior as a separate discipline of biology. In contrast to older definitions of instinct, which already contained the triggering chain "drive-affect-action-goal," Lorenz replaced the concept of instinct with his more precisely defined concept of instinctive action. He criticized animal psychologists for leaving the concept of instinct unexplained and for interpreting every inborn purposeful action as emanating from an instinct. A more careful analysis revealed that there were at least two clearly distinguishable innate patterns of movement: orientation responses or taxis and movements based on inherited coordinations. Although in Lorenz's view no sharp line exists between orientation responses and learned actions, behaviors based on inherited coordinations can be clearly separated from higher psychic capabilities. That is why he counterposed a species' behavior based on inherited coordinations ("instinctive action") to all other forms of behavior that were earlier subsumed under "instinct."

As mentioned above, in 1936 Erich von Holst had called Lorenz's attention to the central nervous system's capacity to spontaneously generate impulses. Holst suggested seeing *Triebhandlung* (literally "drive-action") not as a "complex chain reflex" but as a kind of "automatism"—that is, as the product of physiological systems which became active independent of the environment—and Lorenz integrated this approach into his model of instinctive action. Thus he defined instinctive acts as invariant, inherited movements of an animal species, which from a physiological perspective correspond to the central nervous system automatisms in the sense of Holst and are thus autonomous. The nonautonomous, environmentally conditioned actions that preceded an instinctive movement and were done in search of an impulse to trigger this movement he called appetitive behavior (1937b).[87] As early as 1935, Lorenz—drawing on the work of Uexküll, who had shown that every animal separated out its own sphere of life from the environment and that only a few stimuli of the environment became in each case significant for animal actions—described the stimuli-filtering apparatus of the peripheral and central nervous system as "innate releasing schema" *(angeborenes auslösendes Schema)* (Lorenz 1935). In this work he also described and characterized for the first time the phenomenon of "imprinting" *(Prägung)*.

Lorenz's definition of instinctive action has prevailed within ethology. His thesis of the "species-preserving purposefulness" *(arterhaltende Zweckmäßigkeit)* of instinctive actions, which is not tenable according to newer insights of behavioral biology, is also still found in many German schoolbooks and textbooks on biology. The subsequent development of the model of instinctive action, and the scientific controversies Lorenz carried on especially with American zoologists and psychologists on the validity of this model, are not the subject of this study.[88]

Both in Austria, where Catholicism had a large influence on the politics of science, and initially also in National Socialist Germany, Lorenz had to overcome difficulties before his biological behavior research was recognized and financially supported. Unless otherwise indicated, the following information and quotes come from the DFG file of Konrad Lorenz in the BAK. Following medical school, Lorenz studied zoology and psychology. After obtaining his doctorate in 1933, he became an assistant to Ferdinand Hochstetter at the Anatomical Institute of the University of Vienna; Hochstetter made it possible for Lorenz to pursue his studies in ethology on the side. In 1935 he obtained his *Habilitation* at the University of Vienna in anatomy and animal psychology. When Hochstetter's successors prohibited Lorenz from pursuing his ethological research, he quit his job at the institute and worked there and in Altenberg without salary on the behavior of cichlids and birds.

In Austria, owing to the power of the Catholic Church, Lorenz had no prospects of support for his research, since important parts of it had an evolutionary orientation. Fritz Knoll, professor of botany in Vienna, wrote in this regard to Fritz von Wettstein on October 17, 1937: "Special support is hardly possible for him in Austria, since here, for reasons of the worldview of the ruling circles, biology tends to be unwelcome rather than welcome, most especially the direction in which Lorenz is doing such splendid work."

In 1937 Lorenz submitted his first application to the DFG for financing for his research on specific patterns of movement in geese, which were determined by inherited coordinations. Lorenz stated that the goal of his work, which drew on that of Oskar Heinroth, was to furnish exact proof of the existence of movements that behaved like organs within phylogenic development and inheritance. His application was turned down in 1937, despite an excellent review by Erwin Stresemann, in which he said: "Through his publications, Dr. Lorenz has become known in ornithology and animal psychology circles as a path-breaking scientist of quite exceptional ability." This decision was made "on the basis of an unfavorable evaluation . . . in which questions have been raised in particular about the political attitude and the descent of Dr. Konrad Lorenz."[89]

Wettstein strove to clarify the issues in question. To that end he solicited evaluations of Lorenz from the Viennese professors Fritz Knoll, Otto Antonius, Ferdinand Hochstetter, Eduard Pernkopf, and the lecturer Alexander Pichler. He summarized them for the DFG: "All reviews from Austria agree that the political attitude of Dr. Lorenz is impeccable in every respect. He is not politically active, but in Austria he never made a secret of the fact that he approved of National Socialism . . . Everything is also in order with his Aryan descent."

These assessments of Lorenz reveal the political atmosphere in Austria as well. Thus Hochstetter wrote about his former assistant Lorenz: "Since I was able to trust him completely, I discussed everything with him, including frequently current political questions. For despite the fact that neither of us belonged to a specific party, we were most interested in these questions. We were of one mind in the sharp rejection of clericalism and everything connected with it. We were also always of the same opinion that the fate of us Germans in Austria was most intimately connected with the fate of the Germans in the Reich, and that we could assent only to a politics that took this bond of fate into full consideration, indeed was virtually built upon it."

According to Otto Antonius, Lorenz "never made a secret of his admiration for the new situation in Germany and the accomplishment in all areas, and his attitude toward them is undoubtedly positive." Alexander Pichler knew Lorenz "as a person who, away from politics, lives only for his scientific work. Despite his apolitical attitude, he by no means approved of the conditions currently prevailing in Austria. Lately Dr. Lorenz has repeatedly displayed to me his constantly growing interest for National Socialism and has expressed himself positively about its idea. As far as I am acquainted with his biological studies, they are in keeping with the worldview prevailing in the German Reich."

In December of 1937, Wettstein submitted to the DFG an application for a research grant for Lorenz, and he included these letters. His application was successful, and from 1938 on Lorenz was supported by the DFG with a grant and subsidies. In 1938 Lorenz began working on disturbances in instinctive behavior as a result of domestication in wild geese and crossbreeds of wild geese and domesticated geese.

On September 1, 1940, Lorenz was appointed professor and director of the Institute for Comparative Psychology at the University of Königsberg, "on the intervention of Minister Rust and against the objection of the faculty" (Geuter 1984, p. 131).[90] This objection had a professional basis. According to Hans Thomae, the appointment of Lorenz was professionally not justified from the perspective of experimental psychology.[91] The Reich education minister responded to pertinent inquiries by the executive board of the German Society for Psychology by explaining that Lorenz, through his reference to the "innate forms of experience," was in fact linking up in the best way with the epistemology of German idealism (Thomae 1977, p. 154). The appointment came about with the involvement of the professor of zoology in Königsberg, Otto Koehler, whom Eduard Baumgarten from the faculty of philosophy had contacted regarding Lorenz (Koehler 1963). Koehler was a friend of Lorenz. In 1937 he had brought him in as coeditor of the *Zeitschrift für Tierpsychologie,* and he played a significant part in spreading the ethological approach in Germany. With his counting experiments in birds he had begun to work on questions involving unnamed thinking in the study of animal behavior. Koehler, like Lorenz, was among the biologists who in numerous publications supported the racial political goals of National Socialism.[92] Under the influence of Koehler and his wife, the philosopher Annemarie Koehler, Lorenz began an analysis of Kant's epistemological writings from an ethological perspective.

His teaching at Königsberg ended on October 10, 1941, when he was drafted for military service. According to Otto Koehler, Lorenz was first stationed in Posen as a military psychologist. After military psychology was abolished in May of 1942, he worked as a neurologist and psychiatrist in the reserve hospital in Posen (Koehler 1963, p. 392).[93] In Posen Lorenz participated in psychological examinations that the Reich Foundation for German Eastern European Research (Reichsstiftung für deutsche Ostforschung) was carrying out on "German-Polish half-breeds" and Poles at the request of government officials (see below). In April of 1944, "to gain experience at the front and for his later appointment as medical officer" (Koehler 1963, p. 392), he was transferred to the Soviet Union, where he fell into Russian captivity in June of 1944 (Heinroth 1978). He returned to Altenberg in 1948.

3.2. Lorenz and National Socialist Ideology and Practice

After the initial difficulties I have described, the ideological significance of Lorenz's ethological research was soon recognized in Nazi circles. Lorenz himself saw in National Socialism a sphere of political activity that appealed to him: he joined the NSDAP on June 28, 1938, shortly after it was legally possible in Austria as a result of the Anschluß in March of 1938. In addition, he became a member of the NSDAP's Office of Race Policy with permission to lecture.[94] As early as 1939, he was scheduled as a speaker at an event of a National Socialist organization. The program for the Third Annual Meeting of the German Society for Animal Psychology, which was to take place in Leipzig from September 21 to 23 but was canceled when the war broke out, contained the following announcement for the evening of September 21: "Public lecture in conjunction with the NS-German People's Education Foundation Leipzig: Lecturer Dr. K. Lorenz, Altenberg near Vienna: 'Rise and Decline in Humans and Animals,' with slides" (Kalikow 1980, p. 202). Lorenz published the results of his domestication research in professional zoological and psychological journals and indicated their practical application as follows: "Precisely in the large field of instinctive behavior, humans and animals can be directly compared . . . We confidently venture to predict that these studies will be fruitful for both theoretical as well as practical concerns of race policy" (Lorenz 1941, p. 46).

In his eyes, the homology between characteristics that animals have acquired in the course of their domestication and that humans have acquired through civilizing processes was proved.[95] "As has already been discussed, one can claim with near certainty that all physical and moral manifestations of decay that cause the decline of cultured peoples after they have attained the stage of civilization are identical with the domestication manifestations in domesticated animals" (Lorenz 1943a).

This homologizing of complex human behavior and social developments with animal behavior stood in the tradition of Haeckelian monism (see Kalikow 1980).[96] Theodora J. Kalikow, who has examined the interrelationship between National Socialist ideology and Lorenz's scientific publications to 1943, speaks of a "process of reciprocal legitimation, whereby the Nazis lent political power to ideas which were already part of Lorenz's world view." She believes that this process "may help explain Lorenz's increasing emphasis on animal and human degeneration after 1938; and Lorenz's 'scientific evangelism' may have moved him to try to explain and justify Nazi racial policies ethologically" (1983, p. 56). I concur with this assessment.

Lorenz assumed genetically conditioned degenerative changes in the sphere of instinctive behavior to be the cause of domesticating as well as civilizing processes. Thus he perceived in the "overcivilized big-city man a whole series of hereditary traits that would readily mark any other animal life-form exhibiting them as a typical domesticated animal" (1940a, p. 5). In his view the degenerative manifestations were mainly caused by "the change, the elimination, indeed in some cases the radical reversal of environmental influences of selection and eradication" (1943a, p. 295).[97] As far as we know, Gustav Kramer was the only biologist who criticized Lorenz's theses on domestication at the time, though he expressed this criticism only to friends. Among them was Erich von Holst, who wrote to Kramer on October 27,

1940: "Of course your critique of Konrad's domestication work is also of great interest to me."[98]

Lorenz assessed the results of the domestication research with animals and their application to human societies against the perception—which was gaining ground at the time among racial hygienists,[99] as well as biologists—that the decline of the German Volk was becoming evident.[100] German pessimism about human civilization, which originated in the nineteenth century, was given new nourishment by Oswald Spengler's *Decline of the West* (1920). In this work Spengler described the senescence and death of cultures as a necessary fate. Lorenz, who like many other biologists (including Otto Koehler) frequently referred to Spengler, accepted his view that the cultural decline of all cultured peoples in history was preceded by a phase of civilization, but he rejected Spengler's pessimism. Thus he criticized "Spengler's hopeless pessimism, which denies all possibility of progress; although he believes in the hitherto phylogenetic creation of everything organic, he dogmatically considers any further development impossible for mankind" (Lorenz 1943a, p. 293).

For Lorenz the decline of cultures described by Spengler was a problem of biology. He hoped that humankind would be able to learn from its oldest biological experiment, the domestication of wild animals, "*why* the civilization period of a cultured people is followed by its decline, or—optimistically stated—has so far always followed" (1943a, p. 294). In his opinion the number of "degenerative types" increased in civilized society as a result of domestication: "In a very short time the degenerative types, thanks to their larger reproductive rates and their coarser competitive methods toward the fellow members of the species, pervade the Volk and the state and lead to their downfall, for the same biological reasons that the likewise 'asocial' cells of a cancerous growth destroy the structure of the cellular state" (1943a, p. 294). Biological measures were needed to stop the biologically conditioned degenerative process. Lorenz asserted that a "deliberate, scientifically founded race policy" would prevent the ruin of a cultured people (1943a, p. 302). Science, especially evolutionary biology, should determine the ethical values and goals of this policy: the "racial improvement of Volk and race" (Lorenz 1940b, p. 25).[101]

Lorenz made the assumption that instinctive actions were normally—that is, in wild populations—directed toward the preservation of the species.[102] In human societies, too, social forms of behavior had a species-preserving purpose, as long as they had not degenerated through civilizing processes. As indicated above, in this context he used the terms "species," "race," and "Volk" as synonymous. Thus in the decay of behavior that occurs in the course of "becoming a domesticated animal [*Haustierwerdung*]," he saw "damage that was highly dangerous to the preservation of the species" (1943b, p. 120). Elsewhere (1940a, p. 57) we read: "Our species-specific feeling for the beauty and ugliness of the fellow members of our species is most intimately related to the manifestations of decay that are caused by domestication and threaten our race. One can see in this feeling almost a distinction of species-preserving importance for the elimination of such manifestations of decay."

In his analogy between a cancerous growth and "asocial humans" he described the following danger: "Just as healthy tissue generally treats a tumor as 'identic' [*artgleich*] . . . the healthy volkish body often does not 'notice' how it is being pervaded by elements of decay" (1940a, p. 69). To the charge that with his vigorous dissemination of the evolutionary idea he was advocating a materialism that was also the foundation of Socialism, he responded: "It is true that Socialism and Communism can spring from a half-digested Darwinism, which erroneously conceives of all of humanity as a homogeneous unity of equal value. But already the self-evident correction that not all of humankind but only the race is such a biological unity turns Socialism into National Socialism" (Lorenz 1940b).

According to their contribution to this "biological unity," Lorenz classified people into those of "full value" *(vollwertig)* and those of "inferior value" *(minderwertig)*. The latter included the "defective type" *(Ausfalltypus)* created by the selection conditions of the big city, who, because of the enlarged sphere of releasing schemes, brings a lot of children into the world, much like the "domesticated animal that can be bred in the dirtiest stable and with any sexual partner" (1940a, pp. 67f.). This phenomenon, caused by the absence of the selection conditions of prehistory, in which humans had to prevail against hunger, cold, wild animals, and hostile hordes, led to the already mentioned dangerous increase of "socially inferior human material." Since the species-preserving principle lost effectiveness among humans as civilization caused natural selection to disappear, any measure that led to the "elimination of the ethically inferior" (1940a, p. 66), of "elements afflicted with defects" (1940a, p. 75), could be legitimated.

The principle of species preservation, supposedly the result of natural selection, became, as the principle of the subordination of the individual to the concerns of the Volk and the race, the ethical norm for human life within communities: "To us Volk and race are everything, the individual is virtually nothing" (1940b, p. 32).

Lorenz also saw a biological justification for the value of pure racial stock. In the "mixed heredity of humans, so utterly like that of domesticated animals," he perceived the danger that innate norms or schemes and drives might disintegrate. Thus in his view disturbances occurred also in the "presence of fully valid [*vollwertiger*] schemes" when they were linked with strongly disturbed drives: "Since there is no doubt that one must see in the function of the innate schemes a last anchor for the cultured peoples threatened by the destructive forces of civilization, a reasonably bearable racial balance of a people takes on an entirely new importance in our eyes: it simply is an important precondition for the normal functioning of the schemes" (1943a, p. 314).

The racial hygienic defense he called for against the "members of the Volk who have become asocial through defects" consisted, as in the treatment of a cancerous growth, of the earliest possible "recognition and eradication of the evil" (1940a, p. 69). Lorenz did not name any agency to take over this selection. But he was of the opinion that the "selection as to toughness, heroism, social commitment, and so on" had to be taken over by a social body, and in this regard he considered the situation in Germany exemplary: "The racial idea as the foundation of our form of government has already accomplished a very great deal in this direction. From the very beginning the Nordic movement has been emotionally opposed to this 'domestication' [*Verhaustierung*] of humankind, all its ideals are such as would be destroyed by the biological consequences of civilization and domestication I have discussed" (1940a, pp. 71f.).

In National Socialist Germany, "the sound sentiment of the people" *(das gesunde Volksempfinden)* was considered a serious moral criterion. A quote from *Civilized Man's Eight Deadly Sins* (1980, trans. 1973, pp. 50–51) bears out that in this regard Lorenz remained true to himself later:[103]

It is one of the many dilemmas into which mankind has maneuvered itself that here again, what humane feelings demand for the individual is in opposition to the interests of mankind as a whole. Our sympathy with the asocial defective, whose inferiority might be caused just as well by irreversible injury in early infancy as by hereditary defects, endangers the security of the nondefective. In speaking of human beings, even the words "inferior" or "valuable" cannot be used without arousing the suspicion that one is advocating the gas chamber.

Unquestionably the mysterious "sense of justice" referred to by Peter Sand rests on a system of genetically anchored reactions, causing us to take action against asocial behavior of fellow human beings.

Donald Campbell has discussed the antidemocratic content of Lorenz's publications after 1945 in his essay in Richard I. Evans's *Konrad Lorenz: The Man and His Ideas* (1975). While he appreciates Lorenz's accomplishments in many fields, especially in evolutionary epistemology, he criticizes the right-oriented, potentially racist ideological tendency of some of these publications, above all *Civilized Man's Eight Deadly Sins*. He denounces an omission on the part of Lorenz: "While nuclear weapons are on his list of deadly sins, genocidal nationalism was already deadly sin number one even before the atom and hydrogen bombs" (Campbell 1975, p. 106). Campbell evidently sensed the implications of such publications, which end up in genocide.

The writings quoted up to this point show that the "scientifically underpinned race policy" Lorenz demanded remains incomplete in one central point. Who are "those to be eliminated," and how does one recognize "those of full value"? Since science cannot offer an answer here, Lorenz found other ways of effectively conveying his conviction of the necessity of a state policy of selection and elimination. He used pertinent quotations and changed them to suit himself; for example in his essay "Duch Domestikation verursachte Störungen arteigenen Verhaltens" (Disturbances in species-specific behavior caused by domestication) he wrote:

> Any attempt to reconstruct elements that have fallen out of their relationship to the whole is therefore hopeless. Fortunately their elimination is easier for the people's doctor [*Volksarzt*] and much less dangerous for the supra-individual organism than the operation of a surgeon is for the individual body. The great technical difficulty lies in recognizing them. In this regard the cultivation of our own innate schemes, in other words our emotional reaction to defective phenomena, can be a great help. *A good man, in his dark striving, knows*

full well whether the other person is a scoundrel. This gives rise to a peculiar piece of advice, which may sound a bit strange from the mouth of a scientist concerned with causal analysis: namely that, when it comes to the desirable type of our Volk, we should rely on the reactions of the best of us, reactions that are rooted in the un-analyzed. (1940a, p. 70; emphasis added)

In Goethe (*Faust* I), however, we read the following: "A good man, in his dark striving, still knows well where lies the moral course" ("Ein guter Mensch in seinem dunklen Drange, ist sich des rechten Weges wohl bewußt"). Nor did Lorenz shrink from abusing moral injunctions from the Bible, tearing them completely out of context and twisting their meaning into the opposite. "Love your neighbor as yourself" (Lev. 19.18) is one of the central Jewish commandments. Justice and respect for the individual human being are basic pillars of Jewish teaching. Lorenz (like many theologians then as now) overlooked the origin of the commandment, namely in the Old Testament, and called it "the most essential ethical truth of Christianity." He tried to co-opt this "Christian commandment" for National Socialist ideology by going on to say: "Since race and Volk are everything to us, the individual almost nothing, this commandment is a quite self-evident demand for us."

Another "Christian commandment," however, he considered relevant for the German Volk only in modified form: "And if there is for us an ethical commandment: 'Thou shalt love the future of your Volk above all else,' we thereby merely replace the highest manifestation of our creator who is immanent in his creation for what the Christians call 'God.'" The God of the Bible becomes the "immanent creator" of the German Volk, that is, the God of evolution through mutation and selection. From the two central Jewish commandments of love for God and love for one's fellow human being, which were also adopted by the Christians as the core of their theology, Lorenz derived the "racial care" of the German Volk through the means of eradication. The Christians in Germany took no offense at this.

Lorenz had a tendency to falsify relevant literary or biblical fragments and reinterpret them in the sense of National Socialist racial doctrine; the practice was not a one-time slip-up. He used another falsified quotation to underline his call for racial purity. Thus he wrote: "Our innate scheme of 'the Beautiful and the Good' in our fellow human beings, which is our receptive correlate to very specific characteristics that mark the desirable type, is without a doubt originally completely uniform. The ancient Greeks made virtually no distinction between beautiful and good, the aner kalos k'agathos [beautiful and good person] was to them the quite uniform desirable type of their race. The biblical expression *'Beware of those who are marked'* expresses the same principle. As long as a tribe or a Volk possesses a very high degree of racial uniformity, assessing an individual by his external characteristics alone will be possible and drawing inferences about the full value [*Vollwertigkeit*] of his inner behavioral norms will be justified" (Lorenz 1940a, p. 58, emphasis added).

The theological layperson hearing the term "marked person" thinks of Cain, who after murdering his brother was given a mark by God. If Lorenz was alluding to this instance, he would have been turning the meaning of this mark into its complete opposite: God did not mark Cain to make him publicly recognizable as a criminal. Quite the contrary: he did it "in order that anyone meeting him should not kill him" (Gen. 4.15).

The phrase "Beware of those who are marked" does not appear anywhere in the Bible in this form, either in the Old or the New Testament. The only relevant passage is in the New Testament, 2 Thess. 3.14; the Greek word for "to mark," *semeioomai,* is not used anywhere else.[104] In New Testament literature it can mean both "to make note of" as well as "to remember (someone)." Thus the standard translation of the passage in Thessalonians reads: "If anyone disobeys our instructions given by this letter, *mark him well,* and have no dealing with him until he is ashamed of himself. I do not mean treat him as an enemy, but give him friendly advice, as one of the family" (emphasis added). This admonition occurs after some members of the community in Thessaloniki, in expectation of endtimes, lose any desire to work. Here, too, Lorenz would have been crassly falsifying the biblical passage by turning the instructions "give him friendly advice, as one of the family" into "beware" of people who do not correspond to the uniform "desirable type of their race." "Beware of those who are marked" is a handy formulation that is not a biblical quote but a crass falsification, and at best a half-truth, whose effectiveness, however, is increased by reference to the Bible as its source. Three years later Lorenz used the same quote in the same context (1943a, p. 312): "The old biblical saying 'beware of those who are marked' had always existed for a good reason!" It remains to be noted that from 1933 on it was official policy in Germany to warn of people—primarily the Jews—who did not correspond to the ideal image of the race. And beginning in 1941 there was a public marking (a 'marking' not in the biblical sense) of "racially foreign elements": the Star of David.

Between 1940 and 1943 in Germany, Lorenz called for a "deliberate, scientifically underpinned race policy,"[105] "the improvement of Volk and race,"[106] the "elimination of elements afflicted with defects,"[107] the "elimination" of "members" of the Volk "who have become asocial because of defects"[108] and of "elements who have fallen out of their relationship to the whole,"[109] and "the elimination of ethically inferior people."[110] He did not indicate what sort of concrete measures he associated with the term "elimination" *(Ausmerzung)*. From 1933 on National Socialism incorporated the elimination concept as part of its population policy. The Law on the Prevention of Genetically Diseased Offspring, passed on July 14, 1933, was aimed at eliminating "biologically inferior genetic heritage [*Erbgut*]" through compulsory sterilization (Gütt, Rüdin, and Ruttke 1936, p. 77). The Nuremberg Laws, for example the Law for the Protection of German Blood and German Honor of September 15, 1935, also furthered the idea of elimination and racial purity. The elimination concept culminated in the policy of euthanasia, which was officially begun in 1939, and in the "final solution of the Jewish question," which was launched in 1941.

In an interview with Franz Kreuzer, broadcast by ORF and ZDF (Austrian and German state television, respectively), Lorenz maintained in 1981: "That they meant murder when they said 'elimination' or 'selection' was something I really did not believe at the time. This is how naive, how stupid, how gullible—call it what you will—I was back then." In response to Kreuzer's question "Do you think that you were ever politically active in your life, or did you wish to be politically active?" Lorenz said, "All my life, from childhood on, I wished not to be politically active" (Lorenz and Kreuzer 1988, pp. 95ff.). As described above, Konrad Lorenz was not only a member of the NSDAP but also a member of the party's Office of Race Policy.

Lorenz's activities as psychologist in Posen in 1942 cast doubt on his statement of 1981 about his naïveté when it came to the criminal intentions of the Nazi rulers: during his wartime service, Lorenz participated in Posen as a psychologist in studies carried out by the working group for "suitability research" *(Eignungsforschung)* within the framework of the Reich Foundation for German Eastern European Research (Hippius et al. 1943, p. 12).[111] This working group had been founded in May of 1942 "at the initiative of official places"; a working committee was given the task of conducting "worthiness studies as to psychological suitability and character among German-Polish *Mischlinge* and among Poles, members of groups III and IV of the German Volk-list Posen, as well as among Polish residents of the city of Posen" (Hippius et al. 1943, p. 12). The studies were overseen by Rudolf Hippius, associate professor of psychology at the Reich University in Posen, after December 1942 professor of social and ethnopsychology in Prague. In his book he expressed his gratitude to the Reich Foundation:

> Here I would like to thank the Reichsstiftung für deutsche Ostforschung and its president, Gauleiter and Reichsstatthalter of the Wartheland, Arthur Greiser, for its generous, sympathetic, and farsighted support of the research work. Among its staff, thanks goes especially to the Gauamtsleiter für Volkstumsfragen SS-Sturmbannführer Höppner, the scientific director of the Reichsstiftung Prof. Dr. Carstens, rector of the Reich University of Posen, and the managing director of the Reichsstiftung, University curator Dr. H. Streit. My thanks go also to the Reinhard-Heydrich Foundation in Prague for making possible the quick printing of the report.

Greiser, since 1942 SS general, established the Reich Foundation for German Eastern European Research in 1941 with the primary goal of coordinating research in the Wartheland district (Burleigh 1988, pp. 290, 294–296). The approximately ten working groups of the foundation, all led by professors of the Reich University, had a completely practical orientation. One of these working groups, which supplied material for National Socialist population planning in the East, was the "working team for East Settlement," headed by Peter Carstens, SS colonel and from 1940 on member of the Waffen-SS Bodyguard Regiment Adolf Hitler (Burleigh 1988, p. 292). Hippius's working group for suitability research was part of Carstens's working group.

Unless otherwise indicated, the following account of the studies in Posen is based on the information in Hippius et al. 1943. Carried out from May to September 1942, the studies involved a total of 877 people, who took the tests in small groups, each headed by a psychologist. Like Hippius and Kurt Stavenhagen (professor of philosophy in Posen), Lorenz was one of the psychologists who participated in an honorary capacity. The work proceeded from the assumption that "there are genetic values which are fixed according to peoples, and which undergo specific and regular changes when peoples interbreed" (p. 16). By comparing the accomplishments, emotions, and character of the indicated groups, the studies aimed to "shed light on the psychic backgrounds to national character [*Volkstum,*] namely as a hereditary condition as well as a volkish sentiment" (p. 15). Thus a "Polish genetic substance" was distinguished from a "German" one (p. 46). Two findings of the study shall be mentioned here: researchers determined that the basic structures of Germans and Poles were for the most part mutually exclusive. The German basic structure was characterized by "persistence, dependence, energetic dynamism and aggravated dynamism," whereas the Polish character exhibited "openness to the fullness of life, fear of life, compulsive dynamism, a poverty of vital roots" (p. 64). Analysts of the productivity of German-Polish *Mischlinge* concluded that "the German aptitude for working ability [capacity for professional accomplishment] was largely lost in the interbreeding" (p. 112). They went on to say that "substantial damages in an interbred population mean not only an irksome population difficult to guide, but a considerable defect also in practical and civilized life" (p. 114).

A brief summary of National Socialist population policy in the Wartheland will illuminate the political background to the studies carried out by Hippius. Wartheland comprised the part of western Poland annexed to the German Reich shortly after the attack on Poland; this is where Posen and Litzmannstadt (Lodz) were located. The annexation and Germanification implemented the principles of population policy of the Nazi volkish-racial ideology. The Warthegau was to serve as the settlement area for German immigrants from the Baltic states, the Balkans, and the territory of the Reich. Jews were barred on principle from acquiring German citizenship. As for the Poles, initially, in the opinion of the Reich Interior Ministry, they were to be given the opportunity of becoming German citizens, a position that immediately aroused opposition from the party and the SS (Broszat 1961, pp. 118–130). In Posen, in particular, a method was developed for selecting ethnic Germans. At the suggestion of the Germandom specialist *(Volkstumsspezialist)* of the Security Service in Posen, SS First Lieutenant Dr. Strickner, Greiser set up a central office in Posen in 1939 for keeping a German ethnic list *(Volksliste)*; it reorganized the identification of ethnic Germans according to strict criteria. Strickner also participated in the studies conducted by Hippius in that "he lent a supportive hand . . . when it came to technical questions." Hippius thanked Strickner for "his constant willingness to help and his broad initiative" (Broszat 1961, p. 12). On January 22, 1942, Hippius himself joined the staff of the Ministry for the Occupied Eastern Territories (Geuter 1984, p. 423).

To begin with, the German population of Wartheland was divided into two categories: those who declared themselves Germans *(Bekenntnisdeutsche)* and those of German descent. An expansion of the Germanization of Poles in the district was rejected as undermining the position of Germandom. However, Himmler also wanted the Germanization of "those related to Poles by marriage" and of "certain racially valuable Poles." As a result, at the suggestion of the SD, group 3 (people of German descent absorbed into Polishness and capable of becoming Germanized again), and in January of 1941 group 4 (primarily "ethnic German relapsers who can be re-Germanized") were added to the *Volksliste* (Broszat 1961, p. 123). As indicated above, these two groups were among the target groups of Hippius's "worthiness studies as to psychological suitability and character." Beginning in 1940, members of groups 3 and 4 were sent to the Reich to work and to be "Germanized." Those members of group 4 who were considered "asocial," "of inferior value as to genetic biology," or politically strongly compromised were to be transferred to concentration camps (Broszat 1961, p. 133). All Jews and many of the "Poles incapable of assimilation" were deported from Wartheland. As Reich governor there, Greiser was responsible for the mass deportation and extermination of Jews and Poles. He was hanged in Posen in 1946 (Wistrich 1982, p. 107).

Through his participation in the studies of Rudolf Hippius, Konrad Lorenz had the opportunity to support, by selection measures, the "racial idea as the foundation of our form of government"—which he welcomed (Lorenz 1940a, p. 71), and the call for a "very high degree of racial uniformity" (p. 60). Along the same lines he completed in 1942 in Posen the essay "Die angeborenen Formen möglicher Erfahrung" (The innate forms of possible experience, 1943a).[112] In it, as described above, he attributed degeneration phenomena to "the change, the elimination, indeed in some cases the radical reversal of environmental influences of selection and eradication" (p. 295), and raised as "an absolutely necessary demand" a "conscious, scientifically underpinned racial policy" (p. 302) and a "certain degree of racial harmony [of a people]" (p. 314). The "Germanization" of Wartheland had been a goal of German policy since 1940, and its implementation was in full swing in 1942. In an honorary capacity (without pressure of any kind), Lorenz participated in the research of the working group, which served this policy. Neither he nor anyone else later mentioned his work for the Reich Foundation for German Eastern European Research.[113]

3.3.   The Principle of "Species-Preservation" *(Arterhaltung)*

A brief analysis of the biological basis of Lorenz's argument, which led to the demand for elimination, reveals the following: his assumption that there exist social forms of behavior promoted by natural selection with a view toward the preservation of the species or population has proved questionable not only for humans but also for animals. According to recent findings in evolutionary biology, forms of behavior assessed positively by natural selection are not directed toward the preservation of the species or the general well-being of the Volk. Selection applies to individuals and takes effect via their differential reproduction. Only forms of behavior that serve the so-called inducive fitness of the individual, and not the welfare of a superordinate group or community, are adaptive and thus supported by selection.[114] Both cooperative forms of behavior as well as those that harm the species or the population can be adaptive. Both can therefore be "normal" in the sense of "natural," and it is therefore not possible to use this as a basis for deriving ethical guidelines.[115]

In the opinion of the anthropologist Christian Vogel (1988, esp. pp. 197–202), the notion of a "mysterious principle of species preservation," which is still widespread especially in Germany, has contributed, merely through the linkage of the terms "adapted" and "healthy," to further confusion about moral value judgments such as "good" and "right," and to the derivation from "adaptation" of a typologically understood "normality." Lorenz did not revise this idealistic-typological way of thinking in ethology even later. For example, in 1963 he wrote in *Das sogenannte Böse* (On aggression): "In fact, the only way to assess 'normal' structure and function is by demonstrating that these are the ones which, *under selection pressure on their survival value,* have evolved in this and in no other form . . . However, by normal we understand not the average taken from all the single cases observed, but rather the *type* constructed by evolution, which for obvious reasons is seldom to be found in a pure form; nevertheless we need this purely ideal conception of a type in order to be able to conceive the deviations from it" (Lorenz 1963, p. 194).

Just as one cannot derive from the assumed principle of species preservation the subordination of the interests of the individual to the welfare of the community (however defined) as a moral value, one cannot deduce from modern theories of selection a moral value of egotism or the favoring of kin. Ethical norms for how humans ought to live together cannot be drawn up through scientific analyses.

3.4. Lorenz in the Postwar Period

As I have already mentioned, Lorenz returned to Altenberg from Russian captivity in 1948. His wish in 1950 to be appointed chair for zoology in Graz was not fulfilled. Although Karl von Frisch, who had been professor in Graz from 1946 to 1950 and returned to his old chair in Munich in 1950, worked actively to get Lorenz appointed as his successor in Graz from among the seven candidates, he was unable, as Erich von Holst reported, to do anything against the "strong countercurrent" there. Holst wrote to Lorenz: "The reason why I am telling you all this is my suspicion that your enemies are surely with pleasure gathering everything that proves that you (with the exception of earlier statements that can be used politically) advocate an unethical, pro-Darwinist, materialist conception of human nature and human origins, and especially of the nature of the human spirit."[116]

Lorenz himself had been told, "as reliable inside information," that "the Ministry doesn't give a damn whether you are a Nazi or not, but it doesn't want biology to be served up to the Middle School candidates of Styria in such a pronounced evolutionary form as is the case in the science of comparative behavior."[117] This attitude of Austrian science politicians stood in a line of continuity to earlier years. Lorenz turned down an associate professorship that was promised in Vienna.[118]

In the Federal Republic of Germany, Lorenz initially had difficulty getting appointed to the Max Planck Society, the postwar reconstituted form of the Kaiser Wilhelm Society. However, there is no indication that his trouble was due to his political past. As in 1937, when funding from the DFG was at stake, it was once again colleagues who helped Lorenz, especially Erich von Holst. On November 13, 1950, Holst wrote to Alfred Kühn about his concerns for Lorenz's livelihood and his fears that he might go abroad permanently if nothing were done for him.[119] Along with Boris Rajewsky and Holst, Kühn was part of a commission formed by the MPG to consider the question of Lorenz's association with the Max Planck Institute (MPI) for Marine Biology.[120] Max Hartmann, too, told Holst that he would "support Lorenz's admission to the MPG."[121] At the request of Holst, the president of the MPG, Otto Hahn, worked for Lorenz's admission into the society.[122] In this way Lorenz was given, in 1951, a research position centered in Buldern (Westphalia) in association with the MPI for Marine Biology in Wilhelmshaven, where Holst was head of one department. In 1954 Holst's department and Lorenz's research post were combined into a new Max Planck Institute for Behavioral Physiology, for which a new building was erected in 1955 in Seewiesen. In 1963 Lorenz became director of this MPI and remained in that position until 1973.

Despite his political past, Lorenz won a high reputation as a person and a scientist not only in Germany but internationally as well. The assessment of Lorenz by foreign scientists is reflected in the correspondence between W. C. Allee from the Rockefeller Foundation and European scientists after the war. After his return to Altenberg in 1948, Lorenz had applied to the Rockefeller Foundation for financial support. The man in charge of his application, Allee, asked influential scientists in Europe for evaluation of Lorenz's political activities.

Ethologists in the United States appreciated Lorenz's scientific accomplishments, but his closeness to National Socialism was known. Jewish scientists rejected the idea of American support for Lorenz. As Allee wrote: "I find a strong, or shall I say bitter, attitude on the part of all Jews with whom I have discussed Dr. Lorenz's problems."[123] Other scientists wrote less negative and even exonerative statements. Presumably Lorenz was already so important scientifically that they did not wish to dismiss him; perhaps his personality, which was appreciated by all sides, also played a role. The expert evaluations from Karl von Frisch and Nikolaas Tinbergen go a long way toward explaining Lorenz's international rehabilitation—to the point of winning the Nobel prize in 1973—despite his unambiguous racial-political statements and publications. It is difficult to grasp why these two scientists, who had been persecuted for years during the Nazi period, suffering harassment and repression (Tinbergen in particular), exculpated Lorenz. In his evaluation, Frisch relied on statements from the zoologists Otto Storch, who had been dismissed as director of the Zoological Institute in Vienna in 1938 because he had a Jewish wife, and Otto Koehler, until 1945 full professor for zoology in Königsberg, thereafter in Freiburg. Both had confirmed that Lorenz had never behaved as a Nazi in an unpleasant manner, and that the branch of military psychology where Lorenz worked during the war had been dissolved because it had supposedly aroused the wrath of Himmler.[124]

The division of military psychology was indeed dissolved. Beyond that, it should be noted that Koehler had emerged as one of the explicit Nazi biologists not only through his membership in the SA, but also through his clear statements in support of Hitler and National Socialism. It was Koehler who got the appointment to Königsberg for Lorenz. If he knew anything about Lorenz's work in Posen in an honorary capacity, he would have had every reason to keep quiet about it.

Niko Tinbergen's statement shows, on the one hand, the extent of the destruction and the brutality the Germans brought to the Netherlands and the loss of trust that followed. On the other hand, it reveals his ambivalent relationship with his colleague Lorenz. Did Tinbergen, himself a victim of National Socialism, subordinate his moral and political misgivings about the person Konrad Lorenz to his appreciation for the man's scientific importance? I reproduce in full Tinbergen's letter to Margaret Nice, dated June 23, 1945, as a key document in the history of science in the postwar period.[125]

Dear Mrs. Nice,

I cannot tell you how glad I am to be able to resume contact with my American colleagues, after these five unhappy years. How are you? And how is your family? My family and I are relatively happy and healthy. After an increasingly depressing, terrible nightmare of five years we are free again. It is not possible to express in words what we were feeling, when the Canadian troops arrived, chasing the despised Germans away from our country and with them their whole criminal system of terror, of looting, suppression and murdering. I suppose nobody who not actually has experienced and undergone the German terror, can imagine *how horrid* it was, and how elated we are now, how glorious the world, the simplest privileges of "normal" life are. In spite of undernourishment, of absence of electricity, gas, transport such as streetcars, trains, cars etc., and so many other commodities, of uncertainty about the fate of many friends, of mourning over the death of many others, mostly murdered in incredibly beasty ways by German SS, SD or Wehrmacht, in spite of that all we rejoice, because we can now make a new start in complete moral and spiritual freedom. We will never forget what Americans, Canadians and British men have done to rescue us!

I will try to tell you how things are here, and I want to ask you to renew contact with me and our Dutch ornithology and ethology in general. First a short report over what has happened. The conquering of our country by the Germans was, as you know, a question of five days. Our army, though certainly not lax in spirit, was badly prepared and was soon smashed by the efficient German armies. Moreover, the fate of Rotterdam, the smoke of which could be seen all over Holland, forced us to surrender.—During the first year of occupation the Germans did not interfere so much with our life; I think I wrote you one or more not too sad letters, though it was of course impossible to express our feelings. Soon however the Germans began to register the Jews and to outlaw them, then to influence the schools and the teaching. Interference went on progressively and soon a point was reached where we felt we ought to stiffen and resist as much as possible. Our University was, by accident, the first group of Dutchmen to be tackled by the Germans as a group, and the first to refuse to surrender. The Germans waned to "cleanse" our corps of Jews and of anti Nazi's and proceeded to fire one professor, then another, step by step, on wholly irrelevant grounds. Soon we saw no other way than to resist by refusing to stay in the service of the German-controlled government, and soon after the University had been closed by the Germans because of anti-German "irregularities" 60 of our professors including myself laid down their function. This was at the same time our protest and our means to prevent the Germans to nazificate the University by dismissing only some few of us, for whom they had a nazi-remplacant, and to keep the rest as "flags" to adorn the planned nazi-university. As a reprisal, we were captured (that is to say, some 20 of the supposed leaders of the resistance) and put in a camp as hostages, together with about 1300 other patriots and internationalists. The Germans threatened to shoot a group of these hostages wheneer active resistance or sabotage was threatening some vital part of their war-organization. Twice a group, though a small one, of us were shot in 1942, after that we did not run serious risks, though our families and we lived in uncertainty, which had a more or less depressing influence on most of us. However, the longer the Germans left us alone, the more hopeful we were. Jan Verwey (your remember his Heron-study) and Prof. Van der Klaauw were also among us. In the spring of 43, Van der Klaauw was set at liberty, fall 43 Jan Verwey, but for some reason I was kept as a prisoner til the fall of 1944, when after a short but memorable period in a concentration camp I got my "freedom" again, after an absence of two years. My wife and children were healthy and after the happy reunion we lived for 8 months in the country, where our time was filled by all kinds of practical work, for own household as well as for war-victims, and by doing service as a spy for the intelligence service in our "underground military forces". It was a doubtful kind of "freedom" we enjoyed while the German "Sicherheitsdienst" [Security Service] was hunting men, so that we often had to hide. I won't write you particulars; it was an incredible mode of life. But it belongs to the past; now we are free, ruined but extremely happy.—Kluyver is OK. He prooved to be a *very* reliable friend. He had to leave his house in September 1944 because of the failed British air-landing between Wageningen and Arnhem. His house is damaged but can be repaired; I just heard that he has occupied it again. Junge is allright, also Boschma an De Beaufort. The latter lost a son, who was captured and shot by the Germans in France, when trying to cross to Swiss territory in order to join our armed forces in Britain. My students are gradually returning. Several of them are still in Germany where they were kept as slaves. At least three have died, one was captured and executed as a member of the underground forces. My brother L. Tinbergen is allright, he finished a doctor's thesis on quantitative studies of the relation sparrow-hawk—some of its food-birds. He spent most his time as leader of the Ornithological Station Texel to it. Baerends is allright; he published a very important study of the life history of the digger-wasp Ammophila, his doctor's thesis, of which I am sending you a copy as soon as possible. He lost all he had in a bombing raid on the Hague. By a remarkable coincidence he and his wife and baby were saved. He expected a raid, because of the weather and preceding small attacks, so he took wife and baby in the country for one day. While resting, out in the meadows, they saw a flock of heavy bombers carry out a devastating attack on the part of the The Hague where they lived. Their house and many adjoining blocks were destroyed.—Kortlandt is allright too, and so is Portielje, and Bierens de Haan the psychologist, though both nearly starved to death. Many of us have been emprisoned in some way or another; so that the following joke was circulating: after the war our government will demand from everyone a declaration: "have you been in prison during the war? If not, *why not?*"

During the first years we have been able to continue our scientific work, at least the study-room part of it. Field work was limited. I will send you as soon as allowed the papers of my men and myself that have been published in that time. From 1942 on I have not published anything, but I prepared a little book on animal sociology while in the hostage camp, and I wrote several illustrated books for children of 4–10 years, originally used for my own children as a letter from me during my captivity, one page each week, but forming coherent stories. One of them, the story of a Herring Gull, will be translated into English. I hope you will allow me to ask your advice about it and about the possibility to publish it in the US, so that I may be able to pay the debts I still have in the US.

I know absolutely nothing about our German colleagues. Lorenz was in the Army, Dept. of "Heerespsychologie" [Military Psychology] since 1941. He was rather nazi-infected, though I always considered him a honest and good fellow. But it is impossible for me to resume contact with him or his fellow-countrymen, I mean it is psychologically impossible. The wounds of our soul must heal, and that will take time. Originally, soon after the outbreak of the war, I had planned to personally organize the resuming of cooperation in scientific affairs. Now I see, that we will have to wait at least a year, first to see whether a german science will revive at all, secondly to wait till the Germans give proof of their goodwill in this respect. But the plan as it was, is as follows: I still consider it good, though as yet inopportune. In order to avoid the mistakes from 1918–'26, I did not want, as then, to begin cooperation between allied scientists and leaving the Germans out altogether. It was difficult then, to get them to cooperation again, because they were grieved by the original "lockout". The new plan is based on the following points: 1. Cooperation in science is necessary. 2. The Germans must collaborate, because they have many good scientists. It is, however, absolutely impossible to collaborate with nazi's, SS-men etc., who have murdered and terrorized so many of our dearest relations. This is not a result of a desire for revenge, but we simply cannot bear to see them. They are absolutely spoiled, and still are dreaming of military and political power. "Wir komen zuruck" they cry, and "das nachste Mal werden wir siegen" ["We will be back, and next time we shall be victorious"]. It is not right to think that the atrocities were only committed by a minority of fanatical SS-, SD-, or Gestapo-men. Nearly the whole people is hopelessly poisoned; I can give you many personal experiences! Therefore, collaboration must only be organized with men whom we can trust, whom we know personally, and who have given proof of honest intentions. For instance, SS-man Niethammer will never again be acepted by any of us, or Werner Fischel. But Stresemann, Rensch, Von Frisch and, I hope, Laven are our men. Personally I should regret if Lorenz and Koehler would be expelled. Now to overcome the difficulties of bringing Germans, British, Americans, Dutchmen, etc. together I think the following tactics would be best. I try to organize, next year for instance (so I thought originally but now I see it must be postponed to a much later time) a conference of few (say 10–15) workers in our field (say: behavior-students of birds and animals in general) (or should we split into ornithology and psychology), in which participate only personal friends. For instance, I know Pontus Palmgren, I know Monika Holzapfel from Switzerland, I know Julian Huxley, David Lack, I know Mrs. Nice, Ernst Mayr (I hope he has not been germanized), Bill Vogt, I know Lorenz, Stresemann, Rensch etc. Through the bonds of personal friendship I hoped to persuade them to agree to such a new start towards international cooperation. Should I succeed, then it must be possible to expand this nucleus in a number of successive meetings, until we have our full congress again.—That was my plan. However, I see that we will have to wait in order to see if there will be a German science at all. And I do not like the idea to wait for them; of course we will renew our international connections with USA, Britain, and the other countries as soon as possible. I should very much appreciate to hear your opinion. Anyhow, I hope very much to get the opportunity to visit the US soon, because I fell I *must* renew contact! But the situation here is still so chaotic, and the demands of reconstruction, and of the war against Japan, are so huge, that I realize that opportunity will not come too soon. It is quite possible, also that I will go to the East Indies for some years, but nothing is certain yet.

I am looking forward very much to an answer; I hope you will be able to get some information through about the American friends. How are Bill Vogt, Ernst Mayr, Joe Hickey, Dr. Yerkes, Arthur Allen, and so many other colleagues? How has research been going on? Has Bird Banding been continued? I think I asked you at the outbreak of war to keep my subscription going, has that been possible? Or, anyhow, could you send me the numbers issued during the war? Have you been carrying on the very very valuable excerpting and discussion of internation[al] literature? And do tell me about your own work! And would you kindly tell me the cost of those Bird Banding issues? I hope I will have time to read the many important studies which I am sure have appeared in your journals during the war.

Last but not least: have you ever heard anything from Joost ter Pelkwijk? I hope so much he will come through this war; so much could still be expected from him.

In re-reading this letter, I see it is rather heterogenous and fragmentary. It is difficult to realize what exactly you would want to know about the situation here. But I hope this letter will give you a foundation. I hope to hear from you and/or my other friends soon.

With kind regards,
/s/ N. Tinbergen

4

The Content and Result of Research at Kaiser Wilhelm Institutes

Until 1943, the Kaiser Wilhelm Society published annually a list of the work at all its institutes in the journal *Die Naturwissenschaften;* the publications of the Kaiser Wilhelm Society and of its successor the Max Planck Society, from 1946 to 1951, are listed in *Die Naturwissenschaften* 38 (1951) and 39 (1952).[1] Since 1952, the activity reports of the Max Planck Society have appeared annually in the *Jahrbuch der Max-Planck-Gesellschaft.* As was the case with the universities, the following discussion of the research at KWIs between 1933 and 1945 treats only a small part of all scientific publications.

1. The KWI for Biology, Berlin-Dahlem

All departments of this institute focused on genetics. Max Hartmann, department head since the founding of the institute in 1914, studied sex-determination and genetics in protozoa, invertebrate marine animals, and fish. In 1939 he discovered in sea urchins the first fertilization substance in animals; its chemical composition was analyzed by Richard Kuhn at the KWI for Medical Research (Hartmann and

206

Schartau 1939).[2] Kuhn and Hartmann introduced the term "gamon" for substances that had an activating and chemotactic effect on gametes. In 1944 Hartmann found four gamons in fish (Hartmann 1944). In Hartmann's department, Hans Bauer conducted cytogenetic research on *Drosophila*. He induced chromosome mutations (especially translocations) with X-rays and after 1940 with neutron radiation. His findings confirmed the hypothesis advanced by Lewis Stadler in 1932, according to which a chromosome fracture and not merely a contact between chromosomes was the prerequisite for the production of chromosome mutations through radiation (Bauer 1939). The generation of single fractures was directly proportional to the dose of radiation, and the dose-effect relationship for chromosome mutations could be represented by a curve that lay between a single-hit curve and a hyperbolic curve (Bauer 1942). When Max Hartmann was appointed director of the German-Greek Institute for Biology in Athens-Piraeus (also a Kaiser Wilhelm Institute) in 1943, Bauer became department head at the KWI for Biology.

In 1934 the botanist Fritz von Wettstein succeeded Carl Correns, who died in 1933, as first director of the Kaiser Wilhelm Institute for Biology, and he continued the research begun by Correns in physiological genetics. The majority of research in Wettstein's department fell into three fields: the analysis of the effect of genes in the ontogeny of plants; the study of genetic aspects in the formation of subspecies and species; and experimental mutation research. Mutation research, which included work on polyploidy and its physiological effects, received particularly strong funding. In his applications to the DFG, Wettstein emphasized the great importance polyploidy was likely to take on for plant breeding and for clarifying questions of species formation in evolution. From 1937 on, researchers used almost exclusively colchicine for the production of polyploid plants, which were then experimentally tested in nature as to their fitness under extreme conditions and their capacity for selection. Researchers stressed that polyploids had the significant advantage of markedly greater adaptability to extreme climatic zones as compared to diploids (Wettstein 1941). However, at least for grain plants this later turned out to be a fallacy.

The expectation that the yield and the resistance to parasites were higher in polyploid plants than in their diploid ancestors was also not fulfilled (Straub 1941). Karl Pirschle (1942a) determined that the dry weight of autopolyploid plants was in many cases even less than that of diploid plants. The hopes for better results with allopolyploids were also disappointed. Based on experiments with mosses *(Bryum)*, Wettstein and Josef Straub formulated a hypothesis on species formation through polyploidy in nature. They showed that a fertile tetraploid plant with normal cell size can be produced from an infertile tetraploid starting plant with large cell size through gene mutation and subsequent selection (Wettstein and Straub 1942).

In view of the thesis (increasingly accepted in the thirties) that larger changes in evolution also come about not through large mutations but in small steps through successive small mutations, an important question arose about the way in which useful traits based on polygenic effects, that is, traits and organs produced by many genes and capable of being selected, could take shape in the process of evolution. Starting from the observation that haploid organisms possess a much weaker capacity for organ formation than diploids, Wettstein surmised in 1943 that the significance of the diploid stage lay above all in the fact that recessive mutations, concealed by the dominant allele, can concentrate themselves until they reach favorable combinations.

With good support from the DFG, Hans Stubbe continued in Wettstein's department his research on gene and chromosome mutations in *Antirrhinum* and other plants, which he had begun at the KWI for Breeding Research. After 1938 he studied the mutagenic effect not only of X-ray and UV radiation but also of chemicals. This research bore no results (see Chapter 3, Section 1.2). In 1940 Stubbe and Pirschle were able to explain the nature of the heterosis effect for *Antirrhinum* mutants.[3] In contrast to the widely accepted hypothesis that the effects of dominant genes add up in the hybrid plant, they demonstrated that the heterosis effect depends on the heterozygotic condition. In later generations, too, they found no constancy of effect, as one would have expected—on the basis of the addition hypothesis—owing to the homozygotic presence of dominant genes (Stubbe and Pirschle 1940). After 1945, Stubbe became a leading plant geneticist in the German Democratic Republic.

The zoologist Alfred Kühn became second director of the institute in 1937 as successor to the dismissed Richard Goldschmidt. Under Kühn, research intensified on developmental genetics and genetic questions relating to evolution. During this time he carried out his well-known experiments on gene action in mutants of the flour moth

Ephestia. Kühn had formulated the hypothesis that genes influence visible traits by stimulating the production of diffusible substances (hormones) in the organism. Together with Adolf Butenandt, who had taken over the KWI for Biochemistry in 1937 as successor to the dismissed Carl Neuberg, he tried to clarify the gene action chain for the eye pigment of *Ephestia.* This research led in 1940 to the identification of kynurenine as an intermediate product of the pigment synthesis (Butenandt, Weidel, and Becker 1940). At the time scientists debated whether these "gene hormones" like kynurenine entered into the process of the synthesis of eye pigment as a preliminary stage, or whether they acted as biocatalysts. Kühn and Erich Becker reached the conclusion in 1942 that kynurenine appeared also in insects with great probability as an intermediate product in the pigment formation, which started from tryptophane. The American biochemist Edward L. Tatum had shown this for bacteria in 1939. The genes supposedly provided the enzyme systems that effected these transformations in the substrate chain (Kühn and Becker 1942).

Biologists could not use complex organisms like *Drosophila* and other insects for proving and generalizing this conclusion. For that they had to work with simpler biological objects, in which biosynthetic pathways could be analyzed. Kühn, Butenandt, and their collaborators stuck to their experimental object. The Americans George W. Beadle and Edward L. Tatum, who had also studied the genetics of the synthesis of eye pigment in an insect, *Drosophila,* continued this research in 1940 with the mold *Neurospora.* With this organism they could analyze mutations that deleted the enzymes of simple, known biochemical reactions.[4] They showed that a single gene controlled a single biochemical reaction (Beadle and Tatum 1941), clear proof for the indications that Kühn and Butenandt were getting. The "one gene–one enzyme" hypothesis became one of the fundamental principles of molecular biology. In 1958, Beadle and Tatum received the Nobel prize (jointly with Joshua Lederberg) for their work showing that genes determined single steps in biochemical pathways.

In Germany research with *Neurospora* had begun before Beadle and Tatum collaborated on their experiments.[5] At the KWI for Breeding Research, Helmut Döring conducted mutation research in *Neurospora* in Stubbe's department between 1935 and 1937, and after that at the Kaiser Wilhelm Institute for Biology. Because of his uncertain position at the KWI for Biology, in 1939 he accepted a post as assistant at the

Institute for Botany at the University of Jena, where he continued this work. He tried inducing mutations with radiation and with various chemicals. Among other things, he examined the effect of ethanol, phenol, arsenious acid, iodine, and iodates, achieving sure results only with iodine.[6] To determine possible mutagenic effects of nutritional conditions, he also worked with various synthetic media. In a letter to the DFG, Stubbe expressed his hope that *Neurospora* would become a new experimental object of fundamental importance for genetic research, comparable to *Drosophila*.[7] After Döring's death in 1940, there was nobody who was in a position to continue the *Neurospora* work.[8]

2. The Division for Virus Research of the KWIs for Biology and Biochemistry, Berlin-Dahlem

In the United States in 1935, Wendell Stanley crystallized a virus for the first time, the tobacco mosaic virus (TMV). Following the publication of Stanley's results, the directors of the Kaiser Wilhelm Institutes for Biology and Biochemistry, Alfred Kühn, Fritz von Wettstein, and Adolf Butenandt, agreed in 1937 to work together in the field of virus research. The money for this research was provided by the DFG and especially by Heinrich Hörlein of IG Farben.[9] The research group, which carried out basic research on viruses beginning in 1937, was made up of the botanist Georg Melchers from Wettstein's department, the zoologist Rolf Danneel, guest in Kühn's department, and the chemist Gerhard Schramm from the KWI for Biochemistry. In 1941, at the initiative of Kühn, Wettstein, and Butenandt, the Kaiser Wilhelm Society established the Division *(Arbeitsstätte)* for Virus Research, which included the three research groups plus the zoologist Hans Friedrich-Freksa, then working at the KWI for Biochemistry. In 1942 the Entomological Department under Gernot Bergold was affiliated as another working group; as the head of the entomological branch of the KWG, Bergold at first remained in Oppau.

Another working group existed at the Reich Biological Institute (Biologische Reichsanstalt) for Agricultural Science and Forestry; it carried out basic research on the tobacco mosaic virus and other plant viruses. From 1937 on, Gustav Adolf Kausche, head of the laboratory for experimental virus research of the Reich Biological Institute, together with E. Pfankuch carried out mutation experiments by using radiation on a TMV dry specimen from Stanley, on their own TMV specimen, and on the x-potato virus (see below).[10]

The Division for Virus Research and the KWI for Cultivated Plant Research (see the next section) could not have been set up without strong financial support from the Industrial Bank. In February of 1941, the board of the bank decided to give the Kaiser Wilhelm Society the following financial subsidies, among others: RM 500,000 to continue virus research and to establish an institute planned for that purpose; a one-time grant of up to RM 200,000, and RM 120,000 a year for the next three years to set up an institute for the collection and study of wild and primitive strains of cultivated plants.[11] The yearly support that the Division for Virus Research received from the REM increased from RM 60,000 in 1941 to RM 141,000 in 1944.[12]

The Division for Virus Research (the Kaiser Wilhelm Institute for Virus Research was founded only after the war) specialized exclusively in basic research on viruses, especially the tobacco mosaic virus. Though Stanley had isolated the virus in crystalline form in 1935, and believed it was pure protein, a subsequent chemical analysis showed that the virus was a nucleoprotein. Because TMV could both replicate itself and mutate, research on it was carried out from the beginning with the hope of finding out something about the structure of genes.

Gerhard Schramm did the most interesting basic experiments on the structure and replication of the virus.[13] In 1938 he went for several months to see Arne Tiselius and Theodor Svedberg in Sweden, where he became acquainted with the techniques of electrophoresis for separating proteins and of ultracentrifugation for determining their molecular mass. Thereafter he played an essential role in the construction of an ultracentrifuge, which was not yet available for purchase, in Berlin. Schramm showed in 1941 that the existence of nucleic acid was a prerequisite for the TMV's self-reproduction.

During his attempt to analyze the structure of RNA, Schramm also carried out experiments with yeast RNA.[14] At the time, RNA was believed to be a tetranucleotide or a polymer of tetranucleotides.[15] Schramm determined a molecular weight of 11,000 and concluded: "The nucleic acid of yeast is thus composed of about thirty units of mononucleotides strung together. Analyses, particularly enzymatic ones, have shown that, contrary to the hitherto prevailing notion, the mononucleotides are linked to one another in the same way and that no tetranucleotides are preformed. The structure of yeast nucleic acid thus corresponds to that of other highly polymeric natural substances, as for example cellulose and starch."[16] The RNA Schramm analyzed was not messenger RNA and thus probably did not carry genetic information. But the import of his results, though not clear to Schramm or anyone else at the time, was that RNA in principle could carry genetic specificity.

With respect to the structure of TMV itself, Schramm's analyses led him to determine that it was composed of a large number of similar, if not identical, protein subunits plus RNA (Schramm 1943, 1947).

That laboratory also worked with a tomato mosaic virus that is closely related to the TMV, characterizing it biologically, chemically, and electron-microscopically (Melchers et al. 1940). Melchers also isolated and bred spontaneous mutants of this virus and of TMV, in which the homologous gene had mutated into the same allele (Melchers 1942).

After 1940 the fact that nucleic acids played an important role in the process of self-replication was generally recognized in Germany as a result of the work of Torbjörn Caspersson in Stockholm in 1940 and the theoretical reflections of Hans Friedrich-Freksa in Berlin the same year. Friedrich-Freksa surmised that there were electrostatic attractive forces between the basic components of the proteins and the acidic phosphate groups of the nucleic acids. From this assumption he developed a model conception for the identical duplication of nucleoproteins: the nucleic acid supposedly represented the negative in the replication process by reflecting the charge pattern of the proteins with a negative charge. These negative sites were in turn able to serve as attachment sites for protein molecules with a corresponding positive charge. He thereby anticipated the meaning of complementarity in the replication process.

With this model he contradicted the physicist Pascual Jordan, who called for a resonance attraction (derived from quantum mechanics) between *identical* macromolecules and in this way tried to interpret the attraction between chromosomes in physical terms (Jordan 1938). According to Jordan, the duplication of protein molecules could be understood solely from the structure of these molecules itself. On theoretical grounds he rejected the necessity of participation by other compounds like DNA in the "autocatalysis of protein" (1944; see also Section 5 in this chapter).

Different ideas existed among geneticists and virologists on the question of whether mutations occurred in proteins or nucleic acids. A working group composed of Gustav Adolf Kausche and E. Pfankuch from the Reich Biological Institute, Hans Stubbe from the Kaiser Wilhelm Institute for Biology, and Helmut Ruska from the Charité clinic and the laboratory for electron microscopy of the Siemens-Halske Company, studied sedimentation and electrophoretic migration speeds of five strains of TMV. They attributed the high migration speeds of some mutants to a difference in the free phosphate groups of the nucleic acid component as compared with the wild type and came to the conclusion that "the production of mutations should be moved to the nucleic acid component of the virus molecule, and that quantitative or qualitative changes occurred here" (Pfankuch, Kausche, and Stubbe 1940). Schramm and L. Rebensburg (1942), however, clung to their view that these mutations would also produce changes of the amino acids in the corresponding viruses.

To test the hypothesis that the virus's capacity for hereditary transmission was based on the presence of certain active groups in the protein component, Schramm and his coworkers tried to produce mutations in the TMV through chemical changes of the amino acids. While an acetylation of the amino groups did not show a mutagenic effect, the effect of nitrous acid led to a loss of the biological activity of the virus (Schramm and Müller 1942). Schramm and Hans Müller suspected that the mutagenic effect lay in the deamination of the amino acids. They did not consider a possible effect of nitrous acid on RNA, that is, a deamination of bases. With this experiment they had evidently discovered the mutagenic effect of nitrous acid as early as 1942, but they interpreted the result incorrectly.

The work continued after the war. Despite the finding in 1946 by Friedrich-Freksa, Melchers, and Schramm of minor biological and chemical differences between two mutants of the TMV, which were confirmed years later by further analyses under Melchers (Aach 1957; Kramer 1957), Schramm was not able to determine any analytical difference in the composition of the amino acids between various TMV mutants. Thus, until the mid-fifties he held the view that the mutations did not cause any exchanges of amino acids. The mutations, he said, consisted only in the refolding of the polypeptide chain of the protein component, leading in the mutants to a firmer bonding of the nucleic acid to the protein (Schramm 1948, pp. 320, 322; 1954, p. 481).

The working group of Kausche, Pfankuch, Ruska, and Stubbe also worked with bacteria and bacteriophage. They were able to incorporate radioactive phosphorus into the nucleic acid of bacteria (streptococcus, staphylococcus) and phages.[17] With the help of the electron microscope, Helmut Ruska discovered shapes with a spherical head and a rod-shaped extension, which occurred following a bacterial lysis through phages and which occasionally attached themselves by their extension to the walls of bacteria. These things were in fact phages, which he himself strongly suspected (Ruska 1941). Thomas Anderson and Salvador Luria recognized phage particles in electron micrographs in March 1942.

3. The KWI for Cultivated Plant Research, Tuttenhof

On the decisive initiative of Fritz von Wettstein, the Kaiser Wilhelm Society founded the Institute for Cultivated Plant Research in Tuttenhof, near Vienna in 1943.[18] This institute was to become a central place for the collection and study of the existing wild and primitive strains of cultivated plants. The model for this institute was the world-famous collection of wild and primitive plants of the Soviet geneticist Nikolai Ivanovic Vavilov. Since his theories were of great importance to plant genetics and plant breeding in the thirties and forties, I will briefly introduce them.

Vavilov (1887–1943) became internationally known as a geneticist and plant breeder primarily on the basis of two theories. His "theory of the homologous series of hereditary changes" describes the appearance of similar mutations in related species. Erwin Baur and Reinhold von Sengbusch successfully applied it in the breeding of alkaloid-free lupines through selection. Darwin had already described "parallel variations." Vavilov studied these phenomena in a large collection of cultivated plants and put them on a genetic basis. Homologous series of phenotypes were thus shown to correspond to homologous series of mutations (Kuckuck 1962, pp. 177–196).

The "gene-centers theory for the creation of cultivated plants" that Vavilov developed in 1926 was of great importance for research in plant geography and for plant breeding. According to this theory, the area of the greatest variety of a cultivated plant, its so-called gene center, is also its region of origin, where its wild strains should still exist. Through systematic collecting expeditions over wide areas of the world, including the Near East, the Caucasus, and South America, Vavilov was able to compile a collection of 200,000 cultivated plant seeds and make it available to Soviet biologists and breeders as starting material for basic genetic research and for plant breeding. As head of the Soviet delegation, Vavilov presented this theory at the Fifth International Congress of Geneticists in 1927 in Berlin. In this way he inspired biologists in many countries to conduct additional collecting expeditions to centers of diversity that were important as genetic reserves for agricultural breeding.[19] Vavilov's school, together with all of Soviet genetics, broke up after 1935 through the influence of Trofim Lysenko on Josef Stalin. Vavilov himself was arrested in 1940 and died in 1943 in a Siberian prison.

The director of the KWI for Breeding Research, Wilhelm Rudorf, rejected the establishment of a new Institute for Cultivated Plant Research. He saw in it a competitor with his own research work. Since its founding, the KWI for Breeding Research, at the suggestion of Erwin Baur, had been conducting research in the field of primitive plants. Baur had also undertaken the first collecting expeditions to South America. Rudorf thus felt slighted by Wettstein when the latter suggested establishing the new institute and secured the approval of Konrad Meyer, without first consulting him. It was his opinion that this kind of research should continue to remain in the hands of the already existing plant-breeding institutes, especially his own.[20] Rudorf's criticism, however, had no influence on the decision of the KWG.

At the suggestion of Wettstein, Hans Stubbe was named director of the new institute; he was at the same time head of the Department of Genetics there. Elisabeth Schiemann, who, as described above, had lost her position as associate professor at the University of Berlin in 1940 for political reasons, became head of the Department of Plant Geography. Rudolf Freisleben was to become head of the Department of Applied Research and Cytology, but he died in 1943. Lothar Geitler of Vienna took the post. The Department of Physiology was headed by Karl Pirschle, who also became deputy director of the institute. Work at the experimental estate, called Tuttenhof, and in the laboratories of the Vivarium in Vienna began as early as 1943.[21] The workspace in the Vivarium—a private biological institute in Vienna founded by Hans Przibram, Wilhelm Figdor, and L. Porges von Portheim—had remained unused since 1940, the founders themselves having been expelled following the annexation of Austria in 1938 (see

Chapter 1, Section 2). Geitler and Schiemann remained in Vienna and Berlin, respectively, until the end of the war.

Under Erwin Baur at the Institute for Hereditary and Breeding Research, in Berlin, Schiemann had begun to conduct systematic phylogenetic studies with strawberries and grain. Beginning in 1940, she worked at the Botanical Museum in Berlin with a DFG grant on questions of cultivated plant research; she studied genetic relations in strawberries and carried out research in prehistoric cultivated plants and descent in grain. Her work on the origins and genetics of cultivated plants led to a critique of Vavilov's theory of gene centers. Although she did consider it proved that certain diversity centers existed for the various cultivated plants and that these centers corresponded with the respective gene centers of the plants, she showed in 1943 that the equating of the gene center with the center of origin of cultivated plants did not hold as a general law. Schiemann was criticized by the National Socialists because she had determined that grain originated in the Near East and not in Germany (see, for example, Stokar 1938).

Researchers undertook botanical expeditions to Asia and to eastern and southern European countries to collect wild and primitive plants; Vavilov's theory of gene centers once again provided the theoretical backdrop. The great German Hindu Kush Expedition of 1935, under the leadership of Arnold Scheibe, collected wild and primitive strains (Freisleben 1940). The RFR organized and financed two expeditions to Albania, Greece, and Yugoslavia in 1941 and 1942, and one to northern Spain and southern France in 1944. The high command of the Wehrmacht took on the responsibility of providing military protection for the scientists.[22]

Research on wild and primitive strains of cultivated plants attracted the interest of the SS. Thus the SS expedition to Tibet in 1938–39, led by Ernst Schäfer, aimed at least in part to collect primitive strains, especially cold-resistant strains of grain.[23] Schäfer, too, had plans to establish an institute for collecting and studying wild strains. However, Himmler, who welcomed this plan, instructed him to approach the Kaiser Wilhelm Society to find out about the possibility of a collaboration. Wettstein, Telschow, and Schäfer subsequently decided to establish only *one* institute as a KWI, with Wettstein assuming its overall direction.[24] Schäfer named Heinz Brücher and Konrad von Rauch as members of the institute from the SS. He expressed political reservations about Stubbe. He considered Stubbe suspect because Stubbe was known not to be a National Socialist, an attitude that had led to his dismissal from the KWI for Breeding Research in 1937. Wettstein dismissed these reservations by pointing out that according to Walter Groß and Konrad Meyer these political concerns no longer existed. The institute was founded in 1943.

This being settled, Stubbe, who in the meantime had been installed as head of the institute, and Schäfer worked out the details of the collaboration between the Kaiser Wilhelm Institute for Cultivated Plant Research and the SS. Brücher was to be affiliated with the KWI and function as an intermediary between Stubbe and Schäfer, and Schäfer was recommended by Stubbe as a member of the institute's board of trustees.[25] The Kaiser Wilhelm Society had reason to be confident in 1942 of being given control of the most important biological institutes in those territories of the Soviet Union already in German hands and those still to be conquered.[26] However, because of the German defeat in the winter of 1942 and subsequent military developments, this was increasingly in doubt. In the face of the impending loss of the Soviet institutes that were already within the German sphere, the SS, at the instigation of Schäfer, organized a special command, which in 1943—under Brücher's leadership—stole material at numerous Soviet breeding stations, including parts of the Vavilov collection. Brücher assembled and worked on this material in the Institute for Plant Genetics in Lannach, which was specially set by the SS for this purpose and whose director he became (see Chapter 5, Section 3).

The planned collaboration between the KWI for Cultivated Plant Research and the SS came to an end. When Brücher informed him at the end of 1943 that the SS had founded its own institute for plant genetics, Stubbe was surprised and felt snubbed.[27]

Scientists at the KWI for Cultivated Plant Research also intended to work with collections stolen from the Soviet Union. Thus we read in a research report by Stubbe in 1944: "As part of the development of the institute, additional collections of primitive forms will be acquired, among them a Near Eastern wheat and barley collection and especially the world wheat collection salvaged from Russia, which contains unique types of the utmost value for German breeding research."[28] It is not known from whom Stubbe wanted to acquire these collections and whether he did get them.

4. The KWI for Breeding Research (Erwin Baur Institute), Müncheberg

This institute was founded in 1907 by Erwin Baur, who headed it until his death in December of 1933. He sought to improve plant products through breeding to the point where they could serve as a food, and to make the old cultivated plants resistant to parasites and climatic influences (immunity research). He regarded autarky in food supply as desirable and attainable without problems. Baur introduced cross-breeding (combination breeding) as a complement to selective breeding, which in the twenties was still the method used almost exclusively in agriculture. A conflict between him and Minister of Agriculture Richard Darré led to a drastic cutback of the institute's budget in 1933.[29] Baur suffered an attack of angina pectoris and died that December.

After the first Four-Year Plan—to prepare the economy for war—came into effect in 1936, the DFG and the RFR strongly supported research at the institute; in addition, the institute received large state subsidies. Research on the breeding of cold-, drought-, and parasite-resistant plants, which Baur had already promoted, became a major focus of research at the institute especially after 1937 (Rudorf 1937b; see also the DFG files of scientists of this Kaiser Wilhelm Institute). Rudorf, who became director of the institute in 1936, received considerable financial support from the DFG and the RFR for immunity research and for work on vernalization, which he promoted in Germany. As mentioned in Chapter 3, Section 1.3, vernalization makes use for breeding purposes of the interplay of light and temperature in various developmental stages of plants. Rudorf was working on utilizing vernalization as a method for the early selection of cold-resistant seeds in order to use them for further breeding (Rudorf 1938).

Bernhard Husfeld, who as head of the Department of Vine Breeding at the Kaiser Wilhelm Institute became director of the newly established Kaiser Wilhelm Institute for Vine Breeding in October of 1942, carried out immunity research in vines with substantial funding from the DFG and the RFR. Large sums were spent to support breeding work on grasses and winter-hardy legumes by Walter Hertzsch and lupine breeding by Joachim Hackbarth, both men working at the East Prussian branch (set up in the spring of 1933) of the KWI for Breeding Research. Theodor Gante and Friedrich Gruber did immunity breeding with berries and corn. Hermann Ulrich, the head of the chemical-technical laboratory of the Kaiser Wilhelm Institute, carried out optical-microscope studies on the characterization of the cold resistance of the protoplasm in plant cells. A member of the lab, Paul Schwarze, developed serial methods for determining the protein, oil, fiber, and alkaloids in the plant Kok-saghyz, and after 1944 also the rubber content in these plants (see Chapter 2, Section 10.3). Wild species and primitive forms of cultivated plants, which (as described above) were collected on various botanical expeditions, were used for further breeding. They became an important source for resistant genes (Römer 1941–1943).

The basic research in genetics at this Kaiser Wilhelm Institute consisted primarily of mutation research, which after Stubbe's dismissal was carried on by his successor Edgar Knapp. Knapp induced mutations in mosses through X-ray and UV radiation. During the radiation of moss spermatozoids with monochromatic UV light, he discovered in 1939, together with A. Reuss, Otto Risse, and Hans Schreiber, that the radiation showed the greatest mutagenic effect in the range of the spectrum—265 micrometers—that corresponded to the absorption maximum of DNA (Knapp et al. 1939). As with the corresponding observations by Milislav Demerec and Alexander Hollaender in the United States in the early forties, this result did not lead to the abandonment of the notion that genes were proteins. Knapp and his colleagues merely concluded that thymonucleic acid (the term for DNA at the time) had a decisive importance for genetic and physiological changes through radiation, and that the relationship between "thymonucleic acid and the genetic substance proper" had to be left for further studies. The research by Peter Michaelis on cytoplasmic inheritance was funded on a much smaller scale (see Chapter 3, Section 1.2).

5. The Genetics Department of the KWI for Brain Research, Berlin-Buch

The Genetics Department of the Kaiser Wilhelm Institute for Brain Research was set up in 1925 by Nikolai V. Timoféeff-Ressovsky. In 1925 the then director of the Kaiser Wilhelm Institute, Oskar Vogt, brought him from the Soviet Union to the KWI as a geneticist. Vogt was hoping that a genetics section would clarify questions concerning gene manifestation, that is, the influence that genes or the environment exert on the effect of other genes. This work was to help recognize and influence various forms of hereditary brain disease (Vogt and Vogt 1936, pp. 391ff.). Vogt did not find a geneticist in Germany competent to address these questions. As a result of several years of research work in the Soviet Union he had good contacts among Russian neurologists. For example, in 1924 he was asked by the Soviet government to perform in Moscow an examination of the cellular architecture of Lenin's brain, and under his direction the brain was sliced into a series of sections between 1925 and 1927 (Kirsche 1986, pp. 11–13; Mayr and Provine 1980, pp. 279ff.). Lenin had died in 1924 as a result of arteriosclerotic changes.[30] Thus, through his Soviet contacts Vogt was able to bring Timoféeff-Ressovsky and S. R. Zarapkin as geneticists to the institute in Berlin-Buch. Timoféeff-Ressovsky (soon to be world renowned, as noted in the previous chapter) was a student of the Russian geneticists Nikolai Koltsov and Sergei Chetverikov and at this point nearly unknown outside a small circle of Russian scientists. Timoféeff-Ressovsky's life and personality as well as his scientific accomplishments are detailed in the biography by Daniil Granin (1988) and in the account by Diane Paul and Costas Krimbas (1992). Here I would like to briefly mention his fate after 1945. At the end of World War II the staff of the KWI for Brain Research was evacuated to Göttingen, but Timoféeff-Ressovsky decided to stay in Berlin, where he was arrested when the Soviet troops arrived. He was imprisoned for several months and then transferred to a labor camp in north Kazakhstan, where he nearly died from starvation. After two years he was moved to a military research center near Sverdlovsk, where he organized a radiation biology laboratory. Only in 1955, two years after Stalin's death, did he receive amnesty, but he has still never been officially rehabilitated. In 1988 the Soviet government denied an application for his rehabilitation on the grounds that he had betrayed his motherland by having conducted research that enhanced Fascist military power (Paul and Krimbas 1992).

Chetverikov's population-genetics approach had a strong influence on the direction of Timoféeff-Ressovsky's research and, later, on the synthetic theory of evolution. This theory arose in the late forties and complemented Darwinist explanatory approaches with findings from genetics and population genetics. Incorporating the discovery of pleiotropic genetic effect by the Morgan group in the United States, Chetverikov had developed his concept of the "genotypical milieu" of a gene (Mayr 1982, p. 447).[31] According to this concept, a gene manifests itself differently depending on the complex of other genes in which it is found, that is, it has different effects on the phenotype. Under Chetverikov, Timoféeff-Ressovsky had discovered manifestations of pleiotropy in *Drosophila*. In contrast to Chetverikov's approach, the mathematical population geneticists of the West, especially Ronald A. Fisher and John B. S. Haldane, started from a genetics that left the reciprocal influences of genes largely out of consideration.

With regard to the mechanism of evolution, Chetverikov was one of the first to recognize that animal and plant populations, despite their morphological uniformity, had great genetic variability—which constituted material for the working of selection. Thus he had created a synthesis of Mendelian genetics and classical Darwinism (particularly the theory of selection), much as the Englishmen Fisher and Haldane and the American Sewall Wright had, and at about the same time. Chetverikov presented his findings for the first time at the Fifth International Congress for Genetics in Berlin in 1927. After leaving the Soviet Union, Timoféeff-Ressovsky in Germany and Theodosius Dobzhansky (who had not worked under Chetverikov but had followed the findings of his research) in the United States helped spread the ideas of the Chetverikov school.[32]

From 1925 to 1928 Timoféeff-Ressovsky worked as assistant and then department head at the KWI for Brain Research. Vogt was driven from his post by SA terror in 1935 because of his democratic attitudes. Hugo Spatz succeeded him. In 1936, after the Reich Education Minister had increased the budget of his department, Timoféeff-Ressovsky turned down an offer from the Carnegie Institution to go to Cold Spring Harbor in the United States.[33] Other factors also played a role in his decision to remain in Germany; as reasons for declining the offer, Timoféeff-Ressovsky indicated to his American colleagues that his children had already gone through one major transition, that his coworkers and technical assistants would lose their jobs, and that the social status of a professor was higher in Europe than in the United States.[34]

In his working report to the DFG in 1938, Timoféeff-Ressovsky listed the following as the general research tasks of the department: the phenomenology of gene manifestation, experimental mutation research, and population genetics, with the main focus on the last two fields.[35] Most of the research was done with *Drosophila*. The DFG and RFR grants for research in the departments were given by the sections on medicine, population policy, genetics and racial care, general biology, and physics. The greater part of the money flowed into experimental mutation research, where neutron radiation was increasingly used after 1938, and into research on the biological application of radioactive isotopes. Funds from the RFR and the Philips Company were used in 1938 to set up a biophysical laboratory in the department with apparatuses for radiobiological and radiophysical experiments, among them a powerful neutron generator, a linear accelerator for more effective generation of neutron radiation.[36]

In population genetics, Timoféeff-Ressovsky and his wife, Elena, analyzed the temporal and spatial distribution of individuals of several species of *Drosophila*. They showed that each species had its own "population wave," that is, fluctuation in the number of individuals within a period of time. They believed that the break-up of the species population into smaller, territorially isolated subpopulations was very important for microevolutionary processes, in particular for the fluctuations in the concentration of various alleles and mutations within a population (Timoféeff-Ressovsky and Timoféeff-Ressovsky 1941a–c). The evolutionary factors described here correspond largely to the random factors in the evolutionary changes in small populations, which Sewall Wright called genetic drift. With their work in population genetics the Timoféeff-Ressovskys contributed to the neo-Darwinist or synthetic theory of evolution.

After Hermann Muller's discovery in 1927 of the artificial induction of mutations through X-ray radiation, scientists were hoping to get closer to the nature of the gene by analyzing the mechanisms of the induction of mutations. For a number of years mutation research became the most modern—the most strongly funded and most widely pursued—field of genetic research. Mutations were induced by X-ray, UV, and neutron radiation. The few experiments that were carried out at this institute with chemicals produced no evidence of chemical mutagenesis.[37]

Together with the physicists Max Delbrück and Karl Günter Zimmer, Timoféeff-Ressovsky published "Über die Natur der Genmutation und der Genstruktur" (On the nature of gene mutation and gene structure), the so-called three-man work (Timoféeff-Ressovsky, Zimmer, and Delbrück 1935). This influential piece resulted from discussions within a group of physicists, biochemists, and biologists organized by Delbrück; beginning in 1934, it met privately, mostly in the home of Delbrück's parents. The work of the biochemist Hans Gaffron and the physicist Kurt Wohl (1936) on the primary processes of photosynthesis also resulted from the discussions in the Delbrück home. Delbrück, at the time a theoretical physicist and assistant to Lise Meitner at the KWI for Chemistry, took an increasing interest in questions of biology, prompted above all by the lecture "Light and Life" which Niels Bohr gave in Copenhagen in August 1932. In 1937 Delbrück received a grant from the Rockefeller Foundation to carry out genetic research under Thomas Morgan at the California Institute of Technology. Delbrück moved to the United States and in 1939 decided to stay.[38] Shortly thereafter he shifted to genetic research in phage, and he received a Nobel prize together with Alfred Hershey and Salvador Luria in 1969 for discoveries about replication mechanisms and genetic structure of phage (see Chapter 7, Section 2).

In their work on gene mutation and gene structure, Timoféeff-Ressovsky, Zimmer, and Delbrück applied the target theory of the effect of radiation to the action of radiation in genetics. The target theory was formulated in 1922 by Friedrich Dessauer as a mathematical description of the biological effects of radiation and was given specific form in 1924 by James Crowther.[39] Timoféeff-Ressovsky, Zimmer, and Delbrück reached the conclusion that a mutation represented a one-hit result that came about through a single, radiation-caused ionization in a hit area, the gene. Their notion that the gene was a cluster of atoms for the first time ascribed a certain stability to genes. It was now possible to subject genes, much like molecules, to chemical and physical analyses.[40] In 1970 Delbrück recalled that the finding that genes had a stability comparable to chemical molecules was not a trivial one in the thirties: "Genes at that time were algebraic units of the combinatorial science of genetics, and it was anything but clear that these units were molecules analyzable in terms of structural chemistry" (Delbrück 1970, p. 1312).

The three-man work, especially Delbrück's model of the relative stability of a gene, which he developed on the basis of atomic theory, influenced Erwin Schrödinger in the development of his model of genes as "aperiodical crystals" in his 1944 book *What Is Life* (see Fleming and Bailyn 1969, p. 173). Schrödinger's book, in turn, inspired James Watson to search for the secret of genes. According to Delbrück

(1970), at the time Timoféeff-Ressovsky, along with Zimmer, was doing "by far the best work in the field of quantitative mutation research."

Zimmer became a close collaborator of Timoféeff-Ressovsky's in Berlin-Buch. In the thirties and on through the war, additional research was done to develop the target theory, research that also involved the use of neutron radiation (Zimmer and Timoféeff-Ressovsky 1942). In 1936 Timoféeff-Ressovsky and Delbrück rejected the notion that the size of the hit area would correspond to the size of the gene.

After Delbrück emigrated to the United States in 1937, Timoféeff-Ressovsky continued the interdisciplinary discussion in Berlin-Buch with several physicists. A regular participant at these gatherings was the theoretical physicist Pascual Jordan, who had a strong influence on the analysis of the primary process of the effect of radiation by Timoféeff-Ressovsky and his collaborators. Jordan, like about 60% of the biologists at that time, belonged to the NSDAP and publicly expressed anti-Semitic views. Like Delbrück, Jordan had turned to biological questions during the thirties. Again like Delbrück he had been influenced by Niels Bohr's thoughts on complementarity. In 1932 Bohr had suggested searching for a complementary relationship between life and the laws of physics and chemistry similar to what existed between wave and particle according to the findings of quantum mechanics. However, in contrast to Delbrück, Jordan was not planning any experiments but deduced from quantum mechanics theoretical conclusions for biological problems. Based on quantum mechanics, he posited an attractive power between *identical* macromolecules ("quantum mechanical resonance attraction"), and with it he tried to explain not only the attraction of homologous chromosomes but also the self-replication of "genes and virus molecules." He argued that during replication similar molecules were built, the atoms of which were quantum-mechanically in different states (Jordan 1938).

Delbrück and Linus Pauling sharply criticized Jordan's thesis that quantum-mechanical resonance phenomena would lead to an attraction of identical molecules and to an autocatalytic reproduction of molecules (Pauling and Delbrück 1940; see Fischer 1985, p. 107).[41] There is reason to believe that the intense contact with Jordan contributed to Timoféeff-Ressovsky's decision to make the analysis of the primary process of mutation into one of his main areas of research.

According to Jordan, many biological reactions, especially biological radiation effects (mutations) were microphysical reactions in the quantum-mechanical sense. He saw a phenomenon of quantum mechanics specifically in the movement of the energy released by ionization somewhere in the hit area to the place where it was used for the change of the gene molecule: electrons did not wander in the normal sense of the word but spread over a molecule like a wave (Jordan 1940).[42]

The exclusive concern with questions concerning target theory and energy movement, which in the final analysis did not advance the investigation into the nature of the gene,[43] prevented scientists in Berlin-Buch from pursuing other paths to solve this question. Nikolaus Riehl (the head of the research division of the Auer Society, who collaborated with the scientists in Buch in the field of biophysics), Timoféeff-Ressovsky, and Zimmer reached the conclusion that these questions about energy movement and about the primary processes of radiation effects had to be solved before the nature of elementary biological units and processes made any sense (Riehl, Timoféeff-Ressovsky, and Zimmer 1941; Timoféeff-Ressovsky and Zimmer 1944). For this reason, the radiation experiments that the scientists in Berlin-Buch carried out in the forties with viruses and phages served primarily the purpose of developing the target theory further.[44] This was also the purpose of mutation experiments with neutron radiation conducted by Timoféeff-Ressovsky and Zimmer. In 1942 they confirmed that the formation of an ion from one atom should be seen as the hit effect. A threshold value of the radiation effect did not exist; the mutations produced were directly proportional to the radiation dose right up to saturation level; and the mutation-inducing effect proved to be independent of the wavelength in the range of X-ray and gamma radiation (Zimmer and Timoféeff-Ressovsky 1942). The mutation experiments with UV radiation by Alexander Catsch and Ilse Sell-Beleites, although they used monochromatic radiation of different wavelengths, also served initially the purpose of examining dose proportionality and determining the hit area.[45]

Many experiments to clarify the mechanism of how chromosome mutations were produced by X rays and neutron radiation were carried out by Catsch, Karl Eberhardt, Kanellis, A. N. Panschin and Igor Panschin, P. P. Peyrou, Georg Radu, and Zimmer. Here, too, the question of dose dependency predominated.[46] The Panschins as well as Peyrou tried to refute the notion of the fundamental inseparability of point and chromosome mutations. They showed in mutations of the X chromosome of *Drosophila* that even the smallest deletions as chromosome mutations are two-hit events, which, in contrast to point mutation induced by a one-hit event, show a dose proportionality that follows a nonlinear equation.[47]

Beginning in 1940, Zimmer intensified his research on the development of a neutron dosimetry. The biological effects of fast neutrons proved, with respect to their quantitative effects, greater than those of the corresponding dose of X-ray radiation (Catsch, Zimmer, and Peter 1947). The experiments on dosimetry and on the spatial distribution of radiation-conditioned ionization were also to serve as the basis for the development of radiation therapy. In his DFG applications, Zimmer emphasized the possible importance of neutron radiation for cancer treatment, listing results of his own experiments.[48] At the end of 1942, despite many war projects, he still considered the study of neutron therapy urgent, since according to his own information clinical trials with carcinoma patients had already been carried out with favorable results in the United States using the larger sources of neutrons that were available there. Because of its strong side effects, radiation therapy with neutrons is today used only at a few cancer sites; medical indications are restricted.

Research on biological applications of artificially radioactive isotopes also received strong financial support from the DFG and the RFR; in addition, it was promoted by the Reich Office for Economic Development, a division of the Ministry of Economics. In this work scientists in Berlin-Buch, especially Elena Timoféeff-Ressovsky and the radiochemist Hans Joachim Born, collaborated with P. M. Wolf, the head of the Radiological Department of the Auer Society. The Auer Society in Berlin was part of the Degussa Company and produced, among other things, luminescent substances and purified reduced uranium. The artificially radioactive isotopes were produced through neutron radiation in a biophysical laboratory in Buch, and methods for their isolation and concentration were worked out by Born and Zimmer.

In collaboration with scientists from the division for virus research, researchers were able to label viruses radioactively by marking the tobacco plant with radioactive phosphorous (phosphorous 32) and subsequently infecting it with the tobacco mosaic virus (Born, Lang, and Schramm 1943). Experiments to determine the temporal and spatial distribution of some short-lived isotopes of chlorine, phosphorous, arsenic, and manganese in small mammals formed the basis for the use of radioactive isotopes in medical diagnosis. For example, the liver in rats turned out to be the main storage organ for manganese (Born, Timoféeff-Ressovsky, and Wolf 1943). Joachim Gerlach (1941) carried out experiments in humans to determine the distribution and excretion of radium (radium 224, at the time thorium X).[49] Radium 224 was also injected into the bloodstream of human subjects to determine the circulatory time (Gerlach, Wolf, and Born 1942). According to Diane Paul and Costas Krimbas (1992), the dose of radium 224 used at the time, because of the low half-life, did not cause any major health problems.[50] This conclusion presupposes that there was no contamination with isotopes with a greater half-life. In general, we can note that radioactivity was handled much more carelessly in those days than is the case today.[51] The study of the excretion of thorium in rabbits with the help of the thorium isotope uranium X (today thorium 234) showed that thorium, because of its storage by cells of the reticuloendothelial system, was excreted only very slowly (Wolf, Radu, and Catsch 1944). These experiments were important in view of the use of thorotrast (a thoriumdioxidesol) as an X-ray contrast medium.[52]

6. The KWI for Biophysics, Frankfurt

The Kaiser Wilhelm Institute for Biophysics was founded in 1937 in Frankfurt by Boris Rajewsky, who in 1934 had become Friedrich Dessauer's successor as professor of biophysics at the University of Frankfurt. Dessauer, the founder of the Institute for Biophysics at the University of Frankfurt, had emigrated to Turkey in 1934 because he had been persecuted and imprisoned for his Roman Catholic and political views;[53] in addition, he had one Jewish grandparent. As already mentioned, in 1922 Dessauer formulated for the first time his thesis of the hit theory, which held that the biological effects of radiation on individual elementary processes like the denaturing of proteins could be traced back to specific radiation-sensitive regions and were thus accessible to statistical treatment. At first he developed the theory primarily through radiation experiments with proteins, in the denaturing of which he saw the basic process of the effect of radiation (Rajewsky in *FIAT-Reviews*, vol. 21, pp. 9f.). Rajewsky continued the radiobiological experiments, studying primarily the biological effects of radium, X-ray, and neutron radiation.

Rajewsky and his collaborators discovered that in assessing a case of radium poisoning one had to take into account not only the incorporated radioactive substance but also the special importance attached to the daughter products produced by the decay of radium, called radium emanation at the time, now known to be the gaseous element (*FIAT-Reviews*, vol. 21, p. 162). In 1943 Rajewsky developed a method for determining temporal changes of radon radiation in the animal body: following treatment with radium, animals were killed after a certain time, deep-frozen and thawed, dissected, and examined for the distribution of radioactivity. One finding was a concentration of radium in the bones (p. 169).

In the spring of 1937 Rajewsky also became head of a scientific research section at the radium bath of Oberschlema. The research work aided studies on radium-balneological treatment of rheumatic conditions and radium poisoning in the mines and studies on the effect of radioactive substances on the human body.[54]

At the suggestion of the Reich Committee for Cancer Treatment, Rajewsky examined the importance of radioactive substances as carcinogenic agents. His research included systematic work on the problem of Schneeberg lung cancer. Rajewsky and his collaborators were able to induce lung tumors in white mice through the inhalation of radon and thus to clarify one of the causes of the disease (Rajewsky, Schraub, and Kahlau 1943). The inhaled radon was much more important in causing Schneeberg lung cancer than other factors, for example radioactive dust (Rajewsky 1939). Rajewsky carried out experiments with the aim of determining a tolerance dose for the inhalation of radon. To that end animals were exposed to different kinds and doses of radiation and examined for a variety of factors, for example weight and survival time. In this way he determined a tolerance dose for radon in a situation of continuous inhalation that was about twenty times higher than the one that had previously been published by foreign scientists; young mice were much more sensitive than old ones (Rajewsky, Schraub, and Schraub 1942a, 1942b).

The DFG and the RFR gave Rajewsky RM 375,000 in the years 1940 through 1944 to set up a three-million-volt installation devoted to the production of X-ray, neutron, and electron radiation for radiating animals and people. The money came from the RFR sections on medicine and racial care. However, the installation was not finished by the end of the war.

7.   The Department of Hereditary Pathology of the KWI for Anthropology, Human Genetics, and Eugenics, Berlin-Dahlem: The Example of Hans Nachtsheim

The Kaiser Wilhelm Institute for Anthropology, Human Genetics, and Eugenics was founded in 1927; the directorship was given to the man considered the leading German anthropologist, Eugen Fischer. The establishment of the institute took place on the occasion of the Fifth International Congress for Genetics, in Berlin, and helped to gain international recognition for the study of human inheritance in Germany. Like other sciences, it too had suffered for some years from international isolation owing to the boycott of German scientists after World War I.

When Fischer became emeritus in 1942, his former student Otmar von Verschuer succeeded him. In the Department of Human Inheritance, which he headed, Verschuer carried out the genetic analysis of pathological and normal traits of humans, especially with the help of research on pairs of twins. At that time the institute contained other departments, including hereditary pathology (Kurt Gottschaldt), racial science (Wolfgang Abel), racial hygiene (Fritz Lenz), and experimental hereditary pathology, founded in 1941, with the zoologist and later human geneticist Hans Nachtsheim as its head.

It is known that in important aspects Fischer, von Verschuer, and Lenz supported not only the racial-hygienic but also the anti-Jewish and racial goals and laws of the Nazi regime. Their research programs and their activities in racial politics, along with the extensive collaboration between von Verschuer and Josef Mengele in Auschwitz, are described in Müller-Hill 1988 and Weindling 1989. Nearly all scientists at the KWI for Anthropology identified with the racial political goals of the regime and worked for their practical implementation. For example, members of the institute drew up expert opinions on "racial membership" in connection with the Nuremberg Laws. Because of this close identification with Nazi policies, the institute itself was dissolved by the allies after the war. However, scientists did not have to cease their work. With the exception of Fischer, who was already emeritus, all of them received influential positions in anthropology and human genetics immediately after the war (see Weindling 1989, pp. 566–573). Of course, owing to the fact that human geneticists in Germany had placed themselves at the service of racial political crimes, they were internationally discredited. Many foreign colleagues at first refused to resume their contacts with German human geneticists and anthropologists.

Hans Nachtsheim is considered the only department head of the KWI for Anthropology with a clean political past (Weingart, Kroll, and Bayertz 1988, p. 566). He never belonged to the NSDAP, and geneticists in the United States and Great Britain, among whom he had many friends, were convinced that he was an anti-Nazi; for example, the émigré geneticists Hans Grüneberg and Richard Goldschmidt vouched for his integrity (Weindling 1989, p. 566). He was the only German geneticist to participate in the United Nations Educational, Scientific, and Cultural Organization's declaration of 1949 against racism, the UNESCO Statement on the Nature of Race and Race Differences by Physical Anthropologists and Geneticists (Weingart, Knoll, and Bayertz 1988, p. 605).

In Germany after 1945, Nachtsheim acquired a central role in the establishment of human genetics, successfully prevailing upon the DFG to fund this field of research (Weingart, Knoll, and Bayertz 1988, pp. 583 and 627). In 1955 he was awarded the Federal Service Cross with Star, a decoration of the state. As a member of the Max Planck Society, the Federal Health Council, and the German Society for Population Sciences, he influenced the application of the insights of human genetics to health and population policies. In 1961 Nachtsheim served as an expert adviser to the Restitution Committee (Wiedergutmachungsausschuß) on the question of compensation for people sterilized against their will under National Socialism. He died in 1979. Since 1979, the Society for Anthropology and Human Genetics has been awarding the Hans Nachtsheim Prize for Theoretical Research and the Hans Nachtsheim Prize for Research Oriented toward Practical Application for outstanding scientific accomplishments in human genetics.

If we trace Nachtsheim's career and research during the Nazi period in greater detail, we find confirmation for the widespread belief that Nachtsheim was not one of the anti-Semitic or racist human geneticists of the Third Reich. However, the claim that he behaved with integrity politically and personally and that National Socialist ideology did not enter into his research is not tenable. With this case study of Hans Nachtsheim, I want to shed some light on the practice of human genetic research in a society that had abolished the concept of equal human rights.

7.1. Nachtsheim as Hereditary Scientist and Hereditary Pathologist until 1945

Hans Nachtsheim was born in Koblenz on June 13, 1890. He studied zoology and, following a stint as assistant in the Zoological Institutes in Freiburg and Munich, became a lecturer at the University of Munich in 1919, a lecturer in genetics at the Agricultural College in Berlin (since 1935 part of the university) in 1921, and an associate professor at the college in 1923.

Early on, Nachtsheim took a keen interest in genetics. In his book *Ein halbes Jahrhundert Genetik* (Half a century of genetics, 1951a) he describes how, while working as a military censor during World War I and examining publications from Switzerland, he came into contact with scientific publications from the school of Thomas Hunt Morgan in the journal *Genetics*. These papers reported that scientists had determined the relative positions of hereditary dispositions on specific sites of the chromosomes in *Drosophila*.[55] With the help of Erwin Baur, Nachtsheim was able to procure the Morgan school's existing works on *Drosophila* (Calvin B. Bridges, Hermann J. Muller, and Alfred A. Sturtevant) and, during the war, to publish a review of these works. In 1921 he published a German translation of Morgan's book *The Physical Basis of Heredity*. Beginning in 1921, he did genetic research on *Drosophila*.

From 1925 on he worked in the areas of the genetics and descent of domesticated animals. He spent 1926 in the United States, mainly at Morgan's lab at Columbia University, on a fellowship from the Rockefeller Foundation. Like other geneticists at the time, he advocated sterilization on eugenic grounds starting in the twenties (Nachtsheim 1963) and gave lectures on eugenics, for example at a training course for medical officers that was held January 8–18, 1930, at the KWI for Anthropology under the direction of Fischer.[56] As he was neither Jewish nor politically liberal, Nachtsheim's position was not endangered by the Law for the Restoration of the German Civil Service, implemented on April 7, 1933. But for a short time he was afraid to lose his position as an assistant at the institute because there were plans to eliminate long-term assistant positions. He contacted colleagues in the United States asking for work possibilities, which led to an offer of a position at the University of Wisconsin with financial support from the Rockefeller Foundation. However, when his position in Berlin was extended in October 1933, he decided not to go to Madison.[57] A positive political expert opinion by the faculty body (Dozentenschaft) resulted in another extension in 1937.[58] The opinion stated that Nachtsheim had always acted in favor of the national state and that he was an outstanding geneticist. In 1933 he began to focus his research on hereditary diseases of small mammals, in the process introducing "comparative hereditary pathology" as a new research discipline. In 1937 he wrote the following about his studies of hereditary pathology in rabbits in his funding application to the DFG: "The goal of these studies is to identify, with the help of breeders, the hereditary diseases and hereditary anomalies that occur among rabbits and to submit them to experimental study in order to obtain comparative material for the hereditary diseases and hereditary anomalies of humans, which are not accessible to experimental analysis."[59]

Why did Nachtsheim change the direction of his research in 1933? He wrote to Fischer on April 12, 1948, that the occasion for the beginning of his work in experimental hereditary pathology was the promulgation on July 15, 1933, of the Law on the Prevention of Genetically Diseased Offspring.[60] According to this law, people could be sterilized if they suffered from a disease believed to be hereditary, for example, feeblemindedness, hereditary epilepsy, alcoholism, and schizophrenia. It is estimated that about 350,000 people were sterilized under the law, most against their will. Nachtsheim was speaking from conviction and not merely to secure funding when he justified his grant application by stating: "I believe I can say that with these experiments a new field, that of comparative hereditary pathology, has been successfully launched, a field which is also important for human racial hygiene."[61] After 1933, "important for human racial hygiene" meant that any indication a disease was hereditary could result in the compulsory sterilization of the affected person.

In 1934 Nachtsheim produced a film about a hereditary nervous disease in rabbits, a film that also had a racial-hygienic goal. (The film was made for use at universities.) The disease corresponded to the manifestation of Parkinson's disease in humans and, according to

Nachtsheim, was inherited recessively in a simple Mendelian devolution. In his application of May 17, 1934, to the Prussian Ministry for Science, Art, and People's Education, we read the following: "The film has not only scientific interest but also breeding interest; moreover, in connection with the Law on the Prevention of Genetically Diseased Offspring it also has general interest. After all, it depicts a very clear case of a hereditary ailment that shows how, through healthy parents, who are bearers of the pathological disposition, this trait can be passed on from generation to generation and spread ever more widely, thus seriously harming the race."[62]

Nachtsheim conducted animal experiments on the inheritance and probability of manifestation of epilepsy, Pelger anomaly (an anomaly of the form of the nucleus of white blood cells) gray cataract, certain tooth anomalies, and dwarfism. For epilepsy he showed in rabbits the existence of a recessive allele that increased the disposition to seizures. In electric and low pressure experiments he showed that in principle *every* animal had the capacity to experience seizures. The likelihood, however, depended on an animal's hereditary constitution and age. As the cause of spontaneous seizures he suspected oxygen deficiency in the brain, which could be caused by vasospasms. In the thirties, the question had arisen among psychiatrists whether the intravenous injection of Cardiazol (in the United States: Metrazol), a cramp-inducing drug used in the treatment of schizophrenia, could be used in the diagnosis of hereditary and nonhereditary epilepsy. Some doctors, among them Medizinalrat Dr. Schönmehl, Albrecht Langelüddeke, Georg Stiefler, and F. Langsteiner, believed they had discovered that humans who suffered from hereditary epilepsy would respond with seizures to a lower dose of Cardiazol than those suffering from nonhereditary epilepsy.

Nachtsheim investigated this potential method of diagnosing hereditary epilepsy and wrote in 1942: "The provoked Cardiazol seizure would accordingly provide us with a racial-hygienically important method of gaining insight into the hereditary picture of an individual epileptic." Proof of hereditary epilepsy with the help of Cardiazol would have resulted in the compulsory sterilization of the person concerned (see below). Nachtsheim supplemented the Cardiazol experiments of the doctors with animal experiments, which he justified as follows: "Testing this question sufficiently on human material naturally runs into difficulties. Thus it seems desirable first to study the connection between a disposition to seizures and the genotype by means of the Cardiazol test in animals" (1940b).

The series of experiments with rabbits led to the finding that the Cardiazol test was of only qualified diagnostic value. To justify further development of the method of comparative hereditary pathology in order to apply the results to humans, Nachtsheim had to compare the results of his animal experiments with those of experiments on humans carried out by others. He compared his findings on the latency period of epileptic and non-epileptic rabbits after injection of Cardiazol with the results F. Sal y Rosas obtained in correspondingly extensive tests on humans.[63] Nachtsheim found far-reaching parallels between humans and rabbits. He explained differences in the control group of non-epileptics by maintaining that, unlike the rabbits, the humans tested in this group were unhealthy and suffered from nervous disorders, especially hysteria. Thus he stated: "Had it been possible to use healthy people for the tests with non-epileptics, the correspondence of the curves for rabbits and humans would probably be greater still" (1942, p. 64).

Concerning the effect of Cardiazol injection in humans, the psychiatrist Christel Heinrich Roggenbau wrote in 1937 that in his opinion "the induction of a seizure constitutes in every case a not inconsiderable intervention in the physiological course of the life processes. It seems certain that vasospasms in the brain are in all likelihood the basic process for the epileptic seizure, and that such vasospasms can lead to changes in the brain substance." Nowhere did Nachtsheim protest against the use of Cardiazol in humans, against the fact that humans had to suffer from the induction of seizures. He conducted animal experiments as supplements to these human experiments. What guided him was an interest in diagnosing a disease as a hereditary one, not a desire to counsel sick people and develop a therapy. He sought to prevent them from reproducing in the interest of the "Volk as a whole."

The geneticist Raphael Falk pointed out in 1993 that Cardiazol (or Metrazol) was used in psychiatric diagnosis and therapy not only in Germany but also in Italy, Sweden, the Netherlands, the United States, and Peru. Thus one could see Nachtsheim's Cardiazol experiments as "merely accommodating to the normal conceptions of his time."[64] While I do not deny this aspect of "normality," in my view such experiments must be judged differently depending on where they are being carried out. In his essay of 1993 Falk writes: "The eugenic balance that should be kept between the good of the individual and that of the population was a subject discussed in scientific and juridical circles in Western democracies as much as in Nazi Germany." But according to everything we know this is not what happened. During National Socialism, the good of the whole, the Volk, definitely had priority. Nachtsheim's experiments were part of a medical science that, in keeping with National Socialist goals, oriented itself to the welfare of the Volk and not of the individual.

In the commentary on the 1933 Law on the Prevention of Genetically Diseased Offspring, which made the compulsory sterilization of hereditary epileptics possible, we read the following: "The varied manifestation of the disease makes it necessary that all persons suffering from seizures suspected of being epileptic be reported by those required to do so to those authorized to submit a request [for sterilization]. Those authorized to submit a request as well as the genetic health courts have the sole task of clarifying the diagnosis 'genuine (hereditary) epilepsy.' In every case where this diagnosis is made, sterilization is mandatory" (Gütt, Rüdin, and Ruttke 1936, p. 142).

According to section 3 of the law, a physician with civil-servant status and the head of an institution had the authority to submit a request; in addition, according to section 2, the genetically diseased person himself or herself could request sterilization. The people required to report a suspected hereditary disease to the public health officer included doctors practicing on their own, heads of institutions, and midwives (Gütt, Rüdin, and Ruttke 1936, pp. 211f.).

However, Nachtsheim's Cardiazol experiments were, in the words of Raphael Falk, "only the foam on water."[65] To guarantee scientific exactitude in the service of a higher goal—that is, as Nachtsheim stated, to reduce in the long run the number of diseased persons by preventing the "racial-hygienically" undesirable spread of the "deviant mutations"[66] underlying the disease—he did not shrink in the end from conducting his own experiments with humans. Animal experiments had shown that although oxygen deficiency was the basic cause that induced the epileptic seizure, young and old animals behaved differently. Nachtsheim and his collaborator, Gerhard Ruhenstroth-Bauer, a scientific assistant at the KWI for Biochemistry who obtained his doctorate in 1943, drew the following conclusion from their studies: "The characteristic difference in the behavior of young and old epileptic animals in the face of oxygen deficiency makes it desirable to undertake comparative tests of young and adult epileptics in humans. Gremmler studied only adult epileptics without being able to induce a seizure through hypoxemia. Following the conclusion of our own studies in young epileptics, studies of interest also to clinical practice, we intend to report further details" (Ruhenstroth-Bauer and Nachtsheim 1944).

In his DFG application, Nachtsheim elaborated on this point:

> In humans it has been shown by others that adult epileptics do *not* respond with a seizure to oxygen deficiency. Since our animal experiments had revealed a significant difference in the behavior of young and adult epileptics, we tested epileptic children in a similar manner in a low pressure chamber. However, to date we have been able to use only children eleven to thirteen years of age, in whom no seizure occurred at a low pressure equivalent to a height of 4–6,000 meters. Eleven to thirteen human years are equivalent to 5–6 months in rabbits, an age at which in epileptic rabbits, too, the seizure threshold is no longer low enough to make it possible to induce seizures with a certain regularity in low pressure. To make comparison possible, epileptic children about 5–6 years of age would have to be tested.
>
> A preliminary report about the tests are in print in *Klinische Wochenschrift,* the more detailed paper which is in preparation is to appear in *Zeitschrift für menschliche Vererbungs- und Konstitutionslehre.*[67]

The preliminary report on the experiments with children appeared (Ruhenstroth-Bauer and Nachtsheim 1944), but the publication of the detailed report was preempted by the end of the war. It was not published in the first issue of the *Zeitschrift für menschliche Vererbungs- und Konstitutionslehre* after the war or, as far as we know, in any other journal. According to the bibliography of Hans Nachtsheim's scientific publications in the *Festschrift* for his sixtieth birthday (Grüneberg and Ulrich 1950), he did not publish any more work on epilepsy between 1944 and 1950.

Where did Nachtsheim and Ruhenstroth-Bauer get the children for their low-pressure experiments and what happened to them? Following a hunch that the children may have come from the euthanasia institution in Görden (Brandenburg), Benno Müller-Hill contacted Ruhenstroth-Bauer. Through Ruhenstroth-Bauer's lawyer he was told that "the experiments were not experiments but clinical trials. They were permissible, because the epileptic children came from an orphanage, and nobody had died or was physically harmed. Moreover, Prof. Ruhenstroth had claimed that the results would be of therapeutic and prognostic value" (Müller-Hill 1987).

As it turns out, this information was false. As Nachtsheim himself wrote in a letter (dated September 20, 1943) to the human geneticist Gerhard Koch, the children did in fact come from Görden. What follows is an excerpt from the letter:

Dear Dr. Koch,

Thank you very much for your kind letter of September 10. By the time I received it Dr. Ruhenstroth had already contacted Wuhlgarten and had learned that only adult epileptics are there. Wuhlgarten referred him to Görden, from where, with the kind help of Senior Medical Officer Dr. Brockhausen, we obtained 6 epileptic children (4 genuine, 2 symptomatic). With these children we carried out experiments last Friday in Prof. Strughelt's low pressure chamber. However, the experiments had the same negative results as those of Gremmler on adult epileptics. But at this point it is not possible to say that rabbits and humans behave differently under low pressure, since the children tested by us were between 11 and 13 years old, which corresponds to an age of 5–6 months in rabbits. However, epileptic rabbits 5–6 months old no longer show the same readiness to react under low pressure as animals 2–3 months old, which nearly always had seizures. We would still have to test epileptic children 5–6 years of age, though at the moment that is not possible since this age group is not represented in Görden. (Koch 1993, p. 125)

The low-pressure chamber referred to is the same one in which Sigmund Rascher had earlier carried out his deadly experiments in the Dachau concentration camp. The chamber was subsequently taken to the Academy of Military Medicine, in Berlin. The Psychiatric State Institution Görden, in Brandenburg, was run by Hans Heinze, an expert consultant on euthanasia. As was the case in the "special children's wards" of other psychiatric institutions, the children in the juvenile psychiatry ward in Görden were selected for euthanasia. Some were killed in Görden itself, some in the neighboring extermination center of Brandenburg (Klee 1983, p. 300).[68]

It is not known what happened to the children Nachtsheim and Ruhenstroth-Bauer used for their experiments. A single reference to such a child is found in Knaape 1988, and this child survived the stay in Görden. Dr. Hans-Hinrich Knaape examined medical research during the Nazi period in the Psychiatric State Institution Görden. He reports the case of Hildegard K. (born September 3, 1932), a patient in the juvenile psychiatry ward in Görden. She suffered from epileptic seizures; researchers tried to determine whether she suffered from symptomatic or genuine epilepsy.[69] "The patient suffered from genuine epilepsy with compulsive restlessness and dementia. On September 17, 1943, she was subjected to a diagnostic experiment in the altitude physiology lab of the Academy of Military Medicine in Berlin. Upon the inhalation of the oxygen mixture, which corresponded to an altitude of 4,000 meters, an epileptic seizure did not occur" (Knaape 1988). The age of Hildegard K., who was eleven years old in 1943, as well as the year in which the experiment was conducted agree with Nachtsheim's information about his experiments with epileptic children. Hildegard K. was a patient at the clinic until 1946.

"Inferior" and "worthless" people were available in euthanasia institutions like Görden or in concentration camps as "test material" for medical experiments, a possibility for medical research not available before or after the Nazi period. Nachtsheim made use of this opportunity and to that end cooperated with physicians who themselves carried out euthanasia. After 1945, neither he and his colleagues nor scientists who worked at the Academy of Military Medicine spoke about this; only the publication of Nachtsheim's letter by Gerhard Koch in 1993 provided unequivocal evidence of his connection with Görden.

Nachtsheim made clear his view on the ethical basis of medical research in a dispute with the physician Robert Havemann that took place in 1946. The latter had leveled serious reproaches against Verschuer on account of his collaboration with Mengele. Nachtsheim rejected the charges brought by Havemann and asked him: "Do you in fact regard the use of material from a concentration camp as a crime in every instance?"[70] Nachtsheim compared the situation with the procurement of "material" from prisons, as had happened in Hermann Stieve's anatomical institute: "Corpses of people executed for political reasons, and Prof. Stieve did very valuable studies with such material, studies which would not have been possible any other way. One could blame Verschuer only if it had been known to him that the individuals from whom the eyes came did *not* die a natural death or were even killed in order to get the eyes. One cannot question Verschuer's good faith."[71] One should add that the eyes in question came from six members of one family and from several pairs of twins.

Hermann Stieve became full professor of anatomy at the University of Berlin in 1935. In the forties he did studies on women in an effort to prove that ovulation was possible outside the normal cycle if women were suddenly exposed to mental stress. In this way he sought to refute the gynecologist Hermann Knaus, who denied that such ovulation could occur. Stieve was able to prove the existence of "paracyclic ovulations" by using as his study material the corpses of women who had been sentenced to death; he himself spoke of the "opportunity to examine further women who were healthy and died suddenly" (Stieve 1944). According to Hans Harald Bräutigam (1989),[72] Stieve was able at least in a few cases to arrange for the women to be informed of the day of their execution; the wardens in prisons and concentration camps kept menstrual calendars of women sentenced to die. The women were executed on the day they had been told and their bodies were opened in the anatomical institute and examined for ovulation. It is not known if the women were from concentration camps or prisons.

Nachtsheim's own research on human beings deprived of their rights, and his justification in this regard of Verschuer and Stieve, show that he considered it ethically justifiable and compatible with good science to use this chance for human experiments and to profit scientifically from the murder of humans who provided the "material" for research.

A later letter reveals that Nachtsheim knew full well that the people from Auschwitz whose organs von Verschuer examined had been murdered. On February 14, 1961, Nachtsheim wrote the following remarkable lines to Leslie C. Dunn at Columbia University: "If in the near future the large Auschwitz trial which has already been announced takes place, and if by then Dr. Mengele, who was Verschuer's assistant until the end of the war and at the same time doctor in the concentration camp of Auschwitz, is apprehended, a number of things that are incriminating for Verschuer are likely to come up. I must confess that for me it was the greatest shock I experienced during the entire Nazi period when Mengele one day sent the eyes of a gypsy family interned in the Auschwitz concentration camp."[73]

Nachtsheim was more than a witness to this collaboration with Mengele. To prevent his knowledge and his complicity from becoming known, he went further. In March of 1945, shortly before the allies arrived in Berlin, he wrote a very friendly letter to Verschuer, who had already left the city for a safer place. In the letter he expressed his concern about the incriminating documents that Verschuer had left behind in Berlin instead of destroying them. Nachtsheim offered his help. He now felt responsible for the situation and declared that he would destroy the relevant documents in a timely fashion if the city were occupied.[74]

As I have already indicated, in 1945 Nachtsheim was regarded as the only department head of the KWI for Anthropology, Human Genetics, and Eugenics with a clean political record. He was the only human geneticist who after 1945 maintained or was able to establish many different contacts with foreign geneticists. In addition to his commitment to the rebuilding of human genetics in Germany after the war, Nachtsheim participated in the international campaign against the restriction on scientific freedom in the Soviet Union. What follows is a look at Nachtsheim's fight for scientific freedom in the Soviet Union against the background of his research during the Nazi period.

7.2. Nachtsheim and the Freedom of Science: The Struggle against Lysenkoism

After the war, Nachtsheim was among the many Western European and American geneticists who actively opposed Lysenkoism in the Soviet Union: "A totalitarian political system and the freedom of science cannot coexist. Our struggle goes on" (Nachtsheim 1956). By this time he was full professor of genetics at the Free University of West Berlin. In 1949 he left Humboldt University in East Berlin, which had set up a chair in genetics for him in 1946, to accept an appointment at the Free University, which had been established with American support. By doing so Nachtsheim wanted to protest against his inability, according to his own statements, to work freely as a geneticist in the German Democratic Republic.[75]

The backdrop for the call to political activism against Soviet totalitarianism was the era of Lysenkoism in the Soviet Union, a period that saw the brutal suppression of Soviet scientists, especially geneticists, in the name of a pseudoscience. Trofim Lysenko, an agronomist, had worked on vernalization with the aim of changing the vegetative period of crops in order to grow them in regions where they had never grown before because of an adverse climate. His vernalization attempts drew on experiments by Gustav Gassner in Germany, who had shown in 1918 that the treatment of seeds before or after germination permits under laboratory conditions the shortening of the vegetative period and the growing of winter varieties of grains during the summer (Graham 1974, pp. 202–203).

Inaccurate methods, variable agronomic conditions, and manipulations of results make it difficult to decide whether Lysenko had ever managed to make traditional crops ripen earlier. Because of the genetic heterogeneity of his plant varieties, it cannot be ruled out that he achieved single "successes" of vernalization due to selection processes (Graham 1974, pp. 204–206). He did, however, contribute to the increased attention to vernalization. But Lysenko went further. He claimed that he found proof that plants could undergo a "Socialist reeducation": that winter wheat had been permanently transformed into summer wheat through the change in environmental factors. According to Lysenko, the change in the plants' phenotype was thus genetically transmitted to the next generation. His vernalization trials completely lacked the methodology of scientific research, the most obvious methodological error being the almost total absence of control groups (Graham 1974, p. 202, passim). According to Loren R. Graham, "the Lysenko episode was a chapter in the history of pseudoscience rather than the history of science" (1974, p. 195). David Joravsky refers to Lysenkoism as a rebellion against science altogether (Joravsky 1970, p. vii). It is true that in the early thirties Lysenko's work was supported by several scientists inside and outside the Soviet Union; thus it was publicly acknowledged by the internationally respected plant geneticist Nikolai Vavilov in these years. According to Joravsky, however, Vavilov never believed in the scientific value of Lysenko's work, but in the beginning tried to achieve peaceful coexistence between science and Lysenkoism (Joravsky 1970, pp. 93, 379); in any case, Vavilov became the great opponent of Lysenko a few years later. Step by step Lysenko's doctrine became the only permissible basis for research and teaching in genetics in the USSR. This dominance was justified by claims that it was close to dialectical materialism, which rejected the coincidences that were, in the form of mutations, a firm part of scientific genetics. Joravsky (1970) has explained that Lysenko could exert his influence over decades largely because he promised great agricultural successes in a short period of time. In view of the backward state of agriculture, Stalin had in 1929 made the increase in agricultural production the preeminent concern of the biological sciences.

Scientists who did not accept Lysenko's doctrine were dismissed after 1936. Some were sentenced to labor camps, some were shot. Vavilov, in fact, was arrested in 1940 and exiled to Saratov, where he died in prison in 1943 (Medvedev 1969, p. 74). The geneticists Israel Agol and Solomon Levit were arrested and shot (Zacharow and Surikow 1991). The Institute for Medical Genetics in Moscow, which Levit had headed until 1936, was closed down. After the war, Lysenko intensified his campaign. Following the meeting of the council of the Lenin Academy of Agricultural Sciences in August of 1948, where Lysenko, with Stalin's support, pushed through the final politicization of Soviet biology and condemned all advocates of the teachings of Mendel and Morgan as reactionary and antinational, the last followers of "Western genetics" were expelled from the universities (Medvedev 1969, pp. 83 and 117ff.; see also Joravsky 1970). Political support for Lysenko did not cease with Stalin's death in 1953 but continued under Nikita Khrushchev. Only after Khrushchev's fall in 1964 did Lysenko's suppression of biology in the Soviet Union come to an end.

In the West there were protests against this restriction of scientific freedom. (See for example Judson 1979, pp. 370–372.) In Germany, Hans Nachtsheim wrote: "Today German science is threatened a second time by totalitarianism. In my opinion the indifference and carelessness of German scientists bears a good part of the blame for the fact that National Socialism was able to come to power in 1933. Thanks to the help of the free world, West Germany and West Berlin have been liberated from totalitarianism. East Berlin and the Soviet zone, however, have exchanged the totalitarianism of Hitler for an even more inhuman totalitarian system. It is not only the task of West Berlin to do everything possible for the liberation of the East and its reunification with the West. The scientists of the West bear a heavy responsibility" (1951b).

It is remarkable that Nachtsheim, shortly after the full extent of the Nazi crimes had become known, described the Soviet system, compared with National Socialism, as more inhuman. In this way he relativized the crimes of National Socialism (as occurred in 1986 in the controversy among German historians that has come to be known as the *Historikerstreit*). Moreover, he ignored the Soviet contribution to Germany's liberation from National Socialism. As the quotation shows, Nachtsheim attributed the guilt of German scientists to their indifference and carelessness. In a letter to Fritz Lenz he formulated this self-criticism more concretely: "When the 'Third Reich' had come to an end, the charge was leveled against us geneticists that we accepted the abuse of our science without objections. This reproach falls on us not without good reason. In any case, at that time I swore not to be silent a second time when genetics is abused for political purposes. And today we have the parallel in Lysenkoism."[76] Without wishing to detract from Nachtsheim's commitment to fight Lysenkoism, one might add that in the first case silence (though it was in fact not merely silence but also active participation in the crimes) was very positive for the financial support of genetic research (see below). Nachtsheim made this parallel with the racial theory of Nazism the center of his journalistic activism against Lysenkoism, and he created the impression that Stalinism and National Socialism had, to an equal extent, suppressed genuine sciences and replaced them with pseudo-sciences. For example, he wrote: "'Bourgeois biology,' Lysenko's 'reactionary genetics' remind us of the 'German physics' of the National Socialists. I do not wish to deny that the horticultural and plant-breeding work of Mitschurin and Lysenko had practical significance, their accomplishments are on a similar level as those of the American Luther Burbank. However, the fantastic ideas about heredity developed by Mitschurin and Lysenko belong as much to the field of pseudoscience as does the racial theory of the National Socialists" (1948).

Nachtsheim saw the support that Lysenkoism received from scientists inside and outside the USSR as much more serious than what had occurred in Germany: "The saddest thing about this sad chapter is surely the fact that there were some renowned scientists who, taken in by Communist ideology, actually supported Lysenko" (1956). By contrast, he denied that German scientists had played an active role in National Socialist racial policy. In his opinion the majority of German biologists in 1933 were opposed to Nazi ideology; they kept quiet because they were "muzzled" during the Third Reich.[77] "When the National Socialists propagated their racial theory prior to 1933, we geneticists and biologists, with few exceptions, made fun of their 'Nordic spleen,' their 'race mania.'"[78] As I have repeatedly shown, this statement does not accord with the facts.

While the doctrine of Lysenko was, for political reasons, turned into a science by force, and many geneticists in the USSR rejected it in part openly, in part secretly, in Germany important anthropologists and geneticists contributed their share to National Socialist racial theory and reality.[79] Even if a number of National Socialist politicians were hostile to science, rejected basic research, and demanded a "true" German science, the greater number of those involved with science policy were interested in seeing that good scientists were given good positions and that their research, in particular in the field of genetics, was supported. Scientific freedom, the choice of a research topic and methodology, as well as the possibility of publishing one's findings in scientific journals were a reality for most non-Jewish German scientists, even if it was not officially propagated. However, public criticism of the anti-Jewish and other racial-political measures was not possible without endangering oneself or at least one's job.

The analysis of Nachtsheim's research under National Socialism does not indicate that it was impeded by any significant political constraints. The opposite seems to have been the case. In 1933 Nachtsheim himself decided to change his field of research and to place it at the service of the sterilization law. Nobody forced him to do so, and many of his colleagues continued working with *Antirrhinum* or *Drosophila*. His work enjoyed substantial financial support. Although his application to the DFG for a research grant in 1935 was turned down, beginning in 1937 he received regular funding. After his move to the KWI for Anthropology, this support increased. For example, in 1943 and 1944 Verschuer received RM 40,000 for research in experimental hereditary pathology, funds that were in large measure intended for Nachtsheim. Nachtsheim's productivity during National Socialism also does not indicate constraints on his scientific work. Between 1933 and 1945 he published forty-eight papers and one book.

The use of children from Görden for research shows that Nachtsheim, too, did not shrink from taking advantage of the "freedom" that National Socialism offered to use "inferior human material." By justifying Verschuer's collaboration with Mengele, he contributed to absolving German human genetics from its greatest crime. Did he do this to acquit himself of his own knowledge of the facts and preempt a possible uncovering of his own active role in the abuse of this freedom?

7.3. Nachtsheim as Eugenicist after 1945

In 1945 Nachtsheim became director of the KWI for Anthropology, in Berlin, and remained at that post until the institute was dissolved on the decision of the allies. Nachtsheim's Department of Comparative Hereditary Pathology was the only department of this Kaiser Wilhelm Institute to continue in existence after 1945 (see below). In 1946 he was appointed professor of genetics at Humboldt University in East Berlin. As we have seen, he left that university in 1949 because he was no longer able to work freely and accepted a position at the Free University of Berlin. There he cofounded the Institute of Genetics, which he headed from 1949 until his retirement in 1955, during which period he also held a chair in general biology and genetics. At the same time he was director of the Institute for Comparative Hereditary Biology and Hereditary Pathology of the German Research Institute, which in 1953 was incorporated into the Max Planck Society. He headed this institute, which had emerged out of the Department of Experimental Hereditary Pathology of the KWI for Anthropology, Human Genetics and Eugenics, until 1960.

After 1945, he became the most vigorous champion in the Federal Republic of Germany of voluntary sterilization on eugenic grounds. To him eugenics was still applied genetics; it represented a principle of social order, the controlling of reproduction according to criteria of the biology of heredity (Nachtsheim 1963). When, in 1963, a public debate took place in the Federal Republic on the issue of eugenics, Nachtsheim gave the same justification for it that eugenicists had at the beginning of the century: "The selection that once existed no longer occurs, we are engaged in counterselection, the previously existing balance of the genes is upset, the number of the genetically diseased must invariably rise from generation to generation as a result of the increasing spread of pathological genes."

Since 1946 there had been no secure legal basis for any possibility of sterilization for eugenic reasons. As the most important measure for "putting a stop to rampant hereditary diseases," Nachtsheim therefore called for making it "legally possible to prevent the reproduction of genetically diseased persons by their voluntary sterilization on eugenic grounds." With regard to the sterilization law under National Socialism he noted:

> I believe I can say today that the Law for the Prevention of Genetically Diseased Offspring was not an unjust National Socialist law, nor was it abused for racial and party political purposes, nor did it have anything to do with the crimes of National Socialism. The law came out of the draft of a sterilization law from 1932, which was drawn up by the Prussian State Health Office. Even if the law was naturally given a National Socialist dress in 1933, its core is free of National Socialist ideology. The law served genetic health care through the medical measure of sterilization.

Nachtsheim did not explain what he meant by "National Socialist dress." Neither here nor, as far as is known, in other publications did he criticize, let alone condemn, the use of coercion, which was the hallmark of the National Socialist law. He demanded: "Let us finally free ourselves from the taboo that has been on eugenics in Germany since National Socialism! We are only hurting the genetic health of the German people." Alexander Mitscherlich, psychiatrist and author (together with Fred Mielke) of *Medizin ohne Menschlichkeit: Dokumente Nürnberger Ärzteprozesses* (Medicine without humanity: Documents of the Nuremberg physicians trial, 1962), was one of the experts who contradicted Nachtsheim in this debate.

In response to Nachtsheim, Mitscherlich (1963) emphasized that it is untenable to maintain that eugenics and National Socialist race policy had nothing to do with each other: "Moreover, the extermination of millions of people has an intrinsic conceptual connection to the Law for the Prevention of Genetically Diseased Offspring." In the application by the National Socialists of the law prepared by the Prussian government, he saw "how eugenic thinking that based itself on 'elimination' as the essential measure, got the chance to become, step by step, the official doctrine, then the opinion of the legislator, and finally the ideology of the state."

Mitscherlich did not reject eugenics on principle, but owing to the substantial influence of environmental factors on the manifestation of genetically predisposed diseases, he perceived considerable problems for practical eugenics. In his view, practical eugenics should concentrate on combating disease-causing environmental factors. To him preventing the reproduction of undesirable individuals through interventions in the body was an abuse of eugenic ideas. In the final analysis this abuse led to euthanasia, to the murder of Jews and "Slavic sub-humans." He did not wish to hold "serious scientists in the field of genetics" responsible for this. But what gave him pause was the "monomaniacal fixation of all eugenic practice that seems to regard as the final wisdom the active, mutilating intervention in the sphere of the individual for the sake of the general public (and for the moment it is of no importance whether one defines it as 'race,' 'nation,' or the majority of the 'normal people')."

It is true that the sterilization law in question did not stem from the Nazis; important scientists were involved in creating the draft of the Prussian government, among them the geneticists Erwin Baur and Richard Goldschmidt and the anthropologists Eugen Fischer and Hermann Muckermann. According to Nachtsheim (1962), the text of the laws of 1932 and 1933 differed as follows: "The text of 1932 envisioned voluntary sterilization, the law of 1933 also a compulsory one. On the other hand, the law of 1933 was limited to the sterilization of genetically diseased persons, whereas according to the 1932 draft, people who were outwardly healthy but known to be carriers of pathological traits could also have themselves sterilized." The 1933 law also added alcoholics to those targeted for sterilization.

Characteristic of the National Socialist law was the threat and use of coercion. Norbert Schmacke and Hans-Georg Güse, in their 1984 study of compulsory sterilization in Bremen, wrote:

> The people affected saw themselves from the outset of the proceedings confronted by a cynical system of force. The applicant was given the leaflet "Genetically Diseased Offspring is the death of the Volk," in which it was said: "About 400,000 genetically diseased persons in Germany deprive the same number of healthy people of their chance of living!" Contrary to the true situation, it went on to claim: "The best success of popular education regarding genetic health is the fact that most applications are made voluntarily." (p. 95)

In a memorandum to the Reich Chancellery in 1937, even National Socialist Reich Physicians' Führer Gerhard Wagner expressed his indignation about the irresponsible manner in which the sterilization judgments were being rendered (Müller-Hill 1988, p. 32). He complained that this system had caused an alarming decline in the population's confidence in official physicians. To the Reich Interior Ministry he voiced his criticism that the genetic health courts were applying the law in a "schematic and formalistic" way instead of a "National Socialist and biological manner."[80]

According to official statistics, from 1934 to 1936 between 60,000 and 70,000 people were sterilized each year (Müller-Hill 1988, p. 32). Voluntary sterilizations on other than eugenic or medical grounds were punishable under National Socialism. Sterilizations not permitted by law were to be punished as grievous bodily harm even in cases where they had been performed with the consent of the person or persons concerned (Schmelcher 1965). In 1943 a supplementary paragraph to the Penal Code (section 226b) decreed that "the person who in cases other than those permitted by the law destroys another person's ability to procreate or bear children with that person's consent, shall be punished with a sentence of three months in the penitentiary, in particularly serious cases with a longer prison term, provided that the offense is not subject to a more severe punishment under another statute" (Schmelcher 1965). Decisions about a person's ability to procreate or bear children were made by the National Socialist state. The free will of the individual was legally blocked.

In his postwar calls for eugenic measures, Nachtsheim repeatedly emphasized that sterilization was to be voluntary. Did he move to reestablish the possibility of a freely made decision, something the Nazis had made punishable by law, to give a person suffering from or carrying a hereditary disease the possibility of opting for sterilization by his or her free choice? On this point he stated the following:

> If, after a sterilization law has been passed, it is left to everyone's discretion whether or not he will adhere to the law . . . the end result will be that only the reasonable person will forgo having offspring, the person who in many cases would do it even without sterilization . . . In that case eugenics would be conducted not on the federal level but merely on the familial level. The only ones to benefit from this kind of genetic care would be the families concerned, but the law would have missed its real purpose, that of preserving the genetic health of the nation, an effect on the level of population genetics could not be achieved.   (1962)

The demand that sterilization be voluntary appears to have corresponded more to political opportunism than to his personal conviction. Thus he wrote: "Still, to repeat what has been said: given the current state of affairs one can advocate only voluntary sterilization, with only the one qualification that exists also in other countries, namely that in the case of incapable patients, that is above all the feebleminded and mentally ill, who are a burden on public welfare, sterilization can be ordered by the state" (1952, p. 50).

However, a sterilization law alone was not enough; to render it effective, "the will to eugenics" had to be present "in the people" (Nachtsheim 1952, p. 20). His comments at the thirty-fourth session of the Restitution Committee (Ausschuß für Wiedergutmachung) on April 13, 1961, to which he had been invited as an expert, show that in the final analysis Nachtsheim was thinking about the reintroduction of compulsory sterilization. Not only did he argue for the legality of the sterilization law of 1933 with the old argument about "counter-selection," he also openly declared himself in favor of coercion: "A eugenic law without coercion is, in terms of genetic hygiene, as ineffective as an immunization law without any compulsion" (after Schmacke and Güse 1984, p. 161). In this way Nachtsheim remained true to his ideas: after all, as he had written Eugen Fischer in 1948, the National Socialist sterilization law was the occasion for the beginning of his work in the field of experimental hereditary pathology.

In the medical practice of human genetics in Germany today, which focuses on genetic counseling, there is no legal legitimation for a eugenic orientation, that is, for an approach that concentrates on the total population and would allow for coercive measures. In every case the free decision of those seeking counseling and their individual rights must be respected by the physicians. In other Western European countries and in the United States, as well, practical human genetics is oriented toward the individual. Thus, according to Diane Paul (1992), personal autonomy is generally accepted in the United States as the primary right in genetic counseling. She also calls attention to the serious social problems raised by an ethic of radical individualismin in regard to reproductive decisions, for example when it comes to the selection of nonmedical traits such as the gender of the embryo.

In connection with biomedical developments, in particular the expected results of the Human Genome Project, the danger of a new eugenics has been invoked, at least in Germany, and many critics are demanding that the project be discontinued. Benno Müller-Hill (1993) sees another way of countering potential future problems. He defends the Human Genome Project as "scientifically sound," but he encourages us to think now about possible consequences. These include, for example, genetic tests for diagnosing physical and mental ailments, which in the future are likely to be possible to a much greater extent. Furthermore, with the help of anonymous studies it could become possible to determine genetic differences among ethnic and social groups. To prevent existing genetic discrimination from getting worse, Müller-Hill demands not only that individuals have exclusive right of access to the data of their own genome analysis but also that laws be passed to protect the genetically disadvantaged and provide them with the compensation of social justice.

A look at Nazi Germany and the crimes it committed in the name of genetics and eugenics shows that it was possible to transform a civilized society into one in which the value of equal human rights was abolished and many biomedical scientists lost their humanistic bearings.

5

Scientific Research by the SS

1. The Scientific Interests of Heinrich Himmler

The scientific interests and personality of Heinrich Himmler (1900–1945) largely shaped the goals of the science policy of the SS as well as the practical content of the scientific and medical research it initiated. A brief excursus into Himmler's biography may shed light on some of the underlying factors that influenced his interests (see also Höhne 1988; Kater 1966; Reitlinger 1956).

Himmler grew up in a strict Catholic home. Even as a child he took an interest in warfare and agriculture, especially the goals and possibilities of animal and plant breeding. As a boy he maintained a large plant collection, showing a preference for herb gardens. Early on he also developed a strong interest in alternative forms of medicine, such as homeopathy. This may be one explanation for Himmler's seemingly bizarre order later that every concentration camp plant a wide variety of herbs. The concentration camps supplied cheap labor for the cultivation of medicinal plants, from which he expected better healing success than from pharmaceuticals (recall Chapter 3, Section 1.1). In the field of chemical manufacture, too, Himmler searched for alternative "natural" production methods. For example, he commanded the Agricultural Section of the Auschwitz concentration camp to cultivate the rubber plant Kok-saghyz, which came from the Soviet Union, on a scale competitive with the chemical industry (see Chapter 2, Section 10).[1] IG Farben at the time was producing synthetic rubber in the Buna factories near Auschwitz with the help of forced labor from the camp.

Himmler, a stickler and rigorist, was an insecure and superstitious person. Like his later role model Hitler, he was a vegetarian and teetotaler. At his father's wish,[2] he studied agriculture in Munich and received his diploma in 1922.

After his gradual break with Catholicism he became, in the early twenties, a follower of Adolf Hitler, to whom he submitted with unconditional obedience throughout his life. According to Joachim Fest (in the introduction to Smith and Peterson 1974, p. 14), Himmler was, apart from Alfred Rosenberg, the only leading National Socialist who not only used Nazi ideology as an instrument for seizing and maintaining power but also took its doctrines seriously as a historical mission. Himmler found reflected in this ideology his dislike of urban life and of Western civilization, as well as his firm belief in the superior racial value of the Germanic people and his goal of a Greater Germanic Reich. Out of Hitler's ideas of *Lebensraum* he developed, in pedantic fashion, a program for the Germanification of the East, a program that included the extermination or expulsion of "inferior races." He did more than implement Hitler's fanatical hatred of the Jews by erecting new concentration camps and establishing new branches of the SS, such as the Reich Security Main Office (Reichssicherheitshauptamt), whose tasks included carrying out the destruction of the Jews; in the Institute for Practical Research in Military Science, which he set up within the framework of the SS branch Das Ahnenerbe (Ancestral Heritage), he also wanted to clarify the "Jewish question" anthropologically and scientifically: the Strasbourg anatomy professor August Hirt was allowed to murder one hundred and fifteen Jewish inmates of the Auschwitz concentration camp at his own discretion to establish a typology of Jewish skeletons (Mitscherlich and Mielke 1978, pp. 174f.).

A detailed analysis of SS research using the SS office Das Ahnenerbe as an example can be found in Kater 1966, which shows that Himmler's central concern was the study of the history, threat to, and preservation of the Nordic race, the race he regarded as the bearer of the highest civilization and culture. His ideas on breeding and genetics as well as his exceptional interest in medical questions (Kater emphasizes Himmler's rejection of "academic science" as well as his uncommon interest in natural healing and occult medicine) related to the preservation and further development of the Nordic racial type (Kater 1966, pp. 2–3).

Himmler's letters, which are kept in various archives and a selection of which has been published (Heiber 1968), reveal the extent of his abstruse ideas in the fields of medicine and natural science. His instructions for research projects appear highly contradictory, for we find in them brutality and the will to destroy alongside solicitude and the promotion of natural methods. But what gave Himmler's ideas coherence was his overriding concern for the Nordic race. The following examples will suffice. (Individual research projects in biology are analyzed in greater detail in Sections 2 to 4 of this chapter.)

On October 24, 1942, Himmler commented as follows on Sigmund Rascher's undercooling experiments with inmates at the concentration camp Dachau: "I consider people who today still reject human experiments, and instead allow brave German soldiers to die from the effects of undercooling, to be guilty of high treason and treason against their country" (Heiber 1966, p. 163). Himmler reserved for himself the decision on all experiments with inmates of concentration camps (p. 225). In 1942 he had Professor Ernst-Robert Grawitz test homeopathic remedies in Dachau and Auschwitz for the treatment of sepsis and other illnesses, whereby all cases of sepsis, most of which were artificially induced, resulted in the death of the inmate (pp. 144–145). At Christmas that year Himmler donated an herbarium to the library of the medicinal garden in Dachau (p. 177). Artificial insemination he rejected in most cases as an unnatural method of conception. On March 31, 1943, he wrote to his former teacher Heinz Henseler that in his opinion artificial insemination *could* be used, for example, in animals that were being raised for food. It was his belief, however, that artificial insemination led to the degradation of genetic traits, and he was convinced "that we humans cannot improve upon nature." For that reason he would only buy a pet that had been naturally conceived, and he rejected artificial insemination for human reproduction (pp. 207–208).[3]

The growing array of positions of power Himmler amassed allowed him—in contrast to Alfred Rosenberg, for example—to put most of his ideological and scientific interests into practice. Reichsführer (Reich leader) of the SS since 1929, Himmler became in 1936 also chief of the German police, including the secret police and the security service. In 1939 this personal union between police and SS at the highest level was extended to the middle levels. The police and the security service were combined into the Reich Security Main Office of the SS. The concentration camps stood under the supervision of the SS. In 1939 Himmler was named Reich commissar for the strengthening of Germandom, whose task it was to implement the "Germanification" of the East. That same year he also became president of the research and teaching association Das Ahnenerbe. In 1944 he was appointed chief of the security service and of counterespionage.

Using Das Ahnenerbe as an example, I will look at biological research projects that stood under the direct influence of the SS or Himmler and were conducted outside established scientific institutions and universities.

2. *The SS Research and Teaching Society Das Ahnenerbe*

Himmler played a decisive role in the establishment in 1935 of Das Ahnenerbe as a private association, and he was the first curator of the organization. During the following years his influence grew, and at the beginning of 1939 the society was officially listed as a section of the SS (Kater 1966, p. 50a), whose tasks were set by its president, Himmler. Himmler used the association as a research institute he could charge with scientific tasks of every kind. Even though some suggestions were made to him, in most instances Himmler took the initiative (Kater 1966, p. 46).

The Ahnenerbe was initially devoted to historical, in particular ethnographic, research in the interest of National Socialist ideology, but later it turned increasingly to research in the natural sciences. By the spring of 1938 it already had five sections for the natural sciences. Himmler toyed with the idea of turning the society into an SS academy of sciences, a plan that was thwarted by the outcome of the war (Zierold 1968, p. 187). Wolfram Sievers served as managing director.

Within the framework of the Ahnenerbe, the research in Tibet by the zoologist Ernst Schäfer was given special support. Apart from geographic and anthropological goals, Schäfer's work also had biological aims. Schäfer had participated as a zoologist in two American expeditions to Tibet in 1932–1933 and 1934–1935. The several job offers that he received from the Americans and the Chinese prompted the consul general of Shanghai in January of 1936 to impress upon the DFG the need to support another Tibet expedition that Schäfer had already planned, to prevent the loss of a man who "could one day become the ornament of our long line of German scientists."[4] In addition, the consul general recommended intellectual support for Schäfer in the form of an honorary doctorate or the "bureaucratic facilitation of his doctoral exam."[5] Schäfer was at that time working on his doctoral dissertation under Alfred Kühn. He had intended to take his exam in 1934, prior to the second expedition to Tibet, but Kühn's objections interfered with this plan. Schäfer was awarded his doctorate in 1937, most likely after involvement by Himmler, given that, as the quotation reveals, Schäfer had not made any progress by 1936. The Tibet expedition he had planned and subsequently led received RM 30,000 from the DFG and the RFR in 1938.[6] He conducted the expedition in 1938 as an Ahnenerbe mission, on Himmler's orders. Only members of the SS were allowed to participate; Schäfer himself had been a member since 1933, and he rose to the rank of major in the SS.[7]

With this expedition Schäfer was pursuing not only scientific goals but also aims that were clearly political. He believed that "we could accomplish more as SS men and do much more for the lack of understanding for the new Germany by being up front about who we are, than by traveling under the disguise of an obscure, if neutral scientific academy; after, all, we have a clear conscience."[8] He emphasized publicly that the expedition was an SS undertaking, which led to problems with the British authorities in India.

Among those who participated in the expedition were Bruno Beger as anthropologist, Ernst Krause as entomologist, and Karl Wienert as geographer. Beger, writing his doctoral dissertation under the race anthropologist Hans F. K. Günther, was the anthropologist who in 1943, as August Hirt's assistant, selected the one hundred and fifteen Jews used to establish a collection of Jewish skeletons. His task on the expedition was "to study the current racial-anthropological situation through measurements, trait research, photographing, and molds (using Poller's method), and especially to collect material about the proportion, origins, significance, and development of the Nordic race in this region."[9]

As far as zoology was concerned, since the area represented a "biogeographical juncture of the faunal regions of the Himalayas, India, East Asia, and northern Central Asia," the expedition was to study "the wealth and diversity of forms of animal species distributed over a very small habitat" and collect many animal species.[10] The goals of botanical research corresponded to those described earlier for the botanical expeditions of the RFR: eastern Tibet was considered a region that, "owing to its wealth in original useful plants, has been seen as a gene center and promises a rich yield of new discoveries."[11] Winter-hardy strains of grain were to be collected and examined for their suitability for interbreeding.[12]

In January of 1940 Schäfer became head of the new Ahnenerbe Section for Central Asian Studies and Expeditions. He had the ambitious plan of setting up a large institute as a "germ cell for the rebuilding of German science" (Kater 1966, pp. 120ff.). In October of 1942 the Section for Central Asian Studies and Expeditions became the largest research division of the Ahnenerbe; in 1943 the Sven Hedin Institute for Central Asian Studies, named after the Swedish researcher, was founded in Munich, and Schäfer became its director.[13] Given that the war of destruction in the East and the "final solution" were already in full swing, it might seem strange or even macabre that the openness with which Schäfer disseminated his racism was unacceptable even to Himmler for reasons of political expediency. For example, at the end of 1942 Himmler initially authorized the showing of Schäfer's film *Geheimes Tibet* (Mysterious Tibet) only in a closed screening; not until a report appeared in the press as a result of a slip-up did he order a public showing.[14] In February of 1943 he wrote to Schäfer: "Last night I read the first pages of your book *Geheimes Tibet*. You must stop the publication of the book for now, for the first part of the book is written from such an objectively German point of view toward the English that such a statement cannot be published at this stage in the war. In particular, in the book the native states and peoples are characterized in a benevolent but mocking way. This could do us immeasurable damage in the colored world. The misgivings I had about a public showing of the film have thus been justified after all."[15]

To intensify the support for war-important research, Himmler, on

July 7, 1942, ordered the establishment of the aforementioned Institute for Practical Research in Military Science within the Ahnenerbe. Kater 1966 details some key facts about the institute. It was financed with funds from the Waffen-SS, that is, with state funds (p. 222). Five scientific divisions and two medical divisions were set up, the latter attributable to Himmler's special medical interests and rejection of the "reactionary" practices in conventional medicine and science. With this research Himmler also wanted to meet the needs of Hitler's war machine (p. 4). Wolfram Sievers (Ahnenerbe managing director) became head of the institute; he acted as an intermediary between Hitler and the members of the institute. At the request of his good friend Rudolf Mentzel, Sievers in 1943 also became Mentzel's deputy in the management of the executive advisory board of the RFR (pp. 237f.).

The RFR supported a number of war-important projects at the Ahnenerbe. Those included, in addition to the research of Heinz Brücher and Eduard May described below, the human experiments that have become known through Mitscherlich's and Mielke's work (1978) on the Nuremberg physicians trial of 1946; as part of these experiments, physicians accepted the mass murder of inmates. These crimes cannot be discussed in detail in the present study, but I will mention two examples. As with other applications, the section heads of the RFR decided on the funding for this "research." Thus the grants for the "cold experiments" that Sigmund Rascher was carrying out on inmates of the Dachau concentration camp, in which eighty of about three hundred inmates died, were authorized by the RFR's section head for cancer research and Deputy Reich Physician Führer Kurt Blome. The project, given the urgency classification "SS," was described as follows: "Re-warming following a general cooling of the human body, recovery in cases of partial frostbite, cold adaptation of the human body."[16]

The mustard gas experiments that Hirt, professor of anatomy at the University of Strasbourg, conducted on inmates of the Natzweiler concentration camp, which also received the urgency classification "SS," were, in 1942, within the jurisdiction of the general medicine section head, Professor Ferdinand Sauerbruch, and after 1943 within that of Kurt Blome. The project was described as follows: "The behavior of yellow cross in the living organism."[17] Yellow cross, better known as mustard gas, had been used as a chemical warfare agent during World War I; it is a skin toxin with long-term effect. These experiments, in which countless people were killed, and for which after a while no more "volunteers" could be found with the promise of being released afterward, were intended, among other things, to study various forms of treatment for mustard gas poisonings of different degrees of severity.[18] Today Hirt's best known project was the plan to set up a collection of Jewish skeletons for the anatomical institute in Strasbourg. Hirt committed suicide in 1945.

For his complicity in the death of these inmates, Sievers was executed in Nuremberg in 1947 (Mitscherlich and Mielke 1978, pp. 174f.). The following two sections will examine in greater detail two biological divisions of the Institute for Practical Research in Military Science: the division for plant genetics in Lannach and that for entomology in Dachau.

3. Heinz Brücher at the Ahnenerbe's Institute for Plant Genetics, Lannach

On November 1, 1943, Heinz Brücher, lecturer and SS second lieutenant, was made head of the Division for Plant Genetics of the Institute for Practical Research in Military Science, which was being set up in Lannach, near Graz. Brücher, born January 1, 1915, in Darmstadt, obtained his doctorate at the University of Tübingen in 1938 under Ernst Lehmann with botany as his major field and zoology and anthropology as his minor fields; he wrote his dissertation on cytoplasmic inheritance. Subsequently he became assistant to Karl Astel at the Institute for Human Hereditary Research and Race Policy at the University of Jena. The decision to take this post reflected Brücher's increasing political radicalization. Astel, an anthropologist, had been a member of the NSDAP since 1930 and become SS captain as early as 1934 and lieutenant colonel in 1939; he had been appointed to Jena in 1934 and was among the most prominent SS figures at German universities. He ardently championed the SS idea of human breeding and selection and the racist anti-Semitism that went with it. As a representative of the NSDAP's Office for Race Policy for Thuringia, he worked for the practical implementation of these goals. Already by December of 1934, in his capacity as president of the Thuringian State Office for Racial Matters (Landeswesen für Rassewesen), a central state office for population policy and "genetic and racial care," Astel had compiled files on the hereditary biology of 300,000 people and on the criminal biology of 740 people (Weingart et al. 1988, p. 446). In 1935 he approached Himmler with the suggestion of turning the University of Jena into an SS University, and he was able to push through the appointment of prominent SS professors, for example in 1939 that of zoologist and anthropologist Gerhard Heberer (see Section 5 of this chapter). After the war Astel committed suicide before the allies could put him on trial in Nuremberg.

Brücher obtained his *Habilitation* under Astel in 1940, and in 1941 he was appointed lecturer in botany in Jena. Like Astel, Brücher called for the use of selection as an essential instrument of population policy, which was in his eyes tantamount to racial policy. The fight against drugs was for him also a means of preserving the genetic health of the race. For example, in 1941 he expressed the following warning against the consumption of alcohol and the use of tobacco (Brücher 1941):

> Until now too little attention has been paid to protecting the human genetic material against genetically damaging influences. Known genetic toxins are above all nicotine and alcohol. The extraordinary increase in tobacco addiction in the German people—since about 1934—and the inordinate consumption of the addictive poison nicotine contribute, together with the increased consumption of strong liquors, to the further deterioration of the genetic material. However, the genetic health of the German people and of the Nordic-Germanic race which unites it must be preserved under all circumstances. In fact, through a sensible adherence to the laws of nature, through selection and planned racial care, it can even be improved. The racial superiority attained in this way guarantees our people in the harsh struggle for survival an advantage that will make it invincible.

In his social-darwinist ideas, Brücher drew on Ernst Haeckel, whose biography he wrote in 1936 at Astel's suggestion.[19] In addition, he published several articles in *Nationalsozialistische Monatshefte—Zentrale politische und kulturelle Zeitschrift der NSDAP* (National Socialist monthly—chief political and cultural journal of the NSDAP, published by Alfred Rosenberg), in which he hailed Haeckel as the "pioneer of biological state thinking" and defended him against attacks from the Catholic Church, which condemned Haeckel as an "atheist" and "materialist."[20] Brücher appreciated that Haeckel, at a time when people "regarded the death penalty as a hard-to-justify interference by the state in the human rights of the autonomous person," had the courage to put forth eugenic demands later realized under National Socialism: "The idea, too, of a systematically applied differential reproduction, which is today finally being realized through preventive detention, castration, and frequent death penalties, is already found in one of Haeckel's suggestions for the genetic improvement of the Volk."[21] In regard to Haeckel's question about what benefit it was to mankind that "thousand of cripples" and "people burdened with incurable hereditary ills" were being "artificially kept alive," Brücher contended: "Haeckel's eugenic demands thus go even far beyond the foundations that determine the German idea of sterilization, and so far they have hardly entered into the framework of scientific discussion, especially with regard to euthanasia."[22]

Apart from efforts to spread the idea of selection as the foundation of the National Socialist "policy of elimination," Brücher's main commitment was to research. And here the decisive impulse that led him to the study of wild and primitive strains of cultivated plants, to which he devoted himself to the end of his life, came from an assignment by Himmler to establish the SS's own institute for collecting such plants and doing breeding research on them. As already mentioned, in November of 1943, Brücher was made head of this institute in Lannach.

The institute was intended to become a central site for wild and primitive strains of cultivated plants that were collected during the Tibet expedition and in subsequent expeditions. Such plants were of special importance for breeding, and it was hoped that they would prove particularly suitable for cultivation in the East or as suppliers of strategically important raw materials. As described in Chapter 2, Section 10, the SS had initially considered collaborating with the Kaiser Wilhelm Society in the field of wild and primitive forms of cultivated plants. During the German attack on Russia, the Soviet institutes in which Nikolai I. Vavilov's plant and seed collection was housed and studied were for some time within the German sphere. German scientists saw this development as a unique chance to get their hands on the extensive seed collection. The unanticipated opportunity to give their own research a boost led to the sharp competition (described earlier) between Fritz von Wettstein, director of the Kaiser Wilhelm Institute for Biology, and Wilhelm Rudorf, director of the Kaiser Wilhelm Institute for Breeding Research, over who would control the institutes that lay in the European part of the Soviet Union. The SS eventually established its own institute presumably because the German military defeats and the advance of the Red Army eliminated the possibility of control over Vavilov's collection.

This was the hour of the SS and Heinz Brücher, who succeeded in robbing valuable seed material from the Vavilov institutes with his own commando unit. In a letter from the Ahnenerbe to the head of the General Biology Section, Konrad Meyer, dated September 30, 1943, we read:

> Realizing the importance of wild and primitive strains for plant ge-
> netics, the Reichsführer-SS, as early as 1939, had Dr. E. Schäfer's
> Tibet expedition, which was under his aegis, collect a wide array of
> wild and primitive strains of Asiatic cultivated plants. For the same
> purpose a collecting expedition was conducted, with the help of SS-
> Obergruppenführer Pohl, to the Crimea and southern Russia in the
> summer of 1943. There the expedition was able to secure an exten-
> sive collection of wild, regional, and cultivated strains, in part also
> remnants of Vavilov's valuable world collection. The Reichsführer-
> SS has now decided that these materials, as well as future plant ma-
> terial from the expeditions and collecting trips of the Sven Hedin
> Institute, should be worked on and evaluated in terms of breeding at
> the Institute for Genetics, which has been established for this purpose
> in the research and teaching society Das Ahnenerbe.[23]

The plan for the use of this "Russia collecting commando" of the SS had been worked out by Schäfer and Brücher, with Schäfer appointing Brücher to head it. Supported by the Waffen-SS, Brücher and Konrad von Rauch traversed the Crimea and Ukraine in June of 1943, searching out all breeding stations and Vavilov institutes and stealing all seed material they could lay their hands on.[24] In his report Brücher lamented that during the fighting the material of many Russian breeding institutes had been destroyed by the Germans "out of ignorance" or had been "deliberately damaged," and that some of the German managers resisted the carrying out of his special command. Most German agricultural leaders, however, showed "willing cooperation in handing over the material," as a result of which he was able to make off with a large part of the grain collection.[25] Brücher was therefore able to report back to Sievers: "In the summer and fall of 1943 I secured—at the last minute, so to speak—the seed material of many Russian breeding institutes on orders of the SS Main Office for Economy and Administration [Wirtschafts- und Verwaltungshauptamt]."[26]

Brücher justified the theft of this material with its importance for the National Socialist policy of expanding German *Lebensraum* in the East, as is revealed, for example, in his report on the SS commando:

> The conquest of the East has brought into our possession areas that will be of decisive importance for feeding the German Volk in the future. However, the climatic conditions of these eastern areas place very special demands on cultivated plants. The extremely harsh winters, the short, hot summers, and the aridness of the southern regions demand an extraordinary resilience . . . Falling back on the primitive origins of the cultivated plants that Vavilov collected is all the more important in the current state of plant genetics, as an improvement in the resistance to cold, drought, and pests, and thus a protection of the yields, can be achieved only by interbreeding those genetic traits that have already been selected out in free nature under the harshest selection condition.[27]

We can assume that Brücher did not proceed gingerly when it came to getting hold of the seeds. He had no qualms about robbing the plant material the Russians had collected and using the resistant Soviet genes to enhance the German seed stock in order to improve the food supply of the German people. If the "eastern territories" had to first be "cleansed" of Russians, Poles, and Jews—measures to implement the "final solution" were in full swing by this time—that gave him no reason not to support this policy, since it was being executed for the benefit of the "racially superior." Most important, however, was the valuable seed material. We find no indication of a humanitarian motive for his actions, including efforts to protect the Soviet material from Trofim Lysenko, who was dominating Soviet biology with his neo-Lamarckian ideas and was silencing supporters of "reactionary Western genetics." On the contrary, Brücher attested that the Soviet scientific and breeding efforts were successful: "Even though Bolshevist party doctrine was opposed to the ideas on breeding of men like Darwin, Haeckel, and Mendel, Russian plant genetics in practice did successfully use the science of heredity and the doctrine of selection."[28]

Incidentally, by this time Vavilov himself, for reasons mentioned earlier, had already been stripped of all his posts by Stalin and had been imprisoned; he died in a prison in Saratov in January of 1943.

Brücher brought the stolen seeds to his institute in Lannach with the hope of turning the institute into a central place for wild and primitive strains of cultivated plants. However, since other German institutes were also working on cultivated plant research—apart from the already mentioned KWI for Breeding Research there existed the KWI for Cultivated Plant Research (established in 1943 near Vienna) and the Institute for Plant Breeding in Halle—his chances of realizing that goal were not good. Only the theft of the Soviet seed material and the increasing power of the SS moved his dream of assuming a leading place in the field of cultivated plant research into reach. But the material of the Lannach institute was to be available also to other scientists; close working relations had been established with Rudolf Freisleben (Halle) and Arnold Scheibe (TH Munich, since 1942 German director of the German-Bulgarian Institute for Agricultural Research, a Kaiser Wilhelm Institute).[29]

The Ahnenerbe requested for Brücher from the RFR a research assignment with the urgency classification "SS" to make it possible to expand the institute on the estate the SS had purchased in Lannach and to get exemptions from military service for its members.[30] The application was successful: "The head of the Institute for Plant Genetics in the section 'Ahnenerbe' of the personal staff of Reichsführer-SS, lecturer Dr. rer. nat. habil. Heinz Brücher, has been charged by both the Reich interior minister and Reichsführer-SS, Himmler, and the Reich marshal of the Greater German Reich and president of the Reich Research Council, Hermann Göring, with carrying out studies on the breeding of genetically hardy, frost-resistant, and drought-resistant forms of cultivated plants, and for that purpose he has been given the Wehrmacht's assignment number SS 4891-0331."[31] The research contract was described as follows: "Studies on the breeding of genetically hardy, frost-resistant, and drought-resistant strains of cultivated plants for the eastern region."[32] The Asiatic and stolen Soviet material was of particular importance for this project.[33] In addition, Brücher was working on breeding the oil plant Lafi.[34] Its oil, a sample of which he was able to present to Himmler as early as December of 1944, was said to have a low freezing point and to be well suited for industrial purposes (Kater 1966, p. 206). In February of 1945, Brücher received orders from Sievers to blow up his institute, if necessary, to prevent the "Russian spoils" from falling into enemy hands.[35] We do not know what happened to this institute in Lannach and to the Soviet material. We should note, though, that Brücher's plundering expedition in the Soviet Union did no harm to his scientific contacts either at the time or after the war.

For obvious reasons Brücher did not remain in Germany after the war. In 1948 he went for a short time to Sweden, where he worked for the Swedish seed breeding company Svalöf. In 1949 he accepted an appointment as professor of botany and genetics at the State University of Tucuman in Argentina. (After the war Argentina served as a refuge for many National Socialists.) Brücher became professor at the Universidad Nacional de Cuyo in Mendoza in 1954, head of the agricultural division in Pretoria (South Africa) in 1963, university professor in Caracas (Venezuela) in 1965, and director of the institute Biologia Vegetal in Argentina in 1976. He worked in cultivated plant research in a leading position, conducted extensive collecting expeditions to Africa, South America, and Asia, and supplemented the material with mutants produced by radiation. He received support for his work from the DFG, the International Atomic Energy Commission, and the Argentine Research Council. His more than one hundred publications on the origins and diffusion of cultivated plants—which appeared in many scientific journals in Argentina, the Federal Republic of Germany, Sweden, the United States, and Venezuela, and which included also his monograph of 1977 *Tropische Nutzpflanzen—Ursprung, Evolution und Domestikation* (Tropical useful plants—origins, evolution, and domestication)—were translated into five languages. In 1972 Brücher became UNESCO expert for biology.

Brücher was killed on December 12, 1991, on his farm in Mendoza, Argentina. S. Rehm (1992) believes the drug Mafia was behind his murder, since Brücher was known as their enemy through his longtime fight against the cultivation of coca in the Andean regions of Argentina.

4. Eduard May at the Ahnenerbe's Entomological Institute, Dachau

In a directive of January 29, 1942, concerning the "institute for the study and combating of pests that are annoying and harmful to humans," Himmler laid down the tasks and purpose of this new entomological institute. In contrast to the existing institutes of hygiene, whose efforts, he said, were directed at approaching the insect plague from the position of the physician, the Entomological Institute had the task of studying the "habits of these animals in order to combat them as such and if possible be able to eradicate them."[36] A specialized zoologist was to devote himself especially to the "study of lice, fleas, bedbugs, flies, mosquitoes, and horseflies." Himmler assigned him the following tasks:

1. Precise study of the habits and development of these animals . . .
2. Study of the diseases of these animals, for example the question which disease caused fleas in Europe to die out. Connected with this
3. Study of the question which epidemics and bacteria humans can use to initiate and promote the destruction of these insects that are harmful to us.
4. Study of all possible ways of using large animals, birds, mammals, and snakes to harm these insects or destroy them in their nests, to prevent them from reproducing.
5. Study of all possible ways of destroying the existence of these animals through changes in the composition of the soil and the withdrawal of suitable possibilities of reproduction.
6. Study of all possible ways of destroying these insects through chemical means.[37]

Himmler thus expressed his concern with developing comprehensive chemical and biological methods with which the insects he so hated could be eradicated. When it came to the destruction of insects, as well, he was not content with partial solutions.

The search for a suitable director was not easy in 1942. Eventually Wolfram Sievers hired the then and later scientifically completely insignificant zoologist Eduard May. In February of 1942, May directly began setting up the institute. He was not a member of either the SS or the NSDAP. In his curriculum vitae he indicated that he had studied mathematics, physics, chemistry, botany, zoology, geology, and paleontology, stating that he had abandoned the study of mathematics early on because he was "unable to follow the presentations of the leading mathematicians in Frankfurt at the time, Dehn and Hellinger."[38] (Max W. Dehn and Ernst D. Hellinger were Jewish professors; they were dismissed sometime after 1933 and forced to emigrate.) May obtained his doctorate in zoology in 1928 under Otto zur Straßen in Frankfurt. Owing to a lack of available assistant jobs, he began his career advising the chemical company Borchers A. G. in Goslar on how to set up a division of plant protection. During this time he continued to educate himself in mathematics, physics, and philosophy, "pursuing this all the more intensively as the study of Einstein's theories of relativity and the uncertainties in basic philosophic-scientific questions led me to the conviction that the sphere between philosophy and the natural sciences, neglected and left to each person's arbitrariness, was in need of a rigorous systematic investigation."[39]

He published some essays on the philosophy of science in the *Zeitschrift für die gesamte Naturwissenschaft*, "which is pursuing the same directions as I am with respect to reforming the natural sciences and establishing their foundation on the spirit of Indo-Germanicness."[40] This journal, the organ of the Natural Sciences division of the Reich Student Leadership, was taken over in 1937 by Fritz Kubach, Bruno Thüring, and the biologist Ernst Bergdolt. It was considered the unofficial mouthpiece of the advocates of the anti-Semitic "Aryan physics" (Beyerchen 1977, p. 145). May moved to Munich in 1940, among other reasons to establish a tighter exchange of ideas with this Munich circle, which was close to the journal and also included Hugo Dingler, who did work on the theory of science. In his book of 1941 *Am Abgrund des Relativismus* (At the abyss of relativism), May described Dingler as the real founder of exact science. This book won a prize from the Prussian Academy of Sciences in its competition on the prize topic, "The intrinsic reasons for philosophic relativism and the possibilities of overcoming it"; Eduard Spranger and Nicolai Hartmann judged the competition.[41] In this essay May opposed Karl Popper's demand that scientific standing be accorded only to insights that are in principle falsifiable (*Logik der Forschung*, 1935), and tried to show that Popper's conclusion that neither truth nor probability is attainable through science was untenable.

On the basis of this book, his *Habilitation* thesis, May was awarded the *Habilitation* degree for "natural philosophy and history of science" in the faculty of natural sciences (under dean Wilhelm Müller) at the University of Munich on March 13, 1942, over the opposition of the mathematician Oskar Perron.[42]

May took over the Entomological Institute on the condition that he would be given the opportunity to continue his work on the philosophy of science.[43] In 1943 he joined the editorial staff of the *Zeitschrift für die gesamte Naturwissenschaft*, with the goal that he might eventually become editor-in-chief of the journal.[44] Sievers, despite a heavy and diverse workload from running the Ahnenerbe and the Institute for Practical Research in Military Science, from organizing "human material" for the research of August Hirt, Sigmund Rascher, and others, and from his other functions in the RFR, also had time for philosophical reading: on February 19, 1944, he wrote to May: "By the way, I am not familiar with your essay in the first issue of the new Kant-Studien. Could you make it available to me on temporary loan?"[45]

May began his work at the Ahnenerbe by searching for suitable premises for the institute. His search came to an end with Sievers's suggestion, on April 1, 1942, to use space in the Dachau concentration camp. "The excellent medical installations that exist there could be placed at the service of the study, and it would also greatly facilitate the experiments if observations could be made on inmates."[46] Himmler authorized the construction project. However, because of a scarcity of building materials, the institute could not be finished until the fall of 1943. Until then May worked in a room of the Institute of Medicinal Herbs in Dachau (Kater 1966, p. 225). All he had to do in 1942 was fight an infestation of ladybugs and locusts. At the beginning of 1943, he and Rudolf Schütrumpf, who had been assigned to the Entomological Institute in March of 1943,[47] devoted themselves to combating flies, an issue in which "the Reichsführer-SS takes a very special interest."[48] May wanted to study whether it was possible to fight flies in a "natural-biological way" through infection with the fungus *Empusa*.[49] Himmler, always open to alternative methods of pest control, inquired whether one could breed a parasitic hymenopteran against flies and mosquitoes,[50] and he asked May to pursue whether it might be possible to destroy "flies or fly nests" through some kind of shortwave radiation.[51]

Scientists at the institute also studied the malaria-transmitting anopheles mosquito. On December 10, 1942, a branch of the Entomological Institute called the Southeast Institute opened in Vienna; it focused on the anopheles problem in Styria. In collaboration with Joseph Meixner, studies were carried out on the "dependence of the appearance of the anopheles on soil constitution, plant growth, climate, and so on."[52] Meixner, a member of the NSDAP and the SS, had been professor of zoology at the University of Graz since 1939.[53] In the area of chemical pest control, May wanted to develop new insecticides (larvicides) to combat anopheles. He sought substances to take the place of larvicides that had become scarce (Schweinfurth green, gesarol, thiodiphenylamin). As examples of organic compounds that had so far been determined to be effective, he listed the following:

azobenzene, biphenylamin, biphenyleneoxyde, chlorophenyleneoxyde, biphenylenesulfide, chlorobiphenyl, tetrachlorophenol, phenanthren, phenothioxin, chloronitrobiphenylamin. "These and other compounds must be further investigated and above all tested in large-scale trials . . . The goal is to find substances that, nontoxic to humans and as benign as possible to fauna and flora, are specific to anopheles larvae and can be easily synthesized."[54] On October 4, 1943, May was given a contract from the Reich Research Council entitled "Insects Harmful to Humans" with a military contract number and the urgency classification "SS."[55] This assignment was connected with the research on biological warfare carried out by Kurt Blome and will be described in the following chapter.

According to the testimony of Sievers at the Nuremberg trials, in the spring of 1942 May rejected Himmler's request that he use inmates for his experiments and declined to participate in human experiments (Kater 1966, pp. 497f.). May was acquitted. May knew of Hirt's and Rascher's intentions to kill people in their experiments.[56] He met for discussions with the other staff members of the Institute for Practical Research in Military Science. The following people participated in the meeting in Strasbourg on March 17, 1943: Sievers, Hirt, May, Dr. Wimmer, Prof. Stein, Anton Kiesselbach, Otto Bickenbach, Prof. Holtz, Friedrich Weygand, and Rudolf Fleischmann.[57] Fleischmann (physics), Hirt, and Weygand (chemistry) held chairs at the University of Strasbourg. As will be shown in the following chapter, May also worked in research on biological warfare.

In May of 1945, on orders from the military government, May was dismissed from the University of Munich, where he had become lecturer with civil servant status in November of 1942.[58] On December 19, 1945, the district administrator of Starnberg, at the request of the military government, passed on to him the notification that he had been "found to be politically blameless."[59] May did not fully resume his work in Munich and thus no longer had the status of a lecturer. However, he did give courses at the university and the Volkshochschule (continuing education school) in Munich and continued his philosophical studies. In 1951 he was appointed professor of philosophy at the Free University in Berlin. His collaborator Schütrumpf obtained his *Habilitation* in Cologne and was appointed associate professor in 1970.

As I indicated in Chapter 2, Himmler also tried—not always successfully—to gain influence at the universities. The following section, using the zoologist and geneticist Gerhard Heberer as an example, will show the SS career and research of a professor at the University of Jena whose work in biology and anthropology is internationally recognized.

5. SS Research at the University of Jena: Gerhard Heberer, Human Origins, and the Nordic Race

Among biologists and anthropologists who, as members of the NSDAP, were determined supporters of the ideology of National Socialism, there was no uniform opinion about the extent to which the principles of Darwinism—mutation and selection—also applied to human beings and the origins of the human races. The journal *Zeitschrift für die gesamte Naturwissenschaft* ran many articles by scientists who rejected the "mechanistic principle of selection" and "materialistic monism." This group included a man mentioned several times in this book, the associate professor of botany at the University in Munich Ernst Bergdolt, a member of the NSDAP since 1922.

Bergdolt considered a change of human races through mutation and selection to be impossible. He opposed the notion of the "liberal anthropologist" who sees "in the races merely 'breeding products' of the environment, a view that has been ardently advocated by the Jews, for obvious reasons." Without giving any reasons of his own for the possible causes of the development of human races, he advanced the theory of racial constancy: "The Nordic race, just like any other race, was and is preserved where it keeps itself free from the intermixing of foreign blood" (Bergdolt 1937). The morphologist Wilhelm Troll, professor in Halle, rejected Darwinism as a whole as a "crass expression of the English way of looking at things" (Troll 1941, p. 150). Ernst Krieck, National Socialist theorist of education and professor of philosophy and pedagogy at the University of Heidelberg, also fought Darwinism and called Ernst Haeckel the "shallow epigone of materialism."[60]

At the same time, however, the journal *Der Biologe,* published by the volkish Lehmann publishing house, celebrated Haeckel on the occasion of his one hundredth birthday in 1934 in several essays as a pioneering thinker of National Socialism. The Ernst Haeckel Society, born in 1933 from the Monist League, received support from the district leader of Thuringia, Fritz Sauckel (Gasman 1971, p. 173). And according to Alfred Rosenberg, "representative of the führer for supervising the entire spiritual and ideological schooling and education of the NSDAP" (see Chapter 2), and the Main Office for Science he headed, the "National Socialist idea of achievement and selection" was based on the notion of the development of species and races through mutation and selection.[61] This office sharply rejected Krieck's attacks against Haeckel.[62]

The NSDAP's Office of Race Policy had been founded in 1934 to supervise the racial-political education within the party and its branches. In addition, this office, in cooperation with the Reich Ministry of Education and the Propaganda Ministry, also functioned as a censorship authority with regard to all publications and lectures on topics dealing with population and race policy (Groß 1937). Its head, racial anthropologist Walter Groß, advocated the concept of complete scientific freedom in research on human origins; however, he believed that the publications and public talks of racial scientists should take into consideration their political and psychological effects on the population. His stance is revealed by the notes he made on the use of the work of anthropologist Hans Weinert, a decided Darwinist, in public education courses on race policy. Groß, who considered Weinert the most competent specialist in the field of the theory of human origins, did not initiate any restriction on his research and teaching; in doing so he emphasized, however, that no case is known "in which [Weinert] has directly presented his theory of evolution as part of National Socialist ideas on race. On the other hand, it is my opinion that the political racial doctrine of National Socialism should maintain sufficient distance from the doctrine of human evolution, which is . . . still subject to the changes of scientific theories, is frequently still pervaded with Haeckelian ways of thinking in its basic ideological ideas . . . and is thus publicly considered a part of materialistic, monist ideas."[63]

When it came to making political decisions, National Socialist authorities thus considered possible reactions on the part of the population. (In this context, too, it can be asked whether the implementation of the Civil Service Law, the Nuremberg Laws, and the deportations of Jews would have taken place had there been open or even hidden protest among the population.) The Office of Race Policy did not derive any clear consequences from its maxim. While Groß declined to include Weinert in political training and propaganda work "out of considerations of expediency," Konrad Lorenz, who formulated his hypotheses also by leaning closely on Haeckel's monism, was welcome as a member of the Office of Race Policy with speaking permission (see Chapter 3). Irrespective of these different ways of handling the situation when it came to racial-political training, we must note that during the Nazi period scientific research on human evolution was in no way suppressed or directed from above onto specific ideological paths. Scientists themselves chose to place their work in the service of racial ideology and racial policy.

Among these biologists stood Gerhard Heberer, one of the most ardent champions of the idea that human races developed through mutation and selection. He also carried out research with the goal of proving scientifically the "superior worthiness" of the "Nordic race." The following discussion will examine his career as a university professor, his quarrels with the NSDAP, and his research on the biological origins of the Germanic peoples.

Heberer was born in Halle (Saale) on March 2, 1901. After studying biology, anthropology, and prehistory, he obtained his doctorate in 1924 with a dissertation in the field of genetics. From 1924 to 1927 and from 1934 to 1945 he worked as a racial scientist at the State Institute for Prehistory in Halle, which after 1934 was called State Institute for the Study of Volkdom (Landesanstalt für Volkheitskunde). In 1928 he became assistant at the Zoological Institute of the University of Tübingen, where he obtained his *Habilitation* for work in zoology and comparative anatomy in 1932. The book *Die Evolution der Organismen* (The evolution of organisms), which he edited in 1943, is considered an early contribution to the synthetic theory of evolution (see Mayr 1982, p. 568). In addition to contributions from anthropologists, racial hygienists, and plant and animal breeders, it contains essays by the geneticists Hans Bauer and Nikolai W. Timoféeff-Ressovsky, the zoologists Konrad Lorenz, Wilhelm Ludwig, and Bernhard Rensch, and the botanists Karl Mägdefrau and Walter Zimmermann.

Heberer's essay examined the extent to which the mechanisms of microevolution (phylogenetic differentiation of races, forms, or subspecies) also suffice as a causal explanation for the course of macroevolution (creation of species and higher systematic categories). Most geneticists, including Timoféeff-Ressovsky, believed that the factors of microevolution—mutation, selection, and isolation processes—could be readily transferred to macroevolution. This view is part of the synthetic theory of evolution. In contrast, most paleontologists, among them Otto Schindewolf, maintained that these were two fundamentally different processes, positing spasmodic "typogeneses" for macroevolutionary processes. On the basis of what was at the time modern genetics, Heberer developed a model with which he explained macroevolutionary changes using the mechanisms of microevolution (Heberer 1943, pp. 545–585; see also Heberer 1942). Heberer was and is regarded, not only in Germany, as a leading scientist in the field of evolution research, especially in the area of human evolution.[64]

His work in the fields of racial science, early history, and zoology, as well as his political attitude—he was considered to have uncompromising volkish and National Socialist sentiments—aroused the interest of the SS in 1936.[65] As early as April 12, 1937, Heberer, a member of the SA since 1933 and of the NSDAP since 1937, was named SS-Untersturmführer (second lieutenant) by Himmler and appointed to the staff of the Main Office for Race and Settlement (Rassen- und Siedlungshauptamt, RuSHA); in 1938 he became Obersturmführer (first lieutenant) and in 1942 Hauptsturmführer (captain). The RuSHA, as part of the Germanification policy of the SS in the occupied eastern territories, was responsible for the "racial evaluation" and selection of people.

In contrast to Heberer's rapid political rise, his career as a university professor was less successful. The SS considered it desirable that Heberer be given a chair at a university so that he would be able publish his work on hereditary biology and racial science as a full professor;[66] good prospects for such a position existed in Frankfurt. In the winter semester of 1935–36, Heberer had been given a temporary appointment by that city's university to the chair in zoology, which had become vacant upon the retirement of Otto zur Straßen. In a course on general biology for biologists and medical students, he discussed the German laws on race and heredity in great detail, to the enthusiastic applause of National Socialist students.[67] That he was not given the permanent appointment despite this reception shows that successful protests against Nazi measures were in fact possible. It was Catholic students who opposed Heberer's appointment. What bothered them was not his SS membership or the National Socialist ideology expressed in his lectures; rather, they rejected him because of his views on the theory of the origin of species. The REM therefore decided "that it would not be advisable, given the tensions that have appeared, to leave Dr. Heberer in Frankfurt/Main."[68]

Subsequently, Himmler spoke to the REM in support of appointing Heberer to one of the zoological chairs opening up in Tübingen, Erlangen, and Halle, thus giving him a place to pursue his work in the fields of racial science and hereditary biology.[69] However, the SS was not able to influence the decisions of the universities and the REM. For example, in Tübingen in 1936, Heberer's name did not even come up when twenty-one experts gave suggestions for candidates for the chair in zoology.[70] In a letter to the rector, the dean justified his rejection of the long-time assistant Heberer, in spite of Heberer's good work, by arguing that after such a long period of collaboration the appointment of someone who was much younger would lead to serious discord at the institute.[71]

With renewed support from Himmler and the energetic advocacy of rector Karl Astel, Heberer, in the winter semester of 1938–39, was appointed to the newly created chair for general biology and human evolution (anthropogeny) at the University of Jena as associate professor with civil service status. The University of Jena had thus acquired another eminent champion of the National Socialist racial doctrine, with Heberer joining the likes of Astel, the head of the Thuringian State Office for Race Matters,[72] the jurist Falk Ruttke, who coauthored the commentary on the Law on the Prevention of Genetically Diseased Offspring, and the less well known professor of zoology Viktor Franz, an ardent National Socialist and follower of Haeckel.

At the end of 1941, Heberer's hopes for yet another full professorship were dashed once again. At the University of Strasbourg, newly founded following the occupation of France, a genetic institute was to be established for the first time within the faculty of natural sciences. This institute was to be headed by a "zoological geneticist thoroughly trained in racial science" and by a botanical geneticist;[73] Heberer and Edgar Knapp were the candidates for these positions. In April 1941 Heberer had received a draft agreement on his appointment as full professor of biogenetics, zoological genetics, and hereditary science as well as director of the institute for biogenetics.[74] The party chancellery had already given its approval to his appointment,[75] and Sievers of the SS's Ahnenerbe had already congratulated him.[76]

And yet, to Heberer's great chagrin and for reasons unknown, only Knapp was appointed; Heberer remained in Jena until 1945.

A champion of the neo-Darwinist theory of evolution and a follower of Haeckel, Heberer promoted the idea of evolution with missionary zeal also as a member of the SS—that is, he fought against the attacks on Haeckel and the theory of human evolution that came from Catholic and National Socialist groups. In *Volk und Rasse,* the journal of the German Society for Racial Hygiene, he defended the notion of the creation of human races through mutation and selection against attempts by Catholic circles to present the theory of evolution, especially as it applied to humans, as unproved or utterly false. He could not "cheerfully" ignore the "atrocity propaganda of 'scientific' obscurantists," for in his view "the situation . . . is too serious." Heberer used the arguments from Catholic critics to emphasize the ideological importance of the line of research he was pursuing and thereby strengthen the position of the advocates of human evolution also within the NSDAP: "In the final analysis, the goal [of the critics of the theory of evolution] is to attack racial science in its biological roots, since the theory of evolution is and remains the root soil of racial science" (Heberer 1937).

Heberer got support from the Reich Health Office and from Astel, who urged him "to write about Haeckel in your capacity as a zoologist and anthropologist and to point out his fundamental importance for all of racial science."[77]

The controversy among biologists over human evolution was fought, among other places, in the journal *Der Biologe,* which had been taken over in 1939 by the society Ahnenerbe. In 1936 Heberer became copublisher of the journal with responsibility for the field of phylogeny. Enraged at the appearance, against his will, of Bergdolt's article "On Transformations in Form—Also a Critique of Theories of Species Formation" (1940), he asked the managing editor, Walter Greite, to drop him as copublisher: "It goes too far if polemics against the natural breeding of human races are launched from one's own camp."[78] He withdrew his letter of resignation only at the request of Sievers. Heberer also appeared as an expert for racial science at conferences of the National Socialist Teachers' League,[79] and from 1937 to 1945 copublished the "Annual Courses for Continuing Medical Education."

From 1934 on Heberer focused his research, for which he never

6

Research to Develop Biological Weapons

"Biological warfare is dirty business"—Secretary of War Henry Lewis Stimson wrote in a memorandum (dated April 4, 1942) to President Franklin D. Roosevelt, in which he proposed extensive research for the development of biological weapons by pointing to the threat of biological warfare from the axis powers (Bernstein 1988). The Americans, in cooperation with the British, developed great potential in biological weapons during the World War II (Bernstein 1987, 1988). Although their fears about corresponding preparations by Japan were justified, the Americans and the British had greatly overestimated German "accomplishments" in this field.[1]

In Germany, research into weapons of mass destruction focused on chemical weapons: in 1936, chemists at IG Farben discovered the nerve poison Tabun, which was available for large-scale use beginning in 1942.[2] By contrast, no weapons worth mentioning had been developed in the field of biological warfare. In this chapter I seek to shed light on the goals and content of biological warfare research in Germany on the basis of files on biological warfare research that were confiscated by the allies in 1945 and of documents from the Reich

Research Council and the SS. I offer no analysis of the strategic-military background and the use or refrain from use of biological weapons during World War II.

Two factors impeded biological weapons research in Germany during the war: first, by an order of Hitler all research into offensive biological warfare had been barred since 1942, at the latest; second, there was no central office responsible for all areas of biological warfare research. The essential work in this field was carried out by a committee of the Wehrmacht, the working group "Blitzableiter" headed by Heinrich Kliewe and the Deputy Reich Physicians' Führer Kurt Blome, who was assigned to the committee by Göring and was strongly supported by Himmler.

1. The Working Group Blitzableiter

It appears that prior to 1940 there was no serious thought given in Germany to the use of biological weapons and hardly any research in this area.[3] The impetus for biological warfare research came in 1940 from reports about preparations in this direction by the enemy:

> The enemy powers Russia, France, and England have continued the development of bacteriological weapons in contravention of the agreement of the Geneva Convention of 1934 [sic; actual date 1925]. The scientific representatives in Geneva rejected bacteriological warfare as pointless. Germany therefore did not work in this field after the war. Only when, in the wake of the French campaign, four bacteriological laboratories—in which extensive experiments for bacteriological warfare had been conducted—were found in the Proudrerie Nationale du Bouchet ... did we realize what the enemy believed was possible in this field. Because of this discovery, Prof. Kliewe was transferred to the [Academy of Military Medicine] in Berlin on January 1, 1941, to work on all questions relating to bacteriological warfare; at the same time he was assigned as an adviser to the H S In [presumably, the Army Medical Inspectorate] and the Wa Prüf 9.[4]

Another increase in biological warfare research occurred in 1943, in part in response to reports that the Americans were planning to release Colorado beetles over Germany.[5] The informal working group for the study of biological weapons, established in 1940, was renamed the working group Blitzableiter and given the official responsibility by the high command of the Wehrmacht for research into biological warfare, which by Hitler's order was intended to be only defensive. The working group comprised representatives of various branches of the Wehrmacht. In October of 1941, the areas of biological warfare research were divided up as follows: Army Medical Inspectorate: human bacteriology (protection of the troops and the population from biological weapons); Army Veterinary Inspectorate: veterinary bacteriology (protection of the military's animal stock and of useful animals); General Office of the Wehrmacht, Science Section: use of pests and measures to combat them (protection of vegetation and plant food stocks); and Ordnance Department: delivery methods.[6]

In May of 1942 the high command of the Wehrmacht circulated Hitler's order prohibiting preparations for the offensive use of biological warfare. A corresponding directive was given by General Field Marshal Wilhelm Keitel, chief of the high command, following a conference with Hitler. This directive reiterated the prohibition but also called for research to be carried out into defensive measures for protection against attacks with biological weapons of every kind.[7] A circular from the staff of the high command to the corresponding Wehrmacht offices stated: "With letter AWA Nr. 141/42 g. Kdos., dated 6.24.42, the OKW has decided that a use of biological weapons is not being considered in the attack against England. All preparatory measures must therefore cease. Instead, the führer has demanded that German science work intensively on the combating of biological pests."[8]

Opinions on how strictly the prohibition should be applied diverged widely among members of the working group. Colonel Münch from the WFSt./Org noted at a meeting of the working group on July 21, 1943: "Prof. Blome, deputy reich physicians' führer, has been charged by the Reich Marshal with looking into biological weapons. On the other hand, the führer has prohibited all work dealing with the active use of biological weapons. The utmost effort must therefore be put into working on defensive measures. It is therefore important to know how the enemy could use biological means; for that reason delivery methods have to be tested."[9]

In 1942 the Ordnance Department did not carry out any field tests because it as yet had not received "permission from higher up." For reasons of jurisdiction it also objected to the repeat of the sort of experiments Heinrich Kliewe had carried out in July by dropping *Prodigius* bacteria (harmless red bacteria) on a testing range at the Munster base (officially the testing range of the air force).[10] In contrast to Kliewe, who in September of 1943 no longer considered the large-scale use of biological weapons, in particular human infectious pathogens, by other countries likely, Ministerial Manager Erich Schumann, who headed the Science Section of the Wehrmacht, maintained that such a mass deployment was being planned. Schumann repeatedly advocated offensive biological warfare. He told Kliewe: "Surely the führer is not sufficiently informed, he must be briefed once again. We must not watch heedlessly but must also prepare for the large-scale use of biological weapons. In particular, America must be attacked simultaneously with various human and animal epidemic pathogens as well as plant pests. The führer should be won over for this plan."[11]

The working group was unhappy with the ordered restriction to defensive research: "So far valuable time has been lost for the field testing of possible delivery methods. Reason: 1. So far there has been no order that preparations for the active use of biological weapons should be made. 2. There is no responsible office for those working on specific issues . . . No agreement was reached on the type of bacteria most suitable for biological warfare and on the delivery method. Cooperation became increasingly loose and was dormant for months at a time because field tests were supposedly prohibited."[12]

Which biological weapons, then, did scientists work on? Despite the problems I have mentioned, preparations began for offensive use of biological weapons and such weapons were field tested. The section concerned with weapons against humans seems to have been most strongly impeded in the development of methods for mass use by the orders restricting the work to defensive weapons.[13] According to Kliewe, "plague, typhoid, and paratyphoid bacilli, cholera vibrios, and anthrax spores" were suitable for sabotage and large-scale use of bacteria against humans.[14] Laboratory tests were conducted on the use of these pathogenic agents, and a synthetic medium for the spread of these bacteria was developed. In this medium cholera vibrios remained virulent for eight to ten weeks and anthrax bacteria for more than twelve weeks.[15] The medium was suitable above all for sabotage, and in the assessment of the American secret service it may have been one of the outstanding German "achievements" in the field of biological weapons.[16] The Veterinary and Agricultural Sections were able to combine their war research more easily with their normal activities without attracting attention. The Veterinary Section released pathogenic agents of the hoof-and-mouth disease by plane on an island in Lake Peipus. The test succeeded in producing infections in reindeer and cattle. In the opinion of the veterinarian Dr. Nagel, these pathogens "could be used successfully against England, because the cattle stock there has a weak immunity," whereas German cattle was largely immune.[17]

The following institutions and people, among others, were involved in the research of the Agriculture Section (Wehrmacht Science Division): the University of Posen (Pharmacology Department); the Potato Beetle Research Institute in Kruft, where potato beetles were bred under the direction of Dr. Schwartz, and the KWI for Cultivated Plant Research in Vienna, where work was done, under the direction of Hans Stubbe, on the production of weeds to choke useful plants.[18] Research in the area of plant pests was accorded comparatively high importance. Experiments were carried out with pests of meadows and beets and the possible uses of potato blight. The potato beetle seemed suitable for use against England, but because the beetle could be so easily spread, work with the potato beetle, and thus the breeding of the 20 to 40 million beetles deemed necessary, was considered problematic within the territory of the Reich.[19] Nevertheless, in October of 1943 a field test was carried out near Speyer, in which 1,400 beetles were dropped by plane. Of these, 57 were found again, "the rest will have to be collected next summer during the general search."[20] Chemical methods for the destruction of useful plants were not considered.[21]

The person in charge of the working group Blitzableiter in 1943 was Colonel Walter Hirsch, the head of the Ordnance Testing Office; his deputy was Dr. Kurt Stantien. The working group also included the following people: Heinrich Kliewe, the leading epidemiologist Gerhard Rose as representative of the Air Force, Schulz and his deputy Nagel from the Military Veterinary Inspectorate; Erich Schumann and his deputy Beyer; Denner as counterintelligence expert of the Security Service (SD); Deiters, air force engineer; and Kurt Blome, plenipotentiary for cancer research in the Reich Research Council (Hansen 1990, p. 77). As the following section will show, Blome's plans and activities in particular, introduced a new dimension in the field of biological warfare.

In the assessment of the American secret service, the narrow scope of German biological warfare research is attributable not to the incompetence or unwillingness of scientists but to Hitler's opposition to such research and to the limited support it received from the OKW.[22] A reason for this opposition cannot be found in the documents. At the meetings of scientists involved in biological warfare research, two basic problems, in particular, came up: first, the danger involved if the use of biological weapons could not be contained and if pathogens or pests spread to one's own country; second, the fear of retaliation with the same weapons. Despite the prohibition, research into biological weapons was carried out. It had both a preventive and offensive goal but no "successes" worth mentioning.

2. Biological Warfare Research under Deputy Reich Physician Führer Kurt Blome

Kurt Blome began in the RFR as head of the Genetic Population Policy and Racial Care Section. Beginning in June 1942, he was also plenipotentiary of the RFR for cancer research, that is, he decided on the authorization of DFG funds for cancer research. The chief of the security police and of the SD wrote on September 26, 1944, to the head of the Planning Office of the RFR, Werner Osenberg: "Prof. Blome has been installed as deputy Reich physicians' führer and as plenipotentiary for cancer research. However, these are cover titles, since Blome is carrying out special assignments in an entirely different direction."[23]

During questioning at the Nuremberg physicians trial in 1947, Blome stated that linked to his function as plenipotentiary for cancer research in the RFR "was, as a secret assignment, the study of the defense against biological methods of warfare."[24] In this way Blome was able to support projects of biological warfare research with DFG and RFR funds. From the end of 1943, at the latest, he was also part of the working group Blitzableiter.[25]

On Göring's instruction and with strong support from Himmler, Blome in 1943 set up an institute for the testing of biological weapons on the grounds of the former monastery of Nesselstedt near Posen.[26] The institute, officially part of the Central Institute for Cancer Research (see Chapter 3, Section 2.6), came under Blome's direction. The Ministry for Armament and War Production supported the refitting of the monastery.[27] In addition, the managing director of the Central Institute for Cancer Research (and curator of the Reich University of Posen), Hanns Streit, received RM 206,000 in 1943 for the institute in Nesselstedt,[28] and in January of 1945 the RFR authorized another RM 2.5 million for the Central Institute.[29] Himmler supplied a special unit of the Waffen-SS to provide military protection for this war-essential installation.[30] In June of 1944 a branch of the Central Institute was set up in Birnbaum, Warthegau, and assigned "tasks of the utmost urgency."[31] The following divisions of the institute existed in May of 1944, and according to the REM nearly all were up and running:

1. Physiological-biological division
2. Bacteriological and vaccine division in Nesselstedt
3. Gynecological-surgical division in Nesselstedt
4. X-ray-radiological division in Nesselstedt
5. Division for cancer statistics in Posen (at the university)
6. Pharmacological division in Kleist barracks in Posen
7. Farm Nesselstedt.[32]

From the very beginning Blome had envisioned carrying out human experiments in Nesselstedt to test biological warfare agents. The planning of such experiments was also discussed at meetings of the Blitzableiter. For example, Kliewe wrote as follows about a session on September 24, 1943, at which experiments in human bacteriology were discussed: "Since it was not known whether and under what conditions inhaled aerosols or dispersed droplets of certain pathogenic germs produce illness in humans, Prof. Blome suggested human experiments. The carrying out of such experiments in the laboratories of the MA [Academy of Military Medicine] was rejected. It will be further discussed when Prof. Blome visits my laboratory in the MA."[33]

These plans became more concrete as the building of the institute near Posen progressed. Kliewe noted in 1944:

> [Blome said] that so far he has not conducted any experiments in the field of human medicine. However, such experiments were necessary and he would carry them out. A new institute, which will be under his control, is now being built near Posen, and there biological warfare agents will be studied and tested. General Field Marshal Keitel has authorized its construction, the Reichsführer SS and Generalarzt Brandt has promised him extensive support. At the request of General Field Marshal Keitel, the Wehrmacht is not to participate in a role of responsibility in these experiments, since they will involve experiments on humans. However, Prof. Blome would like me to

participate as an adviser, since, as he said, I have already made many preparations and am in a position to immediately derive from the test results the necessary protective measures for the Wehrmacht . . . Prof. Blome believes . . . that biological weapons . . . could become a very serious threat to us. That is why the field must be worked on to a much greater extent and much more intensively than has hitherto been done. In particular, it would be necessary to test our vaccines, especially the plague vaccine. Corresponding tests would have to be done on humans. In addition, he stated that there are quite erroneous ideas about the effect and maximum doses of certain toxins, which could likewise be cleared up only through human experiments. As soon as Prof. Blome has briefed the Reich Marshal and Generalarzt Prof. Brandt he will send word.[34]

The Wehrmacht was thus not officially involved in the preparations for human experiments with deadly outcomes. However, as this passage shows, it was informed about them at the highest level and was willing to cooperate with Blome in order to make use of the results of these experiments for preventive measures.

The biological weapons studied in Posen included plague bacteria. Research into the use of these bacteria as a warfare agent seems to have been one of Himmler's main interests since 1944.[35] According to investigations by the American secret service, Himmler had ordered that biological warfare research be conducted more offensively and to that end had offered Blome the opportunity of using a concentration camp for studies on the plague.[36] Perhaps this interest in the plague was triggered by a report of the SD to Himmler on October 11, 1944, which stated that the Soviet Union had set up large cultures of plague and cholera bacteria to be dispersed by airplanes. Himmler immediately passed this report on to Blome.[37]

In the area of biological warfare research, Blome collaborated with the SS's Institute for Practical Research in Military Science.[38] The head of the Entomological Division at the institute, Eduard May, participated in Blome's special assignments from the fall of 1943, at the latest. He was given a research contract by the RFR with the Wehrmacht assignment number SS 4891-0330 (1880/8)-III/43.[39] The contract was described as follows: "A study of the habits of insects harmful to humans in order to clarify the question concerning specific uses and heightened defense against them."[40]

May worked on two projects relating to biological warfare. The first involved tests suggested by Blome, intended to develop the capacity to drop mosquitoes infected with malaria over enemy territory.[41] May wrote in one of his reports: "Biological studies are being conducted with anopheles. The aim is to clarify to what extent an *artificial mass transmission of the Malaria parasite to humans* is possible and how one could counter an action aimed at such a mass transmission. The plan is to expand these studies also to other questions that fall under the heading of *biological warfare* and concern the sector of insects harmful to humans."[42]

As part of these studies, May carried out starvation experiments with various anopheles strains. These experiments determined the average survival time of the mosquitoes, the goal being to find strains that could go without food long enough to make possible their transport from the breeding station to the drop site.[43] These experiments were concluded in September of 1944 with the result that all strains were in principle capable of surviving long enough.

May also reported that, on the question of artificial mass infection, studies were being carried out with fluorescent microscopy with the goal of investigating the behavior of the malaria parasite in the esophagus and the salivary glands of the anopheles mosquito.[44] It is not unlikely that in the area of fluorescent microscopy he worked together with Professor August Hirt. Hirt was at this time conducting mustard gas experiments on inmates of the Natzweiler concentration camp and was hoping to use fluorescent microscopy to gain insight into the fluorescent substances that appeared in various organs in cases of poisoning.[45]

The second project in the area of biological weapons that May took part in involved an experimental installation in Dachau for the production of organic-chemical preparations for insect control; it was to be set up in collaboration with the company Borchers A.G. Not only were known substances to be produced there in large quantities; as we learn from a letter May wrote to Himmler, the installation was needed "for the production of *all* organic compounds . . . that have shown promise in tests carried out by Prof. Blome and me."[46] New insecticides, which so far had been synthesized and tested only in a laboratory setting, were to be experimentally produced on a large scale. The installation was to be exclusively devoted to the uses of the SS.[47]

This project changed after the RFR turned down the application with the explanation that it did not fund the building of new production sites.[48] However, a contract was signed between the Ahnenerbe and Borchers A.G. It called for an intensification of collaboration in the field of pest control and pesticides, and Borchers committed itself to providing scientific and technical support to the efforts by the Ahnenerbe to turn its Entomological Institute into the central substance testing site in the sector of medical entomology.[49]

In all likelihood the efforts to intensify the production of insecticides, as preventive measures against retaliation by the enemy, were intended to prepare for the German use of insects harmful to humans and plants. As early as September 30, 1940, Sievers wrote to Blome in regard to the planned large-scale plant for insecticides: "If insects harmful to plants will in fact be dropped on a large scale, which means that similar measures would have to be expected from the enemy, it would be necessary, by way of defense, to provide a quantity of pesticides that would far exceed what [pre-1938 Germany] would have needed if everywhere only the most essential measures for pest control had been taken ... The second problem concerns the storage and transport of the pesticides, for it is clear that the substances must already be at hand at least for the first enemy actions."[50]

As head of the Cancer Research Section, Blome supported other projects in the field of prevention. They included, among others: "Combating the Potato Beetle" (Schwartz, Kruft, Reich Biological Institute), "Discovery of Substances to Combat Agricultural Pests; Decontamination of the Soil" (Seel, Posen), and "Development of an Airplane Scattering Device for Dispersible Insecticides and Fungicides" (Borstell, Weimar-Nohra, "secret!").[51] All these research assignments were given the urgency classification "SS." Professor Schwartz bred the potato beetles dropped in tests from planes. A new dusting apparatus was developed for the large-scale application of insecticides by plane.[52] The Airplane Forest Protection League under Colonel von Borstell was used for both of these tasks.[53] In May of 1943 Blome successfully intervened with the Reich Aviation Ministry to prevent the planned dissolution of the Airplane Forest Protection League. He pointed out that the continuing existence of the league was "vital and decisive for the war," "the tasks of the league are intimately connected with the field of work assigned to me by the Reich marshal."[54] In August of 1944, Sievers noted in his diary under "working questions Entomological Institute: special assignments Prof.

Blome do not permit restrictions of any kind under any circumstances."[55]

The following ("secret") telegram from Sievers to Brandt on Himmler's personal staff (dated August 18, 1944) indicates that Blome, on orders from Himmler, was carrying out other projects as well in the area of biological warfare agents: "Prof. Blome requests appointment after 8.25.44 with Reichsführer-SS for the purpose of giving a report ... Briefing points: 1.) Military use by enemy of insects harmful to humans. 2.) Use of potato beetle by the enemy. 3.) Poison experiments in connection with the report given to Reichsführer-SS on 7.21. Prof. Blome is again being pressured to test the poison now."[56]

Apart from Blome's statement—not very credible—at the Nuremberg trial that the poison in question was Doryl, and that he only made sure that this chemical, which was freely available at the time, would be sold only on prescription after 1944, there is no indication of which poison was being referred to. Not without interest in this connection is the attention that IG Farben, beginning in about 1941, devoted to building the Dyhernfurth/Oder plant as a production site for nerve gas.[57] Together with the planned buna and hydrogenation plant near Auschwitz, this factory was to be part of the great chemical complex in the eastern part of the Reich. In the spring of 1943, Hitler ordered a drastic increase in the production of the nerve toxins Tabun and Sarin in Dyhernfurth.[58] The plant was located less than two hundred kilometers from Posen.

The advance of the Red Army forced Blome and his staff to leave the Posen institute in March of 1945. Walter Schreiber, department head in the OKW with the Military Medical Inspectorate, testified at the Nuremberg trials that Blome had come to him at the Academy of Military Medicine in March of 1945 with the request to find a place for him and his people in the laboratories of the army in Saxony so that they could continue their work. He had to leave the institute in Posen "in a great hurry, he was not even able to blow it up. He was very concerned that the installations for human experiments that were in the institute and recognizable as such, would now be very easily identified by the Russians."[59] According to Schreiber, Blome, with the help of an order from Himmler, was able to keep working for a while longer on his plague cultures in Saxony.

Many questions remain unanswered: between 1943 and 1945, the cancer research institute received RM 2.7 million from the RFR alone.

What was actually done there?[60] Were human experiments with plague bacteria carried out? Were the above-mentioned poison experiments, in contradiction to Blome's testimony at the Nuremberg trials, experiments with Tabun or Sarin? Is it possible that these poisons, too, were tested on humans in Posen, or that there were at least plans to do so? Where are the records concerning the research at this institute?[61]

Kurt Blome was charged by the Nuremberg military tribunal with being personally responsible for as well as participating in malaria, mustard gas, gas, and sulfanilamide experiments; killing Poles suffering from tuberculosis; carrying out the euthanasia program; and participating in freezing experiments and in experiments related to biological warfare. The experiments for the preparation of a campaign of biological warfare were allegedly carried out on Soviet prisoners of war and in the majority of cases led to their deaths.[62] During a secret interrogation by the U.S. Army, Blome admitted that Himmler had charged him in 1943 with carrying out experiments on concentration camp inmates to test plague vaccines.[63] In response Blome suggested to Himmler that his new institute in Posen as a more suitable place for these experiments, since its location was so isolated. Himmler selected an SS physician to work with him.

The Nuremberg tribunal decided that this accusation could not be sustained: "It may well be that defendant Blome was preparing to experiment on human beings in connection with biological warfare, but the record fails to disclose that fact, or that he actually conducted experiments. The charge of the prosecution on this point cannot be sustained."[64]

On the other charges, as well, the tribunal declared that there was no proof, and so Blome was acquitted on all counts. His case provides only one example of the tendency of the Americans, for military-political reasons and against their better moral judgment, to take advantage of the strategically useful knowledge of Nazi scientists and technicians, even at the price of exculpating them. I should also point out that Japanese scientists and physicians, who between 1933 and 1945 under Major Shiro Ishii murdered at least three thousand people in Manchuria in experiments with biological weapons, were also not charged as war criminals by the Americans after the war (Harris 1992, pp. 118–157).[65] Along with Hermann Becker-Freyseng, Siegfried Ruff, and Konrad Schäfer, Blome was among the four defendants at the Nuremberg physicians trial whom the U.S. military recruited as part of the action "paperclip" to work on sensitive military programs (Hunt 1985).[66]

On November 10, 1947, two months after his acquittal, Blome was questioned by four experts at Camp Detrick in Maryland about details of biological warfare research in Germany. Between 1948 and 1951, he worked in his own medical practice in Germany, which he sold after signing a contract on August 21, 1951, to work for "Army Chemical Corps" under "Project 63," a secret American program (Hunt 1985). On his personal data sheet he left out information about his activities in the years 1945 to 1948, that is, his arrest and trial before the Nuremberg tribunal. Nevertheless, the U.S. consul in Frankfurt declared Blome "inadmissible for immigration."[67] Since the U.S. Army and the secret service were afraid that other "paperclip" programs might be in jeopardy if Blome's contract were annulled, he was given a position as physician in the European Command Intelligence Center in Oberursel. No information concerning Blome's subsequent activities could be found.

7

Aftereffects of National Socialism

1. The RFR and the DFG after 1945

The essential institutions for the promotion of scientific research resumed their work or reestablished themselves shortly after the war. The Max Planck Society for the Advancement of Science was founded on February 26, 1948, as the West German successor to the Kaiser Wilhelm Society. The name was changed in response to pressure from allies. The year 1949 saw the creation of two competing research organizations with the revival of the old Notgemeinschaft and the German Research Council. At the joint northwest and southern German university conference in May of 1948, the rectors of the universities and the ministers of culture and education of the various states had agreed unanimously to reestablish the Notgemeinschaft (Zierold 1968, pp. 278f.). This was done in January of 1949; the conference of the ministers had secured the funding of the association in agreement with the finance ministers. Only German universities and scientific colleges as well as the Academies of Science became members of the society. Minister of Culture Christine Teusch did not invite the

290

president of the Max Planck Society to the founding meeting of the Notgemeinschaft on January 11, 1949, in Cologne. The bylaws were largely identical to those of the old Notgemeinschaft of 1920. Its self-governing status was pushed through in part against the opposition of the ministers of culture; the influence of the ministries of culture, despite their role in supplying most of the funds, was reduced to a minimum, and scientists were in the majority in all committees. One change of the old bylaws of 1920 concerned the "monarchical" principle, according to which the decision about the granting of funds rested in the final analysis with the president. In 1949 this power was transferred to the Central Committee.

Two months after the establishment of the Notgemeinschaft, the Max Planck Society and the Academies of Science in Göttingen, Heidelberg, and Munich set up the German Research Council (DFR). It was designed to promote and coordinate scientific research in Germany, provide scientific advice to German governmental agencies, and represent all concerns and demands of German scientific research with agencies of the German government and abroad (Zierold 1968, p. 298). The decisive initiative for the DFR came from the physicist Werner Heisenberg, who became its president. Apart from general support for research wherever it was being carried out in Germany, the DFR considered the funding of large-scale research projects to be particularly important (Eickemeyer 1953, p. 62). The DFR did not turn to the state governments but was oriented toward the government of the recently established Federal Republic. It was supported by Konrad Adenauer, the first chancellor of West Germany. With the disappearance of funds from the Marshall Plan, the DFR considered it particularly urgent that funds be raised from the federal budget. In a memorandum drafted for the federal government, the DFR calculated that DM 200 million were necessary as a one-time research subsidy and DM 100 million yearly in addition to the already existing research budgets (Eickemeyer 1953, p. 65). However, in actuality it never got any support worth mentioning.

The universities and the states opposed the DFR, rejecting in particular the involvement of the federal government in decisions about scientific research. The ministers of culture declared the DFR superfluous and refused to finance it (Zierold 1968, p. 299). University scientists pointed out that 70 to 80% of research in Germany took place at university institutes, and that a committee based only on three academies and the Max Planck Society could not claim to represent German science. Both state administrators and university scientists preferred the concept of the Notgemeinschaft. The physicist Walther Gerlach, vice-president of the Notgemeinschaft, leveled the following reproach at Werner Heisenberg: "You, of all people, who have always advocated the autonomy of research, must understand that we cannot abandon a well-considered plan for the self-organization of scientific research in response to an intervention by the state that lacks a compelling justification" (Nipperdey and Schmugge 1970, p. 77).

The DFR was unsuccessful in its attempt to reorganize responsibility for the funding of research. On August 2, 1951, the Notgemeinschaft and the DFR merged under the name German Research Association (DFG) on the basis of the concept of the Notgemeinschaft; the idea of a Research Council was dropped.

Why did eminent scientists like Heisenberg support the establishment of a new Research Council after the National Socialist Research Council had just been dissolved? Heisenberg presented the connection between politics and scientific planning as the concept of the future in the area of scientific policy, and he thought he could "sense a strongly restorationist element" in the efforts by universities and states to reestablish the Notgemeinschaft (Nipperdey and Schmugge 1970, p. 75).[1]

The results of the present study suggest that the advocates of a new Research Council did not primarily aim to counter restorationist tendencies in science policy. Their arguments resembled those used in 1910 to call for the establishment of the Kaiser Wilhelm Society. Then, the founder and first president of the Kaiser Wilhelm Society, Adolf Harnack, had convinced the emperor of the need to establish the society by pointing out the national-political importance of science in the face of the impending loss of Germany's leading role in the natural sciences: "Military power and science are the twin pillars of Germany's greatness, and their cultivation must never cease or rest" (Glum 1936, p. 6).

After World War I, this argument was developed further. According to a view widely held among German scientists and scholars, the status of a great scientific power would compensate for the lost status as a great military power (Forman 1973). In contrast to the situation after World War I, not only was Germany militarily defeated after 1945, but the danger biologists had been warning of in their grant applications had come to pass: Germany had been overtaken by the Americans in the field of science.[2] The founders of the DFR entertained hopes of reestablishing Germany's greatness through scientific achievements. Heisenberg spoke of "attempts to win back the positions that German science had held within the international world of learning prior to 1939" (Eickemeyer 1953, Preface). The above-mentioned memorandum of the DFR quotes the physicist Max von Laue as saying that "scientific research is today virtually the only form in which Germany can still conduct foreign policy. If we relinquish the struggle in this area, in ten years we will be as uninteresting to the world as some tribe of Bantus" (Eickemeyer 1953, p. 66).

DFR members lamented that "two world wars, emigration, the dismantling program, the often senseless fury of de-Nazification, and the exodus of skilled personnel" had inflicted a heavy blow on Germany (Eickemeyer 1953, p. 64). They voiced the concern that because of the lack of sufficient support for research the most talented young scientists would take advantage of career opportunities in the United States or England. (By contrast, in the report "Forty Years of the Kaiser Wilhelm Society," written at the request of the Max Planck Society, the forced emigration after 1933 went virtually without mention [Haevecker 1951].)[3]

A centralized science policy that would, first of all, support all outstanding German scientists in research institutes was to give the Germans the chance to diminish the dominance of other countries in scientific research. Unlike the universities, the Max Planck Society looked favorably upon the idea of the DFR. Support for large-scale scientific projects was one of the goals of the DFR. To justify generous funding for German science shortly after the war, Heisenberg wrote in 1949: "Under the Third Reich, Germany hardly took part in this development [of scientific research] . . . All that was done was to apply the scientific results that existed in 1933 to armament and to support every type of armament industry, while basic scientific research as a whole was left to whither" (Heisenberg 1953).

This assessment was not accurate, at least as far as biological research at the Kaiser Wilhelm Institutes was concerned. On the contrary, since the Kaiser Wilhelm Society had benefited from the centralized science policy under National Socialism, particularly the National Socialist Research Council, it seems plausible that the Max Planck Society and its scientists supported the new Research Council precisely because they continued to associate it with their hopes for large-scale support. The universities, however, were for obvious reasons not interested in a new Research Council in Germany.

The discussion that follows will demonstrate the isolation of German scientists after 1945, a distinct aftereffect of National Socialism. To examine the issue of international isolation, in the remainder of this chapter I will focus on the most modern field of biology at the time, molecular genetics. The comments so far have shown that the postwar backwardness of scientific research in Germany in comparison with Western countries did not primarily result from National Socialist science policy. At least in the example of molecular genetics it becomes apparent that international isolation, in part also a result of the moral failure of German scientists after 1933, played a major role.

2. The Effects of National Socialism on the Development of Molecular Genetics in Germany

2.1. Early Experimental Objects in Molecular Genetics

A brief summary of important developments in early molecular genetics will serve to highlight particularities of genetic research in Germany after 1945. Here I shall limit myself to genetic and biochemical research on microorganisms and viruses, leaving out of consideration the work of physicists and chemists on the structure of proteins, which was done above all at the Cavendish Laboratory in Cambridge. The history of research in molecular genetics has been described and analyzed in numerous works, including the exhaustive studies of Robert Olby (1974), A. P. Waterson and Lise Wilkinson (1978), Horace Freeland Judson (1979), and Franklin H. Portugal and Jack S. Cohen (1979). Pnina Abir-Arm (1992), for example, analyzed the multidisciplinary creation of molecular biology and showed that the development of molecular genetics was, from the first experiments in the 1930s and during the entire phase of its institutionalization as a new discipline in the 1950s and 1960s, an international enterprise in which formal and informal meetings as well as cooperation across national boundaries played an important role.

Two factors influenced the molecular-biological direction of scientific research around 1940: first, biologists like George Beadle for the first time established a systematic link between genetics and biochemistry; second, many nonbiologists, in particular physicists and chemists, turned toward genetic research with the goal of clarifying the problem of the physical foundations of genetic information (Stent and Calendar 1978, p. 24). For example, the phage group Max Delbrück set up in the United States contributed significantly to the development of molecular genetics (see below). It became possible to provide answers to the two central questions of molecular genetics—that concerning the gene function and the biochemistry on which it was based, and that concerning the physical-chemical nature of the gene, its replication and mutation—because the complex experimental objects of classical genetics were replaced by simpler experimental objects.

"Neurospora" and Bacteria. In 1940 George Beadle and Edward Tatum in the United States continued with *Neurospora* the biochemical-genetic studies they had begun with *Drosophila*. These experiments constituted an important transition to molecular genetics in that they made possible a biochemical explanation of gene functions. *Neurospora* had the following advantages over *Drosophila*:

- the biochemistry of the genetically conditioned sequence of enzymatic reactions for the synthesis of specific amino acids was already known;
- it was readily possible to detect deletion mutants with the help of minimal media;
- the vegetative cells of the *Neurospora* mycelium are haploid, as a result of which the genetic analysis of a mutation is not additionally complicated by the dominance or recessivity of the genes.

The work of Beadle and Tatum and their students on *Neurospora* and other fungi made important contributions to the understanding of the genetic control of cellular metabolism. Beginning in the mid-fifties, however, most such studies were carried out on bacteria, especially *E. coli*. Bacteria have the significant advantage of a short generational period: while *Neurospora* has a twelve-day generational cycle, *E. coli* can double in number in twenty to sixty minutes, depending on the medium. Bacterial genetics was founded by Joshua Lederberg in the United States. But bacteria were perceived to have the disadvantage of an exclusively asexual reproduction. Their importance as genetic research objects thus increased after Lederberg and Tatum discovered the genetic recombination in *E. coli* in 1946 and developed a method for selecting the recombinants. In the early fifties, William Hayes in England and François Jacob and Elie Wollman in France made great gains in clarifying parasexual mechanisms in bacteria. Bacteria were used to provide answers on a molecular level to, among others, questions concerning the dominance or recessivity of gene functions and the regulation of gene activity (Jacob and Monod 1961).

Tobacco Mosaic Virus and Bacteriophages. Much of the early work in molecular genetics was carried out on viruses, especially the tobacco mosaic virus. After Wendell Stanley crystallized the virus in 1935, it became, as the model of "self-reproducing molecules," initially the most extensively studied experimental object for the physical-chemical analysis of biological autoreplication. However, owing to their simpler and faster production, bacteriophages proved more favorable for molecular genetics than tobacco mosaic viruses. For example, to produce sufficient quantities of TMV mutants to allow for their chemical characterization, seventy tobacco plants were infected with a mutant in 1944. These plants were harvested one month after being infected and the viruses were isolated (Friedrich-Freksa, Melchers, and Schramm 1946). Phages can be detected on a bacterial lawn only a few hours after the phage lysate is applied by dropper. In contrast to (DNA) phage, no recombination was found in (RNA) plant viruses, hence they are not suitable for genetic analyses (Melchers 1968).

Bacterial viruses had been the object of biological-medical research since 1915. Frederick Twort in 1915 and Felix H. D'Hérelle in 1917 described them as the material that could be separated out by filters impervious to bacteria and brought about bacterial lysis; it was also D'Hérelle who introduced the term "bacteriophage." This early phase of research into phages was, however, strongly guided by the goal of fighting bacterial infections, a use that was not fulfilled. The second phase began in the thirties, when interest in phages as objects of basic genetic research began to take precedence.

The publication of a paper by Emory Ellis and Max Delbrück in 1939 marked the beginning of this modern phage research. In it they developed an experimental approach to phage replication and showed that the replication of phage occurred in one step. Delbrück, a theoretical physicist, had come to the United States in 1937 on a Rockefeller fellowship to work with Morgan at the California Institute of Technology in the field of *Drosophila* genetics. He had already turned to questions of genetics in Germany, the impulse having come from the lecture "Light and Life" that Niels Bohr gave in Copenhagen in 1932. Delbrück soon realized that studying linkage groups in *Drosophila* would not help him find out more about the capacity of genes for identical replication and the storage of information. *Drosophila* genetics seemed to him much too complex for this problem (P. Fischer 1985, pp. 9ff.). Ellis, a biochemist who had begun to study phages as objects for cancer research, brought Delbrück into contact with phage research. Impressed by the simple methods with which one could, for example, make individual phage particles visible, Delbrück began genetic research on phage particles in 1938 with Ellis. While Ellis, for financial reasons, had to turn to cancer research with mice after a year of their joint work, Delbrück continued to receive funding for phage research. In 1941 he began collaborating with the medical scientist Salvador Luria, an Italian emigrant who had also turned to biophysical problems early on; in 1942 the bacteriologist Alfred Hershey joined the two scientists. The working group of these three scientists is often called the "phage group."

Beginning in 1945, Delbrück and other members of the phage group organized a yearly phage course in Cold Spring Harbor; they sought to make interested scientists familiar with the methods and modes of thinking in phage research. Looking back, Elie Wollman characterized the approach of the phage group, which led to a complete reorientation in the field of bacteria and phage research, as follows: "Taking a strictly Cartesian attitude, Delbrück and Luria had swept away the fact and interpretations accumulated by their predecessors over twenty years and started anew. Within a few years they, and the small group of other workers they had attracted by the simplicity, the precision, and the elegance of their new departure, had made enormous advances" (Wollman 1972, p. 212).

However, this departure also meant that other approaches were for now ignored or even rejected. For example, for some years the phage group did not take seriously the problem of lysogeny, on which Wollman's parents had already done preliminary work in the thirties, because the T-phages they primarily used did not exist in a lysogenic condition. That is, there were no bacteria in whose DNA T-phages-DNA was integrated. In contrast, the genetics of lysogenic phages or bacteria became a focal point of research at the Institut Pasteur. André Lwoff and his collaborators furnished proof in 1948 and 1950 that lysogenic bacteria did in fact contain the genetic information of phages, and that the occasionally occurring phage replication did not constitute contaminations with phage particles, as the phage group in the United States was maintaining. In this way research into lysogeny, too, became an internationally recognized field of research.

"Neurospora," Bacteria, and Phage as Experimental Objects in Germany to 1945. Detailed information on experimental genetic objects in Germany may be found in Chapter 4, Sections 1, 2, and 5. While *Neurospora* did serve as an experimental genetic object, it was not used for research into physiological genetics.[4] By contrast, geneticists conducted intensive research on TMV. Mutation research on TMV and research on the chemical analysis of the virus had been carried out since 1937 by an interdisciplinary working group at the Kaiser Wilhelm Institutes for Biology and Biochemistry, after 1941 at the Division for Virus Research at both of these Kaiser Wilhelm Institutes. The experiments of Gerhard Schramm, in particular, were of fundamental importance in clarifying the structure and replication of the virus.

Beginning in 1940, biologists used bacteria and phages as experimental objects at the Reich Biological Institute, the KWI for Biology, and the Genetics Department of the KWI for Brain Research. A part of these experiments was devoted to elaborating the hit theory or hit-area theory. Moreover, genetic experiments were prepared by labeling the nucleic acid of bacteria and phages with radioactive phosphorous.

2.2. Research in Molecular Genetics in Germany (FRG), 1946–1965

The following account is limited to molecular-genetic research conducted with *Neurospora*, bacteria, and phages and concerned with the structure and replication of nucleic acids and with protein synthesis. It is not always possible to set this research clearly apart from other research on plant and animal viruses. Moreover, in analyzing DFG support, only the projects themselves and not the amount of funding could be considered (see below). Despite these difficulties with regard to a quantitative evaluation, the method used here seems suitable for assessing to what extent the new ideas for molecular-genetic research, which came initially from the United States and then also from France and England, were picked up in Germany.

The Continuation of Research by the Division for Virus Research after 1945. In 1943 the Division for Virus Research along with a part of the Kaiser Wilhelm Institute for Biochemistry was moved to a number of university institutes. In 1945 it was reorganized and incorporated into the KWI for Biochemistry in Tübingen (which became the Max Planck Institute for Biochemistry in 1948, following the founding of the Max Planck Society) as the Department for Virus Research. This department was given its own building in 1950. Biochemical and genetic work on TMV and other plant viruses continued after the war both in this department, under the direction of Gerhard Schramm, and at the Kaiser Wilhelm Institute/Max Planck Institute for Biology in the department of Georg Melchers. Gernot Bergold, the one-time head of the Department for Insect Viruses of the Division for Virus Research, went to Canada in 1948. After his departure Werner Schäfer headed the work in the field of animal viruses. In 1954 the Department for Virus Research was transformed into the Max Planck Institute for Virus Research. Hans Friedrich-Freksa became director of the institute, Schramm head and in 1956 director of the Department of Biochemistry, and Schäfer director of the Department for Animal Pathogenic Viruses.

The Number of Publications in Molecular Genetics from Max Planck Institutes, 1947–1965. The first relevant work in molecular genetics from one of the Max Planck Institutes was published in 1951.[5] This was the work of Wolfhard Weidel in Melchers's department in the Max Planck Institute for Biology on the cause of phage resistance in *E. coli B*. Eight of the thirteen papers in molecular genetics that were published up to 1956 came from Weidel, who studied T-phage receptor systems of *E. coli* (he later made decisive contributions to clarifying the structure of the cell wall of bacteria). The number of publications in molecular genetics from MPIs show a stronger increase only beginning in 1957 (Figure 7.1).[6]

Funding for Molecular-Genetic Research from the DFG, 1949–1965. The following information comes from the yearly reports between 1949 and 1965 of the Notgemeinschaft (reestablished in 1949) and

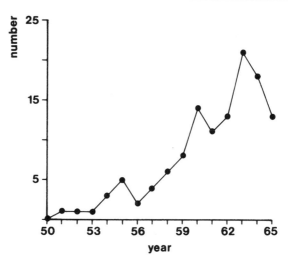

Figure 7.1   Number of publications in molecular genetics from the Max Planck Institutes for Biochemistry, Biology, Comparative Heredity and Hereditary Pathology, and Virology.

after 1951 of the DFG, which was created by the fusion of the Not-gemeinschaft and the German Research Council. The reports list the topics but not the amount of individual grants. Beginning in 1952 the DFG funded research in the field of molecular genetics through modest material subsidies within the framework of the Regular Program for Biology as well as through larger subsidies within the framework of the following four focal programs *(Schwerpunktprogramme):* virus research 1952–1957; genetics 1953–1963; biochemistry 1956–1964; molecular biology from 1964 on. The focal programs had been set up by the Notgemeinschaft in 1951 to provide special funding for areas that lagged behind in research. The applications within the focal program for genetics were submitted primarily by botanists and zoologists, almost none by chemists and biochemists. Traditional genetic research predominated within this focal program (Figure 7.2). The percentage share of projects in molecular genetics rose from 11% in 1960 to 19% in 1961 and 35% in 1962. In the years 1962 and 1963, 50% of the projects in molecular genetics that were supported within the framework of the focal program for genetics were carried out at the Institute for Genetics (founded in 1961) at the University of Cologne.

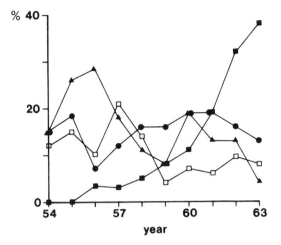

-●- Cytoplasmic Inheritance
-▲- Developmental Genetics
-■- Molecular Genetics
-□- Mutation and Radiation Research
  in Complex Organisms

Figure 7.2 Percentage distribution of projects funded by the DFG under the general category of genetics.

The number of projects in molecular genetics supported within the framework of all four above-mentioned focal programs and the Regular Program for Biology increased noticeably from two projects in 1955 to twelve projects in 1956; after the focal program for molecular biology had been set up in 1964, it rose once again from eight (1963) to twenty-seven (1964) (Figure 7.3).

Between 1952 and 1954, all funded projects in molecular genetics were carried out by scientists at MPIs; between 1955 and 1958 the majority of such projects were. Thereafter the DFG funded such projects largely at universities. The following list gives examples of projects supported by the DFG in the early period:

Max Planck Institute for Biology, Department Melchers, Tübingen
Wolfhard Weidel: receptor of the phage T5 from *E. coli*, chemical structure of bacterial cell walls, beginning in 1952.

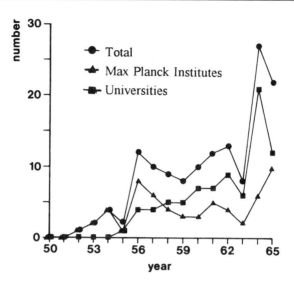

Figure 7.3 Number of projects in molecular genetics funded by the DFG at Max Planck Institutes and universities between 1950 and 1965.

Georg Melchers: studies on bacteriophages, in particular on of first contact between virus and host cell, 1954.

Max Planck Institute for Virus Research, Tübingen
Hans Friedrich-Freksa: relationship of nucleic acids to protein synthesis; gene-based syntheses-chains in bacteria, from 1956 on.
Gerhard Schramm: biochemistry of nucleic acids, from 1955 on.
Fritz Kaudewitz: studies on the one gene–one enzyme hypothesis in microorganisms, from 1957 on.

Max Planck Institute for Breeding Research, Voldagsen
Reinhard Kaplan: mutation process in bacteria and phages, from 1952 on.

Institute for Microbiology, University of Frankfurt
Reinhard Kaplan: mutation and deactivation of bacteria and phages, from 1955 on.

Institute for Organic Chemistry, University of Frankfurt
Theodor Wieland: studies on the biological synthesis of polypeptides, from 1956 on.

Institute for Botany, University of Cologne
Carsten Bresch: gene recombination in microorganisms, from 1957 on.

Institute for Physiological Chemistry, University of Munich
Fritz Turba: reaction mechanism of protein synthesis in living cells, from 1956 on.

On the whole we can note that research in molecular genetics was slow to develop in the Federal Republic and was carried out on a larger scale at university institutes and institutes of the Max Planck Society only from the end of the fifties on. A survey of the state of German research, done at the request of the DFG and published in 1964, stated about biological research: "As for the rest, microbiology, including technical microbiology, is judged to be backward. In the modern and widely overlapping field of molecular biology, in particular, considerable catching-up needs to be done with other countries, despite some important beginnings. The fields of molecular genetics, botanical and zoological genetics, human genetics, and developmental physiology are also strongly deserving of funding" (Clausen 1964, p. 12).

As previously argued, neither National Socialist research policy nor the expulsion of Jewish scientists, despite the significant losses they entailed, can be used as the sole explanation for the lag in molecular genetics research in the Federal Republic until about 1970. The closing of some universities during the war, the moving of a number of Kaiser Wilhelm Institutes from Berlin at the end of the war because of bombing raids, and especially the war-related destruction and lack of money after 1945 made research much more difficult, but these things, alone, also fail to fully explain the lag. Incidentally, the situation in other European countries—England and France, for example—was no different.[7] In Germany the support for research by the federal government and the states resumed soon after the war.[8] We find the following statement with regard to the financing of the Kaiser Wilhelm Institutes (Max Planck Institutes) in the 1951 report of the Max Planck Society (*Naturwissenschaften* 38, p. 362):

Following the collapse, the funding of the institutes located in the West came from the eleven state governments, which in this way have secured the existence and continued operation of the institutes. In the first period after the end of the war, the state governments could fall back on an "iron fund" in the amount of one million Reichs marks which had been made available to the General Administration in 1944 by the Donors Association of German Industry. It played an essential role in allowing the institutes to meet their financial obligations and keep together their staff of experienced scientists and trained personnel. An additional sum of one million marks, made available at the end of 1944 also by the Donors Association via Mr. Hermann v. Siemens, gave another financial boost.

This was the reason why the Kaiser Wilhelm Society (Max Planck Society) was able to set up two new institutes between 1946 and 1951.

In looking for the causes for the delayed development of research in molecular genetics in Germany, one must not overlook the marked difference between genetic research at the university institutes and that at the KWIs or MPIs. For example, after the war, as well, university biology institutes failed to pick up the research on TMV that had been conducted at the Kaiser Wilhelm Institutes for Biology and Chemistry since 1937. Complex organisms were used exclusively as genetic research objects at these institutes right up to the mid-fifties. Melchers attributed this fact not only to a lack of funds and a lack of cooperation among biologists and chemists, but also to the "lack of insight into the importance of this field." In his view it was "not rare to hear that the concern with such a specialized object as the causative agent of a single plant disease did not really go well with the 'universitas litterarum'" (1960b, pp. 111, 112). A number of cases illustrate the ignorance of (botanical and zoological) geneticists with regard to the newest developments or their unwillingness to grapple with them.[9] According to Kurt Wallenfels, since 1953 associate professor of chemistry in Freiburg, as late as 1957 that university's geneticist (and botanist) Friedrich Oehlkers reacted with disbelief to the announcement of a lecture by Wallenfels on the chemical basis of inheritance. In Oehlkers's opinion genes had nothing to do with chemistry; he still regarded them as abstract entities.[10] Wallenfels also remembered that the Freiburg plant geneticist Hans Marquardt, Oehlkers's student, was very skeptical about the lecture.

The inability to recognize the importance of research on *Neurospora,* bacteria, and phage was not due to a lack of opportunities to get information about it. For instance, according to Marquardt, "the molecular genetic results of Beadle, Tatum, and the phage people" since 1945 were certainly known in Freiburg. However, he maintained that the genetics of phage and bacteria could not contribute anything to the understanding of higher organisms, since the significance of genetic findings in prokaryotes for eukaryotes had not been clarified at the time.[11] Similar arguments were put forth by Siegfried Strugger, professor in Münster, when he questioned, as late as 1958, whether the results of phage research, according to which the DNA alone represented the genetic substance, were transferable to the cell; in his view most the geneticists at the time believed that proteins, too, had an important role (Melchers 1958, p. 52).

The molecular genetic research at the universities that was supported by the DFG from the mid-fifties on was initiated not by scientists who held chairs in botany and zoology, most of whom had returned to their positions after the war, but by younger scientists. The example of Carsten Bresch, described below, shows that the possibility of working in the field of molecular biology with modest means had already existed shortly after the war.

Although modern genetic TMV research went on at the Division for Virus research, and the development of research in molecular biology abroad could be followed in Germany, the innovations, in particular those coming from the United States, were picked up late at Max Planck Institutes relative to comparable institutes in England and France. In what follows I will examine the extent to which a politically conditioned lack of international exchange after the war contributed to the fact that science in Germany initially did not catch up with the international developments in genetics.

2.3.   The Limited International Scientific Exchange after 1945

Many biologists who were active scientists in Germany prior to 1945 maintain that German scientists were not isolated after World War II; by way of contrast they point to the international isolation of Germany after World War I.[12] The Versailles Treaty included provisions that essentially dissolved all existing international conventions with Germany in the field of science. A memorandum (dated January 29, 1925) from the Reich Central Office for Scientific Reporting stated: "German societies and scientists and scholars were excluded from international associations or were no longer invited to congresses, German members were crossed off the membership lists of international commissions . . . German journals, in particular those of international renown, were boycotted for a long time . . . the German language was explicitly rejected as a spoken language at congresses and as a written language in journals, in particular multilingual international ones."[13]

Fifty-one of fifty-seven congresses in the humanities and natural sciences between 1922 and 1924 took place to the exclusion of German scholars and scientists.[14] The Congress of Geneticists in Berlin in 1927 was one of the first international biological congresses in Germany after 1918.

In view of this experience, it is likely that German scientists after World War II had expected the reaction of foreign scientists and scientific associations to be far more severe than it in fact was. To be sure, the situation was in any case different, since German had been replaced by English in the 1930s as one of the major scientific languages. But an across-the-board condemnation of German scientists and the exclusion of all German participants from international conferences did not take place.

Nevertheless, the following two examples show that the distrust of foreign scientists often failed to register among Germans. A Mr. Knoll from the British military government stated at the meeting on the future of German universities in Marburg in 1946 that German professors probably had no idea of the mistrust they encountered on the other side of the border.[15] At the same meeting Professor Citroen from Holland said, of a possible student exchange:

> Please do not get any wrong ideas from the fact that I have appeared here, that I am, in a sense, "willing to exchange."—Perhaps it is not sufficiently known what the situation in our countries is—and here I am speaking not only of Holland, but also of the other ruined countries, Belgium, France, Norway, and especially all of southeastern Europe. There the destruction is so horrible, the devastation wrought by the German armies so complete, that one probably could not go out into the street and speak German in any of these countries without risking serious harassment. An exchange with German students will surely be impossible for a very long time to come in countries where one cannot walk in any city without coming upon graves on the sidewalk in which lie the slaughtered victims of Nazi terror.[16]

In contrast to the situation after World War I, on a political level the Cold War played a role in the attempt to integrate Germany as quickly as possible into the West, not only politically and economically but also scientifically. But even a partial isolation of German scientists, which I shall presently examine, could be of importance for the development of genetics.

To what extent political reasons played a role in international scientific exchange, for example the invitation of German scientists to international conferences or the choice of a site for a conference, is a question to which we can give a direct answer in some cases and a conjectured one in others. These examples show, or raise the likelihood, that the National Socialist past remained a factor in contact with foreign scientists even more than ten years after the end of World War II.

According to H. F. Linskens (1962, pp. 36f.), the relationship of Dutch biologists to German biology was very strained after 1933 and especially after 1940. Prior to this time, Dutch science had been strongly oriented toward Germany. For example, 70% of the published papers in Dutch botanical and zoological journals in 1932 were written in German; that figure dropped drastically to about 10% in 1943 and 0 in 1947 (and it remained under 5% until 1960).

I have analyzed the development of German participation after 1945 by using the Cold Spring Harbor (CSH) Symposia on Quantitative Biology as a test case. These symposia made a significant contribution to the international exchange in the field of molecular genetics. The first symposium after 1942 took place in the summer of 1946 on the topic "Heredity and Variation in Microorganisms" and was of great importance for molecular genetics in a number of respects.[17] In the view of Milislav Demerec, director of the Genetics Division of the Carnegie Institution from 1941 to 1960 and organizer of the CSH Symposia until 1952, its purpose was to allow for the exchange of the latest results from Europe (especially France) and the United States in the field of the genetics and physiology of microorganisms.[18] At the 1944 symposium Oswald Avery reported on his transformation experiments on pneumococci, experiments which demonstrated that DNA—and not, as previously assumed, proteins or nucleoproteins—constituted the genetic material. The discovery of the sexual reproduction of bacteria and viruses, which was of extraordinary importance for subsequent genetic research, was also announced for the first time at this meeting (see Stent and Calendar 1978, p. 141). The symposium provided a stimulus to many geneticists, among them André Lwoff and Jacques Monod, to pursue phage research.

The large majority of passive and active participants (70–94%) be-

tween 1946 and 1965 came from the United States. Between 5% and
19% of the participants were Europeans. Their numbers fluctuated
between nine and twenty-eight at each symposium, with the exception
of 1963 (fifty-three Europeans). A comparison of the participation of
German scientists at the symposia with that of scientists from France,
Great Britain, and Sweden (that is, countries with a comparable scientific standard) shows the following: relative to population density,
the number of German scientists (exclusively scientists from the FRG)
until 1958 was noticeably below the average number of participants
for the other three states (Figure 7.4). German scientists were completely absent from seven of the thirteen symposia up to 1958. The
highest number of participants until 1958 came in 1953 (three).

Only after 1959 were German scientists represented at every symposium; from 1951 on, their number relative to population was sometimes greater than and sometimes less than the sum of the three comparison states. The symposium with the greatest number of German
scientists was that in 1960 on the topic "Biological Clocks."

In the case of these symposia we can only conjecture that German
geneticists were not invited, or did not wish to participate, for political
reasons. However, the correspondence between geneticists reveals that
after the war many of the foreign scientists, Jews and non-Jews, Europeans and Americans, were unwilling for some years to meet with
German colleagues, at least those from the older generation. Thus
Demerec, known to be politically moderate, said in 1947 that in general Germans should not be invited to congresses, with the exception
of "those few individuals who either had made important discoveries
or had suffered under the Nazis."[19]

Hence for the International Genetics Congress in Stockholm in July
of 1948 no general invitation was issued to German scientists. Rather,
individual invitations were sent to those German geneticists who were
considered to be unburdened by their political past.[20] Among the scientists working in the field of TMV research, Friedrich-Freksa and
Melchers were invited.

The discussions over the site of the 1963 International Genetics
Congress reveal that Germany was not a neutral location even eighteen years after the war: back in 1958, when the decision was made
to hold the next Congress on Human Genetics and the next Genetics
Congress in Berlin in 1963,[21] most of the participants initially accepted
the German invitation. But in response to an anti-Semitic act in the

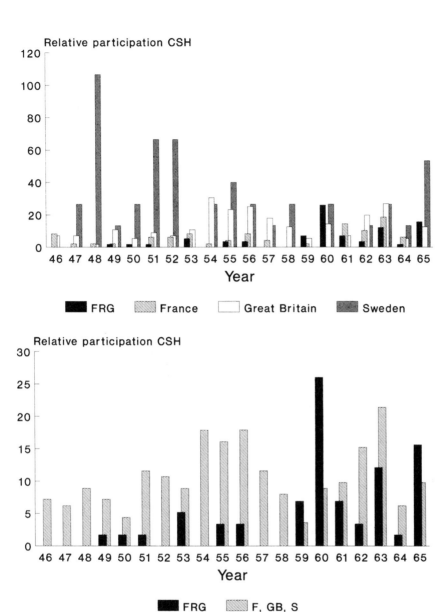

Figure 7.4 Relative participation by scientists from four Western European countries at the Cold Spring Harbor Symposia, 1946–1965. The number of participants is calculated relative to population size, using population figures of 1963 (in millions): Germany (FRG), 57.6; France, 48.5; Great Britain, 55.7; Sweden, 7.5. *Top:* Participation numbers for the four countries, listed individually. *Bottom:* A comparison of the number of German participants with the total number of participants from France, Great Britain, and Sweden, relative to the population size of the three countries.

Federal Republic in 1960, the discussion about the location of the congresses was reopened among American geneticists, especially at the initiative of Walter Landauer and Leslie C. Dunn.[22] In the course of the controversy it became clear that, regardless of the immediate cause, not only Israeli but also a number of American and European geneticists were not yet ready to go to a congress in Germany. The rejection on the part of some British human geneticists, for example, had to do largely with the person of Otmar von Verschuer, who had once again attained an influential position in Germany after 1945, despite the fact that he had placed genetics in the service of Nazi ideology and racial practice.[23] Even among those who supported the idea of a congress in Germany, the news that Verschuer would probably be involved in a significant way in the organization of the congresses caused considerable irritation.[24] The Israeli geneticist Jacob Wahrman summed up the reservations of many colleagues, which had already been voiced in 1958:

> I tried at the time to explain the position of many colleagues who feel that the choice of Germany as a meeting place for geneticists is rather unfortunate in view of the short time that has elapsed since the Nazi regime used a pseudogenetic argumentation for the destruction of human life on an unprecedented scale with apparently very little opposition from German scientists. As is well known, many of the Nazis who were responsible for such atrocities are once again occupying prominent positions. You may know that the German invitation for the Congress of Human Genetics was not accepted and I can hardly understand why a different policy should be adopted for the Genetics Congress.[25]

As a result of these controversies, and owing also to disputes between geneticists in the Federal Republic of Germany and the German Democratic Republic, the site of the congresses was eventually changed.[26]

These events show—regardless of how much the opinions of American geneticists diverged as far as the assessment of the political situation in Germany was concerned—that a single anti-Semitic incident in that country was enough to remind the rest of the world of the Nazi past. Surely the treatment of many eminent Jewish geneticists under the Germans could not be forgotten.

Otto L. Mohr, the director of the Genetics Institute in Oslo and a friend of Thomas Hunt Morgan, was arrested by the Gestapo in September of 1941. After seven months in jail and a concentration camp, he was released but strictly forbidden to live in Oslo and to visit the university or to have contact with students. His books were blacklisted, and the remaining copies of his book *Genetics and Pathology* were destroyed. He was officially discharged in 1943. All education work at the university stopped when 2,200 Norwegian students were either arrested or sent to concentration camps in Germany in December 1943. In June 1945, Mohr wrote to Demerec concerning meeting German geneticists at congresses: "We do not hate the Germans. But it will take some time before we feel like meeting them again."[27]

Ernst Caspari, an assistant to Alfred Kühn in Göttingen was expelled in 1933. He emigrated first to Istanbul and then to the United States, where he became professor at a small university (Wesleyan) in Connecticut. His father, Wilhelm Caspari, for many years head of the Cancer Department of the Institute for Experimental Therapy in Frankfurt, died in the concentration camp in Lodz in 1944. His mother, exiled with her husband, was deported from the camp in 1942 and was never heard of again. The sister of Caspari's wife and her daughter were murdered in concentration camps.

In France, nearly all scientists at the Institut Pasteur, as well as Boris Ephrussi, had had terrible experiences with the Germans during the war. Jacques Monod went underground and joined the armed resistance movement led by the Communist Party. André Lwoff joined another underground network. François Jacob fought in de Gaulle's French Free Forces against the Nazis; almost all of his friends were killed in the war, and he himself had to give up his career as a physician because of permanent injuries.[28] In 1954 he was invited by Max Delbrück to give a talk in Cologne but went back home after one day. In an interview he said that he could not stand it because he could not prevent himself from thinking of the war and of what all the people around him in Cologne might have done during that time.[29]

Elie Wollman is the son of Elisabeth and Eugène Wollman, who had also been scientists at the Institut Pasteur. In December 1943, they were arrested in occupied Paris and soon afterward deported to Auschwitz; they were never heard of again.

At least until the mid-fifties, no scientists from the Institut Pasteur went to Germany for any length of time, and no German scientists made more than brief visits to the institute. On the German side, Georg Melchers voiced his concerns in 1984 about meeting colleagues whose families had been affected by the Nazi murders: "Fortunately I have never received an invitation to Israel. Possibly, my refusal would be interpreted quite wrongly. The fact is that I would find it very difficult to face meeting the survivors, whose close relatives were among the millions systematically exterminated, without any active opposition on our part. But many colleagues of my generation think differently."[30]

Hermann Muller described the reservations that American geneticists had about German colleagues after the war: hearing that Delbrück was about to visit Germany, Muller asked him to inform the Preparatory Committee of the International Genetics Congress "which German geneticists have remained clearly above reproach, so far as voluntarily having helped the Hitler movement at any time is concerned. If you could get such information while in Germany, it would be very helpful to our Committee on Aid to Geneticists Abroad, because many of the members who sent contributions to us did not want to have their money used to help people who had taken part *in the prostitution of science* or in the movement that would have destroyed the values for which civilization should stand."[31]

German geneticists who, shortly after the war, went to the United States as research assistants or participants in the phage course, or to a congress in France or the United States, were almost exclusively young men who had not yet worked in leading scientific positions during the Nazi period.

The first geneticist to travel to the United States was Wolfhard Weidel, who went in 1949 for a year to get experience working with phage genetics under Delbrück in Pasadena. Delbrück himself had suggested the visit.[32] In 1953 Reinhard Kaplan went to Columbia University for one year as a research assistant. Fritz Kaudewitz was a guest of the Carnegie Institution in Cold Spring Harbor in 1954. Young scientists from the Max Planck Institutes for Virus Research (Werner Schäfer and Klaus Munk) and Biology (Weidel) participated at the Cold Spring Harbor Symposium ("Viruses") in 1953. Carsten Bresch and Weidel participated in the phage conference held in 1952 in Royaumont near Paris. Bresch already ran the phage course in Cold Spring Harbor. Among the researchers at the Former Division for Virus Research, Melchers was early on invited to international congresses and lectures.[33] He had never been a member of the NSDAP or any of its associated organizations.

As indicated above, the research in the department of Gerhard Schramm, at the time the leading virus geneticist in West Germany, was for a long time carried out exclusively on plant and animal viruses.[34] It was only after 1954 that occasional papers on phage genetics were published from this department.[35] As at the Cold Spring Harbor Symposia, active scientific collaboration at the Max Planck Institute for Biochemistry with foreign scientists began in the mid-fifties. Beginning in 1956, foreign scientists, Americans in particular, spent their sabbaticals at the institute.

According to Schäfer, the institute became vastly more attractive to foreign, especially American, colleagues after he presented a detailed report about the work being done there at a CIBA Symposium in London in 1956 and a short time later in New York before an "illustrious audience (including Watson and Crick, Lwoff, Burnet, Dulbecco, Henle, Isaacs, Stoker, Enders, Stanley)."[36] His comments indicate that the institute had been at least partially isolated scientifically prior to this time. James Watson later described the attitude of American geneticists and virus researchers to the German work. In his view, Schramm's experiments of the forties were in fact known in the United States, but "virtually no one outside Germany . . . thought that Schramm's story was right. This was because of the war. It was inconceivable to most people that the German beasts would have permitted the extensive experiments underlying his claims to be routinely carried out during the last years of a war they were so badly losing" (Watson 1968, p. 112). According to the biochemist Peter Karlson, the relationship with foreign countries after the war was difficult at first. He had been told by both Butenandt and Schramm that one was judged according to the contacts one had had during the Nazi period. The exchange got easier in the mid-fifties.[37] Schramm had been a member of the NSDAP as well as the SS.[38]

Seymour S. Cohen, who had come to work with Wendell Stanley in 1941 as a postdoctoral fellow, refutes Watson's notion that Schramm's papers had not been taken seriously in the United States during the war: "I tried very hard to reproduce his 1941 paper."[39] However, he was not able to reproduce the work in which Schramm allegedly removed the nucleic acid from TMV and was left with an intact virus protein. He published his findings together with Stanley in 1942.[40] Schramm never reacted to this paper.

According to Georg Melchers, Schramm's reputation in the United

States suffered considerably from his scientific carelessness and his failure to react to the refutation of his results. Melchers recalled that in the spring of 1941, shortly after Schramm had published his experiment on the production of TMV free of nucleic acid (Schramm 1941), Schramm received a letter from Stanley, in which he informed Schramm of his failure to reproduce the experiment. At that time Schramm already knew that his protein, too, was not free of nucleic acid. Melchers advised Schramm to respond immediately:

> I advised Schramm to write to Stanley immediately. Schramm merely laughed and said he would write a footnote about it sometime. I don't know if he ever did that. Until the United States entered the war there would still have been time for a written note. Because he did not explicitly correct his mistake to Stanley, Stanley and his colleagues continued to mistrust Schramm's results on many occasions. One can hardly expect that Heinz Fraenkel-Conrat, as a German-Jewish émigré (his father was gynecologist and professor in Breslau), was very happy about the fact that Schramm was a member of the SS.[41]

2.4. Max Delbrück's Influence on Genetics in Germany

Special ties to Germany were evidently one prerequisite for establishing more intensive contact with German colleagues after 1945. In these circumstances Max Delbrück, after 1945 the first and for a long time only person in the outside world who talked to German scientists in the field of molecular genetics, was of outstanding importance for the development of genetics in Germany. As mentioned above, Delbrück had emigrated to the United States in 1937 on a Rockefeller fellowship. He was not Jewish, and his position—assistant to Lise Meitner at the Kaiser Wilhelm Institute for Chemistry—had never been challenged on political grounds. But a university career would not have been open to him at that time, since his political commitment was unclear. His application in November of 1936 for teaching credentials in theoretical physics had been turned down on the grounds that "Delbrück does not yet meet the political requirements demanded of a National Socialist professor."[42]

Delbrück had family ties and other personal links to Germany. After the war he saw it as his task to participate in building up modern biology in Germany.[43] As early as 1947 he gave two talks in Berlin on new problems in biology, which had been organized by Otto Warburg in the Harnack House in Berlin. About two hundred German scientists attended.[44] During a visit to Tübingen in 1947, Delbrück arranged with Melchers the above-mentioned one-year fellowship for a young German scientist (Weidel) to Pasadena. His talks brought phage genetics to the attention of Bresch and Wolfgang Eckart, for example, and prompted them to conduct their own research in this field. Bresch, today professor of genetics in Freiburg, was carrying out his experiments at the Robert Koch Institute in Berlin with very primitive equipment. The necessary bacteria and phages were sent to him by Delbrück, who was also for many years the only person he could talk to about his research.[45] Through Delbrück's good offices, Bresch, after obtaining his doctorate, was given a position at the Max Planck Institute for Physical Chemistry in Göttingen, where he could continue his phage research (the head of the institute was Karl-Friedrich Bonhoeffer, Delbrück's brother-in-law), and later an associate professorship for microbiology at the University of Cologne.

Beginning in 1947, Delbrück came to the Federal Republic for a seminar or a lecture nearly every year; he conducted the first phage course in Germany in 1956. He played a crucial role in the founding of several institutes for modern biological research, of which the Institute for Genetics in Cologne should be singled out for special mention. Its founding in 1961 came at the initiative of Josef Straub, at the time professor of botany in Cologne, whose efforts to establish molecular genetics in Germany made him an exception among full professors. The institute in Cologne was the first genetics institute at a West German university. At Straub's request, Delbrück was involved in the planning from 1954 on, and after repeated urging he agreed to serve as head of the institute from 1961 to 1963. In 1956 the Max Planck Society was also trying hard to get Delbrück to succeed Alfred Kühn as director of the Max Planck Society for Biology in Tübingen. In a letter to Karl-Friedrich Bonhoeffer, Delbrück explained why the institute in Cologne interested him more than the offer from Tübingen:

What attracts me to the situation in Cologne is not the size of the institute or the salary or the budget, but the opportunity to make a clear break with the organizational rigidification of academic biology, in collaboration with Straub, in whose judgment, skill, and energy I have great confidence. I think that if we succeed in doing this, it will perform a service to German biology ten times greater than could a small research institute on a hill outside of Tübingen, to which no student will ever stray, and whose influence on the academic routine the established interests of the faculty would unanimously oppose.[46]

After his return to the United States, Delbrück was also frequently consulted on appointments, for example in Cologne, Freiburg, and Konstanz. He became a member of the founding committee of the University of Konstanz and adviser to the faculty of natural sciences. His initiatives led not only to invitations for young German scientists to the United States, but also to connections with European colleagues. The contact François Jacob had with the few German scientists researching the genetics of bacteria and phages had come about through Delbrück's efforts.[47] Delbrück's influence can also be seen in the characteristic absence of research on lysogenic phages and bacteria in the Federal Republic: up to 1965, only one work in this field was published from the Max Planck Institute for Virus Research (Cavallo and Schramm 1954). As mentioned above, Delbrück long refused to believe in the existence of the phenomenon of lysogeny (see Judson 1979, p. 373; Wollman 1972).[48] Moreover, German research caught up with bacteria genetics, which Delbrück initially regarded with skepticism, much later than it did with phage genetics.

According to Bresch, Delbrück and Gunther Stent (who also had close ties to Germany) advanced molecular genetics in Germany by making possible the access of German scientists to the "club" of molecular geneticists:

> In the phage world, Weidel and I were for a long time the only Germans, later Wittmann joined us. This was truly a club; people helped each other, sent strains and so on. If it hadn't been for Delbrück and Stent, it would have hardly been possible for us to get access. Of great importance were also the conferences . . . Royaumont was extraordinarily important for phage research, especially European phage research. Many contacts with American scientists were initiated there. If you were a German, trust would be extended to you if Stent, for example, introduced you . . . Until 1955, at least, the situation had not normalized for Germans.[49]

In explaining the goals of the focal program for genetics, the DFG wrote retrospectively in 1959: "For political reasons genetics in Germany had lost touch with scientific developments abroad. The Research Association tried to remedy this situation by supporting young researchers in this field and funding their work within the framework of the focal program."[50]

Clearly, the absence of necessary exchange and contact with foreign colleagues in this field, which rapidly developed after the war, contributed far more to the failure of German scientists to keep up with advances in molecular genetics than did the flaws in National Socialist science policy. Molecular genetics was born from a synthesis of various disciplines. Apart from genetics, microbiology, and virology, which we have examined, biochemistry and X-ray crystallography played an important role. The decisive advances were made in the forties and fifties in the United States, France, and England. For a long time German scientists did not participate in the international discussions that were indispensable for the development of this research. Foreign scientists—many leading geneticists came from countries attacked by Germany or were Jews—could not forget within a few years that German colleagues had been involved in some way (even if only by their silence) in the Nazi murder of millions of people.

Another consideration follows from this observation. The almost unopposed acceptance of the expulsion of Jewish scientists (and the swift filling of the vacancies) on the part of German colleagues was not only an offense against moral principles. The expulsion and halt or substantial reduction of contacts with Jews, steps that already during the Nazi period made scientific exchange at international congresses difficult, was also an offense against a fundamental scientific principle, that of internationality. One wonders whether this step may not have seduced non-Jewish German scientists into taking a more casual attitude also to other scientific principles. Was the readiness for carelessness (as in the case of Gerhard Schramm, who did not write a letter of clarification to Wendell Stanley) or fraud (as with Franz Möwus) greater than it might otherwise have been? The "experiment" of Nazism was much too short (twelve years) to answer this question; nor is it possible to make any predictions about how scientific life would have been influenced in the long run. But it was not too short to have an effect on the science of the postwar period and to create a situation in which it took German science many years to catch up to international science.

Conclusion

"Butenandt spoke at some length about the freedom he had been given as a scientist, and the freedom given to the Kaiser Wilhelm Institutes, from World War I through the Nazi time. Only recently, since the occupation [by the allies], has this freedom been interfered with."[1]

"When Hitler was preparing his seizure of power, he figured the German scientists into the equation as a *quantité negligeable,* and unfortunately he was right. I cannot shake the tormenting thought that it would have been possible to prevent much if, at the first moment Hitler attacked freedom and justice, a group of German scientists had protested" (A. Kühn in *Wissenschaft und Freiheit* 1954, p. 269).

Until now the history of biology under National Socialism has never been examined. In this book I have sought to answer the following questions for the academic discipline of biology: what was the impact of the expulsion of Jewish scientists after 1933? What sort of influence did National Socialist ideology and politics come to exert on the freedom and achievement of biological research in Germany? In what way did biologists during this time participate in the crimes of National

318

Socialism? What impact did the twelve years of Nazi rule have on the development of biology after 1945?

My examination reveals a certain "normalcy" during this period, as scientific activity continued under the Hitler dictatorship. Some surprising findings are worth briefly summarizing here: the main impact of National Socialist policy on biology was the expulsion of primarily Jewish scientists. Despite the serious qualitative loss in terms of personnel, no biological discipline disappeared as such. Beyond that, the influence of ideology and politics on biology in Germany was negligible; financial support for basic biological research, some of which had outstanding results, increased after 1933 right up to the end of the war. The Kaiser Wilhelm Institutes profited far more from the research funding than the universities. The accommodating attitude of nearly all remaining scientists and their readiness to cooperate with the state and the party helped them to obtain funding for their work.

After 1945 it became clear that biological research in Germany was lagging behind international research in Western countries. The present study has shown that this lag did not—as has frequently been assumed—primarily result from National Socialist science policy following the expulsion of the Jews. Instead, it is attributable above all to the moral failure of German scientists after 1933 and the later international isolation that ensued.

The following discussion will put some of the most important results of this study against the background of a number of works on related fields of science.

1. Expulsion and Emigration

All existing studies on the forced emigration in the biosciences between 1933 and 1939 have focused on the great importance of Jewish émigrés for modern fields of research, in particular molecular biology and biochemistry. Donald Fleming (1969), for example, examined the role of émigré physicists in the development of molecular biology during the forties and fifties and reached the conclusion that "one of the most remarkable by-products of the European Diaspora of the 1930's was the profound stimulus given by refugee physicists to the revolution in biology" (p. 152). David Nachmansohn (1979) pointed to the scientific accomplishments of eminent Jewish organic chemists and biochemists to show "how world biochemistry benefited as a result of the Nazi persecution" (p. 366). In contrast to my own approach, these studies focused only on certain groups of mostly eminent scientific émigrés. The focus on elite groups among the émigrés puts the total phenomenon of emigration in a lopsided perspective, however. Moreover, scholars have in general relied on a figure of émigré scientists that is in all likelihood too large. In physics, too, it does not appear that 20 to 30% of scientists emigrated (a figure given in Beyerchen 1977, for example); according to Klaus Fischer (1988), 15.5% of university teachers of physics had to emigrate.[2]

My study of emigration in biology as a whole (that is, botany, zoology, and genetics) has shown that 13% of scientists active in 1932 were dismissed and 10% subsequently emigrated, with Jews and persons of Jewish descent making up 80% of this group. The émigrés were not more strongly represented in the modern fields of research than their non-Jewish colleagues. Outstanding achievements by émigrés in biology came primarily in experimental embryology and genetics. Most of the émigrés who were able to obtain new research positions abroad—about half of those driven from Germany—hardly changed the focus of their research. In physics it was primarily developments within the field that caused some émigrés as well as American and British physicists to switch into the new field of molecular biology.

Even if there was no difference between biologists who left and those who stayed in terms of their fields of research, the two groups did differ with respect to their average achievement. If we quantify that achievement with the help of the Science Citation Index, we find that it was about three times greater among the émigrés, in spite of the often difficult circumstances after their forced emigration. On the whole we have seen that the forced emigration in biology did not lead to the disappearance of entire fields of research. Despite what were sometimes serious losses, emigration alone is not a sufficient explanation for the lag in biology in Germany after World War II and the virtual absence of research in molecular biology until the sixties. Another serious loss to science in Germany can only be briefly indicated here. It was the forced emigration of children and students who later became biologists. Though their accomplishments, some of them significant, are known in individual cases, no comprehensive analysis of their achievement exists.

2. *Science in Nazi Germany: Ideology and Scientific Reality*

Two important American postwar books on the role of individual sciences and scientists in the crimes committed under Nazi rule were never translated into German. Max Weinreich, in *Hitler's Professors: The Part of Scholarship in Germany's Crimes against the Jewish People* (1946), brought to light the responsibility of anthropologists and jurists in the German crimes against the Jews. Samuel Goudsmit, in *Alsos* (1947), attributed the failure of German science, in particular the uranium project under Werner Heisenberg, to serious flaws in the Nazi organizational structure of science, to mistakes by the participating scientists, and to their complacency. Goudsmit rejected the self-justification by Heisenberg and Carl Friedrich von Weizsäcker that they had renounced the building of an atomic bomb for moral reasons. This explanation by German physicists—convincingly refuted by the British government's release in 1992 of transcripts of secret recordings of conversations among German nuclear physicists interned in Farm Hall after the war—has hardly ever been publicly challenged in Germany (in political camps on the right as well as the left). Mark Walker (1989, German translation 1990) provided the first detailed analysis of the scientific, technical, economic, and military backgrounds of the German atomic bomb project and its failure.

I would like to point out another early publication, namely the account by Mengele's Jewish "slave doctor" Miklós Nyiszli. Originally published in Hungarian,[3] and later translated into many languages, it revealed Mengele's crimes in Auschwitz as well as his scientific relationship with the Kaiser Wilhelm Institute for Anthropology under Verschuer. Not until 1992 did this book appear in a German edition.

A new discussion about the relationship between science and National Socialism in Germany was initiated by Alan Beyerchen (1977) in his pioneering work on physics in Nazi Germany. Since that time numerous publications have treated individual sciences and their practical application during National Socialism. Most of these more recent studies have examined the fields of physics and medicine; in the latter field the overwhelming majority have looked at the causes and consequences of racial hygiene.

This is one reason why I decided not to discuss in detail racial hygiene and its causes in the present study. In addition, although meas-

ures of racial hygiene were also propagated by biologists, research that
was clearly directed at racial hygiene was not carried out by biologists,
with one exception described in the following section. The research
and practical application of racial hygiene remained a matter for physicians and jurists.

It must be noted, though, that biologists—botanists, zoologists, and
geneticists—played an important role in initiating and propagating
eugenic and racial-hygienic concepts within the racial-hygienic movement (which had already began in the last century). Paul Weindling,
among others, pointed this out in his detailed study (1989), in which
he called attention to the growing scientification and professionalization in the racial-hygienic movement and practice in Germany after
1870. However, his thesis that great efforts were made to create a
National Socialist "action-oriented new biology" (p. 506), either to
furnish scientific evidence for the racial ideology or to legitimate the
racial-hygienic policy of removing "useless elements" from human society, has not been confirmed by my findings. Even if there were a
number of biologists who supported the racial ideology, we must note
that at no time was there a National Socialist biology with a uniform
ideological objective. The work of most biologists remained committed to scientific criteria; in those cases where it was aimed at confirm-
ing the racial doctrine, for example in the work of Gerhard Heberer,
no external pressure can be documented. Weindling's approach of
seeing every professional biological activity at the time as supportive
of the racial ideology bears the danger of leveling the significant political differences of opinion among scientists and the differing consequences of biomedical research and its application. It lets those off
the hook whose research under Hitler was indeed murderous or profited from the murder of others.

The murderous consequences of human genetics research at the
time, mainly in anthropology and psychiatry, were spelled out clearly
for the first time by Benno Müller-Hill (1984, English translation
1988). "Racial evaluations," drawn up by scientists of the Kaiser Wilhelm Institute for Anthropology, for example, could, if they confirmed
the Jewish descent of a person, lead to his or her death. The director
of this institute, Otmar von Verschuer, did not shrink from cooper-
ating with Josef Mengele, who sent him from Auschwitz "material"
from murdered people for research purposes. My study has shown
that the only department head of this institute who was until now

considered above reproach politically, the zoologist and later human geneticist Hans Nachtsheim, carried out low-pressure experiments on epileptic children obtained from a euthanasia center. With these experiments he tried to clarify the question—important for racial hygiene—of how hereditary epilepsy might be diagnosed. Nachtsheim, whose work on hereditary pathology and human genetics is internationally recognized, was also not pressured or forced to work in the field of racial hygiene and to carry out experiments on human beings who had been deprived of their rights.

In contrast to Weindling, Robert Proctor, in his detailed book *Racial Hygiene* (1988), has concentrated on the biomedical profession during the Nazi period. He emphasizes the racial-political initiative of medical researchers and physicians, whom he describes as helping to originate race policy. Against the notion (widely held at least in Germany) that the roots of National Socialist race policy are to be found exclusively in irrational racist ideological currents of the nineteenth century, Proctor shows that German biomedical scientists, mainly physicians, "played an active, even leading role in the initiation, administration, and execution of Nazi racial programs" (p. 6).

I, too, was able to ascertain an active role of scientists themselves in regard to Nazi race policy in the case of prominent (and nonprominent) biologists. The theoretical biologist Ludwig von Bertalanffy for one wanted to make a contribution to the "organismic age" with his "organismic biology," since the "hope that the atomizing conception of state and society will be followed by a biological one that recognizes the holistic nature of life and of the Volk" had now been fulfilled (Bertalanffy 1941). The botanist Ernst Lehmann, a committed anti-Semite, tried (unsuccessfully) to found an institute for "German biology." The zoologist and later Nobel prize winner Konrad Lorenz joined the staff of the NSDAP's Office of Race Policy and called for the "elimination" of supposedly "genetically inferior people." In addition, he participated in the "racial" selection of German-Polish offspring in Posen. Gerhard Heberer, a zoologist and anthropologist of international renown, attempted to establish the superiority of the "Nordic" race on the basis of population genetics and anthropology. Wilhelm Rudorf and his collaborators at the Kaiser Wilhelm Institute for Breeding Research worked on breeding specially adapted plants for use in the "Germanification of the East." They carried out one project in cooperation with the authorities running the Auschwitz concentration camp, who made inmates available as a labor force. Nevertheless, the influence of ideology on basic research itself, its methods and content, remained, as already said, surprisingly insignificant.

To illustrate this last point, I shall summarize the results of my study with respect to how biological research was funded. An analysis of the files of the Notgemeinschaft/German Research Association (DFG) and the Reich Research Council (RFR) has shown that their funding for basic biological research increased steadily from 1933 to the end of the war. Membership in the NSDAP was no prerequisite for receiving financial support; statistically seen it did not even amount to an advantage. Over time the funding shifted to an ever increasing degree from the university institutes to the Kaiser Wilhelm Institutes. During the war, many university institutes got into great financial difficulties as well as personnel difficulties as younger scientists were drafted into military service. By contrast, the work at the Kaiser Wilhelm Institutes was not impeded by a lack of funds, and large numbers of their scientists were exempted from military service.

The greatest financial support from the DFG and the RFR went to research in genetics and radiobiology. Traditional genetic research, for example the genetics of experimental embryology, constituted a primary area of research also during the period of National Socialism. However, far greater support was given to what was at that time modern experimental mutation research. In addition, beginning in 1937, pioneering molecular research on the tobacco mosaic virus was carried out with growing support from the DFG, the Kaiser Wilhelm Society, and industry. The findings thus indicate that there was by and large continuity in biological research after 1933; at the KWIs it even increased in intensity.

Kristie Macrakis (1993) has also emphasized the continuity of basic biological research at the Kaiser Wilhelm Institutes. She explains this finding of her nonquantitative study of scientific research at KWIs under National Socialism by emphasizing the semiprivate nature of the Kaiser Wilhelm Society. By means of accommodation and passive opposition, she argues, the society succeeded in preserving a good deal of its autonomy in the Third Reich, and many first-rate scientists remained in Germany in order to save German science. Macrakis adopts almost verbatim and uncritically the official version of the Kaiser Wilhelm Society and the Max Planck Society after 1945, in which a veil has been thrown over the active role of leading scientists of the society in the dismissal of Jewish colleagues and the implementation of the Nazi policy of race, conquest, and *Lebensraum*. Although Macrakis addresses the continuing cultural imperialism of the society under National Socialism and the close relationship of some of its institute directors to state and party authorities (as mentioned below), she does not see the connection between these elements and the good opportunities for scientific research at these institutes. In her view some of the institutes became "islands of freedom in a sea of propaganda and used camouflage techniques to continue work undisturbed."

I disagree with this interpretation. The results of my study show, first, that good basic research was done not only at Kaiser Wilhelm Institutes but also at universities. Second, it bears out that the KWIs, in contrast to the universities or even at their expense, benefited greatly from National Socialism, at least in financial terms. There is considerable documentary evidence that leading Nazi politicians took an interest in appointing eminent scientists as directors of institutes and supporting their research financially. For example, it was Reich Education Minister Rust, and not the Kaiser Wilhelm Society, who in 1934 urged Fritz von Wettstein, known as the leading German botanist and as a non–party member, to head the KWI for Biology. At this point I must emphasize once again that scientists, including those who were not members of the party, helped to get funding for their work through their modified behavior and direct cooperation with the state and the party. This cooperation by influential biologists extended from the above-mentioned legitimization of racial-hygienic measures to the acceptance of and support for the expansionist plans of National Socialism and the goal of a "new European order" under German, that is, National Socialist, hegemony. Following the successful war of conquest, the Kaiser Wilhelm Society would take over scientific leadership in this "new European order."

Personalities as different as Fritz von Wettstein and Wilhelm Rudorf, who openly professed his support for National Socialism and belonged to the party, tried to take advantage for science of the offensive war and the territorial conquests in the Soviet Union: with backing from state authorities, the most important Soviet biological institutes were to come under the control of the Kaiser Wilhelm Society, a plan whose realization was thwarted only by the outcome of the war. Some university biologists and the director of the KWI for Breeding Research, Rudorf, were ready to pursue scientific cooperation with the authorities running the Auschwitz concentration camp. This sort of behavior on the part of scientists amounted to encouragement of Nazi policy, which in these particular cases was rewarded with financial support and freedom for research. Such an assessment does not exclude the possibility that in individual situations other motivations may have played a role, and that scientists could hold contrary views in regard to the racial ideology of National Socialism. But these views do not change the fact of support for Nazi policy, no matter whether it was prompted by opportunistic motives or agreement with some or all of the goals of the regime. Those goals included in particular also nationalism and anti-Communism.

Mark Walker has characterized the relationship between physics and National Socialism as one of compromise and collaboration (1989, p. 100). Much the same was true in biology. The attitude of leading biologists—neutral at best, in many cases approving and actively supportive—made possible by and large a continuity in biological research after 1933 (leaving aside, of course, the dismissals). That continuity went hand in hand with the freedom granted most scientists to pursue their own research.

3. *Continuity and Freedom of Research under National Socialism and in the Soviet Union under Stalin*

Many National Socialist politicians, philosophers, and scientists attacked the ideal of the freedom of scientific research. In 1941 Joseph Needham observed, the Nazi writers "are all at one in stating that 'science is a product of blood' (Rosenberg), and that 'We do not know science, but only that science which is valid for us Nazis,' or 'It becomes more and more urgent to lay the foundations of a race-bound [*artgemäß*] scientific knowledge' (Schulze-Soelde)" (1941, p. 18).

Despite these antiscientific currents, which were strong especially at the outset of Hitler's rule and were, for some time, able to exert a not inconsiderable influence in physics in the form of "Aryan physics," ideologues and politicians hostile to science did not impose ideology on biological research and teaching to an appreciable extent for any length of time or shut it down. In physics, too, the influence of Philipp Lenard and Johannes Stark was limited and came to an end in 1941, at the latest.

In this regard the science policy of Nazi Germany differed from that of the Soviet Union under Stalin. A brief comparison reveals a fundamental difference between the two dictatorships both in the nature of their rule and in their policy on scientific research, at least with respect to biology and agronomy. Stalin used terror and mass murder to implement political but never racial goals (Bullock 1991, p. 1245). His terror could fall on anybody; with the exception of the kulaks it was not limited to a specific group. Moreover, beginning in the 1930s, science, in particular biology and agronomy, was increasingly politicized as a result of the influence of the agronomist and scientific charlatan Trofim Lysenko, and the autonomy of science as propagated by Lenin and Trotsky was completely abolished in 1948.

David Joravsky, among others, has analyzed this development in his book *The Lysenko Affair* (1970). He shows that Lysenko's school did not grow out of a moribund scientific tradition but was a rebellion against science as such (p. vii). The influence of this fraudulent school lasted decades because its practitioners promised quick successes in agriculture. In view of the utterly backward state of Soviet agriculture, Stalin in 1929 had introduced the primacy of practical application—"practical success in agriculture is the ultimate criterion of truth in biological science"—as the binding scientific criterion (pp. 63, 95). While this approach created infinite opportunities in agricultural policy for charlatans, the biological sciences in the Soviet Union were increasingly suppressed.

Lysenko won Stalin's confidence through initial practical successes in the vernalization of grain plants.[4] With Stalin's help he was able to climb to the top of Soviet agriculture and to wield power over the entire field of biology. Lysenko claimed, for example, that he was able to "reeducate" winter wheat into summer wheat in one generation, and he fought against the principles of Mendelian and Morganian genetics as well as the principle of natural selection as outgrowths of Western bourgeois science. With disregard for all scientific principles, Lysenko promised great successes in a short time; even after repeated setbacks, the hope that his promises would come true lead the Communist leadership to stick with Lysenko: "genuine science proved incompatible with the hope of getting a lot for a little in a very short time" (p. 149).

In 1935 Soviet geneticists were subordinated to the charlatans. For another thirteen years they were able to continue working with steadily dwindling support, and then in 1948 their right of existence was taken from them. Beginning in 1936, Lysenko's crusade was supported by state terror. Thus as early as the end of the thirties, seventy-seven geneticists and other agricultural specialists were arrested, shot or subjected to other forms of repression; in keeping with the arbitrary nature of Stalin's rule, six Lysenkoists also suffered from coercive measures. Among the victims of the thirties and forties were the developmental biologists Nikolai K. Koltsov and I. I. Schmalhausen as well as the geneticists Nikolai P. Dubinin, Joseph Rapoport, and Nikolai V. Timoféeff-Ressovsky. The human geneticist Solomon Levit was shot (Zacharov and Surikov 1991). The well-known plant geneticist Nikolai I. Vavilov died in prison under obscure circumstances. Most of the geneticists who were still active in 1948 were forced to take on different work (Joravsky 1970, pp. 116–117). The protocols of the conference of the Lenin Academy in August of 1948 make clear that from that point on science and the freedom of research had ceased to exist.[5] Lysenko's power remained intact even after Stalin's death in 1953, only from about 1964 on did real scientists regain control over biological and agricultural research (Joravsky 1970, p. 63).

By contrast, in Nazi Germany, charlatans—like Julius Streicher, who claimed that a single act of intercourse between a Jewish man and an "Aryan woman" was enough to poison her blood with his "race-foreign protein" so that she could never again bear "pure-blooded Aryan" children (see Müller-Hill 1984, p. 81)—did not at any time acquire substantial influence over the biological sciences. The majority of non-Jewish biologists were able to work in relative safety. The National Socialists based their race policy and later their murderous measures on racial-volkish ideologies, in particular the racial doctrines of the teacher and anthropological autodidact Hans F. K. Günther. They also invoked—without opposition and with support from renowned anthropologists, medical researchers, and biologists— genetics and evolutionary biology. These fields supplied the framework of modern, genetically grounded anti-Semitism (Müller-Hill 1984). This anti-Semitism and the evaluation and classification of peoples and individuals with racial criteria formed the foundation of National Socialist ideology.

Michael Burleigh and Wolfgang Wippermann (1991) have also emphasized that the main object of National Socialist social policy "remained the creation of a hierarchical racial new order . . . The Third

Reich was intended to be a racial rather than a class society" (p. 306). In keeping with this goal, the terror was directed primarily against Jews and other "elements" that were "inferior" in terms of "race" and "race hygiene." Thus the great majority of those dismissed in the field of biology in the thirties were Jews or people of Jewish descent. They were dismissed regardless of their political conviction and the fields of research they represented. The styles of research described by Jonathan Harwood (1993) also had no influence on the dismissals. Harwood divides German geneticists into those who pursued a "pragmatic" and those who pursued a "comprehensive" style of thought. The later dismissals are evenly distributed among both groups. Among the few non-Jews dismissed for political reasons were some who were able to continue their work elsewhere in Germany, even if—as in the case of Elisabeth Schiemann—they were deprived of their teaching credentials.

Among the biologists who remained in Germany after this wave of dismissals and subsequent emigration, I know of only one, Walther Arndt, who later became a victim of the regime. A zoologist, Arndt was denounced in 1944 because he had expressed doubts about the final German victory in the presence of witnesses. He was sentenced to death by the People's Court (Volksgerichtshof) and executed.

The old elites of Germany helped in the consolidation of National Socialism and put up with the persecution of the Jews because, as Arno J. Mayer (1988) emphasizes, they mixed anti-Semitism and anti-Communism. According to Mayer (p. 16), "under the banner of anti-Bolshevism" these elites also supported "the military offensive to conquer a boundless 'Lebensraum' in Eastern Europe and closed their eyes to the barbarian abuse and slaughter of the Jews" and large parts of the peoples of Eastern Europe, who were regarded as "sub-humans."

At this point I want to make very clear that I am not trying to trivialize the crimes of Nazi Germany through this comparison with the Soviet Union or question their uniqueness. This comparison has nothing to do with the attempt by German historians—for example, Ernst Nolte—to relativize the singularity of Nazi crimes by invoking Bolshevist practices or even to declare Auschwitz the reaction to earlier horror models in the Soviet Union. In this controversy about the historical assessment of the recent German past, which began in July of 1986 and is known as the *Historikerstreit*, Nolte (1986) even went so far as to ask: "Was the Bolshevik murder of an entire class [the bourgeoisie and the peasantry] not the logical and factual antecedent of the 'racial murder' of National Socialism?"

This same attitude of relativizing and even justifying the crimes of National Socialism with the help of Stalinism is also clearly revealed in Nolte's biography of Martin Heidegger (1992). Here Nolte emphasizes (p. 195) that until 1941 Hitler's concept of destruction was "far less comprehensive than the Leninist one." According to Nolte, Hitler at first wanted only to remove the Jews from German life (without systematically killing them), whereas Socialism was in its essence a "doctrine of destruction" in the wake of which the bourgeoisie as a class would be abolished (p. 149). For that reason Nolte considers the decision by Heidegger and other Germans to support Hitler in 1933 to have been correct: "All those who at that time strove for a 'German Socialism' [National Socialism] should be considered rehabilitated, even if one considers this solution, too, to have been a failure"; "To the extent that he opposed the 'attempt at a great solution' [the Communist attempt], Heidegger—like many others—had historical right on his side" (pp. 151, and 296).[6] A discussion of the primarily political implications of what was on the surface only a scholarly dispute in the *Historikerstreit*, and which includes basic questions about the self-conception of the Federal Republic, can be found in Wehler 1988, for example.

Alan Bullock (1991, pp. 1250–1257) has shown that it is entirely possible to compare the terror systems of Hitler and Stalin without questioning the singularity of the Nazi (or the Stalinist) regime or relativizing it through references to historical precedents. Thus there was under Stalin no equivalent to the Holocaust, the central element of Nazi terror, the systematic annihilation of European Jewry. The statement that there was continuity in the field of biological research also has nothing to do with the normalization, or even trivialization, of the picture of the Third Reich that was called for by some in the *Historikerstreit*. It merely says that the effects of Stalinism and National Socialism on biological research were very different. And here we should note as an additional point that the suppression of biological sciences in the Soviet Union lasted thirty-five years, according to Joravsky (1970, p. 62), while the "experiment" of National Socialism was over after twelve years.

To return to the starting point of this comparison: if geneticists and other biologists had protested publicly against the racial-hygienic propaganda and practice, which was often enough carried out also in the name of genetics, it might have meant the end of their lives, their careers, or at least their research funds. The scientists chose not to use the opportunity to publicly criticize the application and abuse of their work, and they then enjoyed substantial freedom of research and financial support. Freedom of research means that the content and method of research can be freely chosen, that the decision about the necessary material support depends primarily on professional prerequisites, and that the results can be published in scientific journals— though here with the caveat that joint publications together with Jews were no longer possible. As far as human geneticists and medical scientists at the time are concerned, the option of carrying out experiments with people in psychiatric institutions and concentration camps, people who had been deprived of their rights, gave them a "freedom" they had not had before and would not have after National Socialism. The fact that some of them made use of this option reveals the abyss of a science without a humane orientation, an orientation that cannot come from science itself.

My results have shown that freedom of research does not depend on peace and democracy, as American intellectuals assumed in their 1937 "Statement Condemning Nazi Attack on Intellectual Freedom."[7] Needham (1941), too, gives the impression that scientific progress is possible only in a democracy. However, if we include the period after 1945, we must ask if Needham was not right, after all, in a different sense. This is how he described the increasing attack on the basic principles of science in Nazi Germany (p. 20): "Science is thus, and always will be, international, rational, impartial, autonomous, independent, and truly totalitarian. Against this fundamental truth the nazis foam with all the weapons of insane nationalism." All scientists in Germany began to violate this principle of scientific internationalism when they accepted the expulsion of their Jewish colleagues without any major protests; the violation continued when they submitted to the prohibition of any further contacts and collaboration with Jews living in other countries. This was the reason why Adolf Butenandt, for example, canceled a planned collaborative project with Rudolf Schönheimer, who had been forced to emigrate to the United States in 1933 (Karlson 1990, p. 85).

National Socialism did not last long enough for us to be able to say what long-term effects this violation of the principle of internationalism might have had on science. Predictions about the long-term consequences of the National Socialist scientific system are also not possible. The period was not too short, however, to have a lasting effect on the science of the postwar period. In contrast to 1933, we can see here a clear break, at least in the field of genetics. The new directions of research being developed above all in the United States were taken up in Germany only after a long delay. And while German scientists were not officially isolated on the international level, as had been the case after World War I, scientific exchange with colleagues of countries formerly occupied by the Germans and with Jewish colleagues was very difficult or nonexistent for a long time. In many cases the dialogue was broken off and never reestablished. These colleagues could not forget in a few years the active contribution of Germany scientists to, or their silence about, the National Socialist policy of expansion and the murder of millions. By contrast, many German scientists were eager to "forget" this contribution. They saw themselves as victims—victims of a failed policy, a lost war, an unjustified de-Nazification, and an unfair allocation of blame.[8] There was never any attempt made to abolish the Law on the Restoration of the Professional Civil Service (which had decreed the dismissals of 1933) as an unjust law, and to recall all the living scientists who had been affected by it to their old or comparable new positions or compensate them appropriately.

"When men like James Franck, Albert Einstein, or Thomas Mann may no longer continue their work, whether the reason is race, creed, or belief, all mankind suffers the loss"—this was the view of American intellectuals as far back as 1937, before the murder of disenfranchised people had begun. The physicist Lise Meitner, driven from Germany in 1939, made a particularly powerful statement on this immeasurable loss to mankind and the responsibility of scientists in a letter to her friend and former colleague Otto Hahn. The final words shall therefore be hers.

Epilogue

On June 27, 1945, Lise Meitner wrote the following letter to Otto Hahn in Berlin. Later she noted by hand on a copy of the first page (in the Archives Centre of Churchill College in the University of Cambridge):[1] "A letter that never reached him."

"Dear Otto,
"Your last letter is dated 3/25, you can imagine how I yearn to receive news from you. I have been following the events very closely in the English reporting on the war and think I can assume that the area in which you and Laue are has been occupied without fighting. Therefore I hope with all my heart that you personally did not have to suffer. Of course things will be very difficult for you now, but that was inevitable. However, I am very concerned about the Plancks, since there was fierce fighting in their area. Do you know anything about them and the Berlin friends? An American will take this letter along, he is going to pick it up shortly so I am writing in great haste, though I have so much to say that is close to my heart. Please keep that in mind and read it with confidence in my unshakable friendship.

"During these months I have written many letters to you and Laue in my mind, because it was clear to me that even people like you and Laue did not grasp the real situation. One of the things that made me realize this so clearly was when Laue wrote to me on the occasion of Wettstein's death that his passing was also a loss in the wider sense, since Wettstein, with his diplomatic talents, could have been very useful at the end of the war. How could a man who had never opposed the crimes of the last years be of use to Germany? This is, indeed, the misfortune of Germany, that all of you lost the measure of justice and fairness. You yourself related to me in March of 1938 that Hörlein told you that terrible things were being done to the Jews. He therefore knew of all the crimes that were planned and later carried out, and in spite of this he was a member of the party, and in spite of it you regarded him as a decent man and let yourself be influenced by him in your conduct toward your best friend.

"All of you also worked for Nazi Germany and never tried to engage even in passive resistance. To be sure, to soothe your conscience you helped someone in trouble now and then, but you let millions of innocent humans be murdered and no protest was raised.

"I must write this to you, for so much depends for you and Germany on your recognition of what you allowed to happen. Here in neutral Sweden people were already discussing long before the end of the war what should be done with the German scholars and scientists when the war is over. And what might the English and Americans think about this? I and many with me think that one way for you would be to make a public declaration that you are aware that you bear responsibility for what happened because of your passivity, and that you feel the need to participate in the effort to make good what was done, to the extent it is possible to do so. But many others think that it is too late for this. They say that you betrayed first your friends, then your men and children by letting them risk their lives in a criminal war, and finally you betrayed Germany herself because when the war was already quite hopeless you did not even resist the senseless destruction of Germany.

"This sounds pitiless, and yet you must believe me that it is truest friendship which prompts me to write this to you.—You cannot really expect that the rest of the world will feel sorry for Germany. What one hears these days about the incomprehensible horrors in the concentration camps surpasses everything one feared. After hearing on the English radio a very factual report by the English and the Americans about Belsen and Buchenwald, I started to weep loudly and was unable to sleep the entire night. And if you could have seen the people who came here from the camps. One should force a man like Heisenberg and many millions along with him to take at look at these camps and these tortured people. Heisenberg's conduct in Denmark in 1941 cannot be forgotten.

"You may remember what I often told you when I was still in Germany (and today I know that it was not only stupid but a great injustice that I did not leave immediately): as long as only we are having sleepless nights and not you, the situation in Germany won't get better. But you never did have sleepless nights. You didn't want to see, it was too inconvenient. I could prove it to you with many examples big and small. I ask you to believe that everything I am writing here is an attempt to help you.

"With very kind regards to all.
"Your Lise."

Appendix A
Career Information

This appendix primarily lists German biologists. Also included are some biologists from other European countries, as well as some medical, physical, and chemical scientists and politicians.

Academic degrees and titles used in this name list:

Dr. med.: Doctorate in medicine, a prerequisite for a career in medicine but not necessary for the practice of medicine.

Dr. phil.: Doctorate in the Philosophical Faculty, including the natural sciences. Only in the 1930s did some universities established faculties of the natural sciences, the doctorate being Dr. rer. nat. or Dr. phil. nat. or Dr. sc. nat.

Some doctorates given as h.c.: Honorary doctorate *(honoris causa).*

Qualifying for a university teaching appointment required a second dissertation, the *Habilitationsschrift,* and oral examination. Those who succeeded became *Dozent* or *Privatdozent* (if they taught only on an unpaid basis), here referred to as *Dozent.*

A university professor who was head of an institute was usually *ordentlicher Professor* (o. Prof.); other titles were those of *außerordentlicher Professor* or *außerplanmäßiger Professor,* here referred to as ao. Prof.

Aach, Hans Günther, Dr. rer. nat. Prof., born 1919, Dozent Cologne 61, ao. Prof. Aachen 62, o. Prof. 65

Abel, Othenio, Dr. phil. Prof., 1875–1946, o. Prof. Vienna 17–34, Göttingen 35–40

Agol, Israel I., 1891–1937, executed, Soviet geneticist and medical scientist

Ahrens, Willi, born 1906, Dozent Jena 35, Erlangen 38

Alverdes, Friedrich, Dr. phil. Prof., 1906–1952, o. Prof. Marburg 28

André, Hans, Dr. phil. Prof., 1891–1966, o. Prof. State Academy Braunsberg/East Prussia 29, retired 45

Ankel, Wulf-Emmo, Dr. phil. Prof., 1897–1983, Dozent Gießen 30, ao. Prof. 37, Technical College Darmstadt 39, o. Prof. 52

Antonius, Otto, Dr. Prof., 1885–1945, Dozent of zoology at the College for Soil Culture Vienna 21, ao. Prof. 31, Director of the Schönbrunn Zoo 24–45

Apstein, Carl H., Prof., 1862–1950, Dozent of zoology Kiel 98–11, scientific civil servant at the Academy of Sciences in Berlin

Arens, Karl, Dr. phil., born 1902, Dozent Cologne 35, emigration to Brazil 36, Prof. Rio de Janeiro 38

Arndt, Walther, Dr. med. and phil. Prof., 1891–1944, Kustos at the Zoological Museum Berlin 25, executed 44 at the penitentiary of Brandenburg after being denounced for publicly doubting the German final victory

Astel, Karl, Dr. med. Prof., 1898–1945, suicide, SS Colonel, president of the State Office for Race Policy of Thuringia, o. Prof. Jena 34, Rector Univ. Jena 39

Auerbach, Charlotte, Ph.D. Prof., born 1899, KWI for Biology 31–33, emigration 33, Univ. Edinburgh 33, Ph.D., lecturer and prof., discovered chemical mutagenesis 41

Auerbach, Max, Dr. phil. Prof., 1879–1968, Dozent and ao. Prof. Technical College Karlsruhe till 35, then at State Collections for Natural Science in Karlsruhe

Autrum, Hansjochem, Dr. phil. Prof., born 1907, Dozent Berlin 39, ao. Prof. Göttingen 48, o. Prof. Würzburg 52, Munich 58, Vice-President of the DFG 61

Backe, Herbert, 1896–1947, studied agriculture, NSDAP member 23, Secretary of State in the Reich Ministry for Nutrition and Agriculture, head of this ministry 42, Minister and chief of the Farmers' League 43, Reich Nutrition Minister 44, Vice-President of the KWG 41–45, suicide in the Nuremberg prison

Barthelmeß, Alfred, Dr. phil., 1910–1987, Dozent Munich 41, ao. Prof. 57

Appendix A: Career Information

Bateson, William, 1861–1926, Prof. Univ. Cambridge 08, Director John Innes Horticultural Inst. (Merton) 10–26

Bauch, Robert, Dr. phil. Prof., 1897–1957, ao. Prof. Rostock 30, Prof. Greifswald 47

Bauer, Hans, Dr. phil. Prof., born 1904, Asst. KWI for Biology Berlin 32, head of dept. 42, MPI for Marine Biology Wilhelmshaven, later Tübingen 49, Prof. 50, Vice-Director 55

Baur, Erwin, Dr. phil. and med. Prof., 1875–1933, ao. Prof. Berlin 04, o. Prof. 11, Agricultural College Berlin, Dir. KWI for Breeding Research Müncheberg 27

Becker, Erich, Dr. phil., colleague of Alfred Kühn, KWI for Biology Berlin, died 1941

Becker, Karl E., Dr.-Ing. Prof., 1879–1940, suicide, General-Lieutenant in the Reich War Ministry, o. Prof. Technical College Berlin, president of the RFR 37–40, member of the senate of the KWG 33–40

Beger, Bruno, Dr. phil., born 1911, anthropologist, SS-Captain

Beleites (Sell-Bel.), Ilse, Dr. phil. 1911–1944, Genetics Dept. KWI for Brain Research Berlin-Buch

Berenblum, Isaac, M. D., born 1903, pathology and experimental biology, Leeds Univ. 29–36, Oxford Univ. 36–48, Nat. Cancer Inst. (U.S.) 48–50, Weizmann Inst. of Science (Israel) 50

Bergdolt, Ernst, Dr. Prof., 1902–1948, Dozent Munich 35, ao. Prof. 40–45

Bergold, Gernot, born 1911, head of Dept. of Entomology of the KWG in Oppau and since 40 member of the society's Research Division for virology (Canada) 48

Bertalanffy, Ludwig von, Dr. phil. Prof., 1901–1972, Dozent Vienna 34, ao. Prof. Vienna 41, Prof. Univ. Ottawa 49, Prof. of Theoretical Biology Univ. Alberta 61

Beurlen, Karl, Dr. phil. Prof., 1901–1985, Dozent Königsberg 27, o. Prof. Kiel 34, Munich 41–45 (paleontology, theory of evolution), geologist Rio de Janeiro 50, Prof. Curso d. Geologia Recife (Brazil), Rio de Janeiro 59

Beutler, Ruth, Dr. phil. Prof., 1897–1959, Dozent Munich 30, ao. Prof. 37

Bickenbach, Otto, Dr. med. Prof., carried out deadly experiments with phosgene on inmates of the Natzweiler concentration camp

Biebl, Richard, Dr. phil., 1908–1974, Dozent Vienna 39, ao. Prof. 44

Blome, Kurt, Dr. med. Prof., 1894–?, medical District Leader of the state of Mecklenburg 33, deputy chief of the Reich Physicians League, SA (Group Leader,) acquitted at Nuremberg physicians trial 47, employed by U.S. Army

Boas, Friedrich, Dr. phil. Prof., 1886–1960, o. Prof. Technical College Munich 28

Bodenstein, Dietrich, Dr. phil. Prof., born 1908, Research Asst. KWI for Biology Berlin 28, dismissal and emigration 33, German-Italian Institute for Marine Research Rovigno (Italy) 33, Res. Assoc. Stanford Univ. 34–41, Guggenheim Fel. Columbia Univ. 41–43, Asst. Agricult. Station New Haven 44, insect physiologist Army Chem. Cent, 45–57, Dr. phil. Freiburg 53, embryologist Nat. Heart Inst. 58–60, Prof. Univ. Virginia 60

Boehm, Hermann, Dr. med., 1884–?, Honorary Prof. Univ. Rostock 38, Commissioner of the Reich Physicians League for the racial-hygienic education of German physicians, head of the Institute for Genetics of the Leadership School of German Physicians at Alt-Rehse near Neubrandenburg

Böhme, R. Werner, Dr., born 1903, KWI for Breeding Research Müncheberg, SS First Lieutenant 43, Commissioner of the Reichsführer SS for rubber plant at Auschwitz

Born, Hans Joachim, born 1909, Dr. phil. Prof., Researcher Auer Society and Genetics Dept. KWI for Brain Research Berlin-Buch, scientific internment camp USSR 45–55, ao. Prof. Technical College Munich 57

Borriss, Heinrich, Dr. phil. Prof., 1909–1985, Dozent Bonn 40, Prof. Greifswald 49

Borst, Max, Dr. med. Prof., 1868–1946, o. Prof. Univ. Munich and Dir. of the Institute for Pathology since 10

Bosch, Carl, 1874–1940, Nobel prize 1931, Chairman of the supervisory board of the IG Farben Industry, President of the DFG 37–40

Brandt, Karl, Dr. med. Prof., 1904–1947, General Reich Commissioner for Medicine and Public Health, SS Lieutenant General sentenced in Nuremberg physicians trial 47 and executed

Branscheidt, Paul, Prof., 1893–1942, Dozent Würzburg 30, ao. Prof. 34, o. Prof. 35

Brauner, Leo, Dr. phil. Prof., 1898–1974, ao. Prof. Jena, dismissal and emigration 33, Prof. Istanbul 33, Munich 55

Bredemann, Gustav, Dr. phil. Prof., 1880–1960, o. Prof. Hamburg 27

Breider, Hans, Dr. phil., born 1908, head of dept. KWI for Vine Breeding Müncheberg 36–45, Dozent Münster/Westfalia 47, Prof. Mendoza (Argentina) 48

Bresch, Carsten, Dr. rer. nat. Prof., born 1921, Dozent Göttingen 57, ao. Prof. Cologne 57, o. Prof. Freiburg 63

Bresslau, Ernst, Dr. phil. and med. Prof., 1877–1935, o. Prof. Cologne 25, dismissal and emigration 33, Prof. São Paulo (Brazil) 33

Brieger, Friedrich, Dr. phil., 1900–1985, Dozent Berlin 28, dismissal and emigration 34, Prof. São Paulo (Brazil)

Brock, Friedrich, Dr. phil., 1898–1958, Dozent Kiel 38, ao. Prof. 43

Brücher, Heinz, Dr. rer. nat. Prof., 1915–1991, Dozent Jena 41, KWI for Breeding Research 41, Dir. SS Institute for Plant Genetics Lannach 43, Swedish Seed Company Svalöf 48, Prof. Tucuman (Argentina) 49, Mendoza 54, Pretoria 63, Caracas 65

Buchner, Paul, Dr. phil. Prof., 1886–1978, o. Prof. Breslau 27, Leipzig 34–43, since 43 private researcher in Italy

Buddenbrock-Hettersdorf, Wolfgang, Dr. phil. Prof., 1884–1964, o. Prof. Kiel 22, Halle/Saale 36, Vienna 42–45, Mainz 46

Buder, Johannes, Dr. phil. nat. Prof., 1884–1966, o. Prof. Greifswald 22, Breslau 28–45, Prof. Halle 47

Bukatsch, Franz, Dr. phil., born 1909, Dozent Munich 41, ao. Prof. 57

Bünning, Erwin, Dr. phil. Prof., born 1906, Dozent Jena 31, Königsberg 35, ao. Prof. 38, Strasbourg 42, o. Prof. Cologne 45, Tübingen 46

Burgeff, Hans, Dr. Prof., 1883–1976, o. Prof. Göttingen 23, Würzburg 25

Butenandt, Adolf, Dr. phil. Prof., born 1903, o. Prof. Danzig 33, Dir. KWI for Biochemistry 36, Nobel prize for chemistry 39 (with L. Ruzicka), President of the MPG 60

Butterfaß, Theodor, Dr. rer. nat. Prof., born 1926, o. Prof. Frankfurt 73

Carstens, Peter, Dr. Prof., 1903–1945, o. Prof. Posen and Rector of the Reich University 41–45

Caspari, Ernst W., Dr. phil. Prof., 1909–?, Asst. Göttingen 34–35, dismissal 33, emigration, Asst. Istanbul 35–38, U.S. 38, Prof. Wesleyan Univ. 49–60

Caspersson, Tobjörn, Dr. med. Prof., born 1915, Prof. of medical genetics Stockholm 44

Catsch, Alexander, Dr. med. Prof., 1903–?, Researcher Auer Company and Genetics Dept. KWI for Brain Research Berlin-Buch, scientific internment camp USSR 45–55, Dozent Heidelberg 58, o. Prof. Technical College Karlsruhe 62, radiation biologist

Chetverikov, Sergei S., 1880–1959, Soviet zoologist (entomologist), Dozent Univ. Moscow, founder and head of the Genetics Dept. of the Inst. for Experimental Biology (Dir. Koltsov), head of the Dept. for Genetics Univ. Gorky 35–48

Christiansen-Weniger, Friedrich, Dr. phil. Prof., 1897–?, ao. Prof. Breslau 28, Ankara 28–40, Consultant to the Ministry of Agriculture 33–38, Dir. Inst. for Plant Cultivation and Breeding, Agricultural College Ankara, head of the Agricultural Research Station of the General Government, Pulawy 40–45

Cook, James W. Sc.D. and Ph.D. Prof., 1900–?, chemist Univ. of London 32, Royal Cancer Hospital London 35, Glasgow Univ. 39

Correns, Carl E., Dr. phil. Prof., 1864–1933, one of the rediscoverers of Mendel's laws, Dozent Tübingen 1862, ao. Prof. Leipzig 1902, o. Prof. Münster 09, Dir. KWI for Biology 14

Cramer, Heinrich, Dr. med. Prof., 1890–1960, Dozent Berlin 28, ao. Prof. 30, Medical Dir. Rudolf Virchow Hospital Berlin and Dir. General Inst. against the Canceral Diseases

Czaja, Alphons Theodor, Dr. phil. Prof., born 1894, Dozent Berlin 26, ao. Prof. 35, Dir. Botanical Inst. Technical College Aachen 36

Czurda, Viktor, Prof., 1897–1945, Dozent German Univ. Prague 28, ao. Prof. 34, o. Prof. 40, since 41 called himself V. Denk

Danneel, Rolf, Dr. phil., 1901–1982, Dozent Königsberg/Prussia 36, ao. Prof. 41, head of dept. KWI for Biology Berlin 41, o. Prof. Bonn 49

Dannenberg, Heinz, Dr.-Ing. Prof., 1912–1975, Dozent Tübingen 51, Munich 56, ao. Prof. 58, head of dept. MPI for Biochemistry Munich 49, Deputy Dir. 60

Darré, Richard W., 1895–1953, chief of Farmers' League 33–45, Reich Minister for Nutrition and Agriculture 33–42, ideologist of blood and soil, SS Lieutenant General

Daumann, Erich, Dr. phil., 1905–1978, Dozent German Univ. Prague 36, dismissed 39, 42–45 in concentration camp, 45–60 in Czech Ministry for Education, 60 Prof. at Karls-University in Prague

De Crinis, Max, Dr. med. Prof., 1889–1945, suicide, psychiatrist, o. Prof. Cologne 34, Berlin 38

Deegener, Paul, Dr. phil. Prof., 1875–?, ao. Prof. Berlin 22

Dehn, Max Wilhelm, Dr. phil. Prof., 1878–1952, o. Prof. Technical College Breslau 13, Frankfurt 21, dismissed 35, imprisoned 38, emigrated to Norway 39, U.S. 40, Prof. Black Mountain Coll. 45

Delbrück, Max, Dr. Prof., 1906–1981, Asst. Physics Dept. KWI for Chemistry 32–37, Habilitation 36, but denied the right to teach at a university for political reasons, left for U.S. with a scholarship from the Rockefeller Foundation, *Drosophila* and phage research at Cal. Tech., Instructor of Physics Vanderbilt Univ. 40, phage research in cooperation with Alfred Hershey and Salvador Luria 40–44, Prof. Cal. Tech. 47, Dir. Institute for Genetics Univ.of Cologne 61–63, started research in Phycomyces 65, Nobel prize for medicine 69 (with A. Hershey and S. Luria)

Demoll, Reinhard, Dr. phil. and med. Prof., 1882–1960, o. Prof. Munich 18

Denk, Viktor, Prof., 1900–?, o. Prof. Prag 44–45

Dessauer, Friedrich, Dr. phil. nat. Prof., 1881–?, o. Prof. Frankfurt 22, dismissed 33, emigration, Prof. Istanbul 34, Fribourg (Switzerland) 37, Frankfurt 50

D'Hérelle, Felix H., 1873–1949, from Montreal, microbiologist, Paris 09–21, discovered phage against dysentery bacteria, Dir. Bacteriological Service Egypt 23–27, Prof. Yale Univ. 28–34, since 35 Paris

Diebschlag, Emil, 1911–1942, Dozent Marburg 38

Dingler, Hugo, Dr. phil. Prof., 1881–1954, o. Prof. Technical College Darmstadt 32, pensioned 34, lecturer for philosophy, history, and scientific theory Munich 35 and 39

Dobzhansky, Theodosius, Prof., 1900–1975, geneticist in Leningrad 24, with scholarship from the Rockefeller Foundation at Columbia Univ. New York (lab. of Morgan) 27, Cal. Tech. 29–40, Prof. Columbia Univ. 40–62, Rockefeller Inst. 62–71

Döderlein, Ludwig, Dr. phil. nat. Dr. med. h.c., 1855–1936, ao. Prof. Strasbourg 1891, Honorary Prof. Univ. Munich 1919

Domagk, Gerhard, Dr. med. Prof., 1895–1964, Prof. Greifswald 24, Münster 28, Dir. research lab for experimental pathology and bacteriology Bayer Company, Wuppertal-Elberfeld since 27, Nobel prize for medicine 39

Döpp, Walter, Dr. phil., 1901–1963, Dozent Marburg/Lahn 32, ao. Prof. Technical College Dresden 42–45

Döring, Helmut, Dr. phil., 1909–1940, KWI for Breeding Research, Müncheberg 35–37, research fellow KWI for Biology 37–39, Asst. Jena 39–40

Driesch, Hans, Dr. phil. Prof., 1867–1941, o. Prof. Cologne 20, Prof. emeritus 33

Dubinin, Nikolai Petrovic, born 1907, Prof. of zoology and genetics Moscow 35, after 48 transferred to "practical work" in a forestry station

Duspiva, Franz, Dr. phil. nat. Prof., born 1907, Dozent Heidelberg 40, ao. Prof. Freiburg 55, o. Prof. Heidelberg 59

Eberhardt, Karl, 1913–1944, DFG scholar Genetics Dept. KWI for Brain Research 41–44

Eckardt, Theo, Dr. sc. nat., 1910–1977, Dozent Halle/Saale 40, Prof. Potsdam 51, Freie Univ. Berlin 55, o. Prof. 59

Eckert, Erich, died 1960, 44 in concentration camp, after 45 Institute for Physiology, Czech Academy of Sciences

Egle, Karl, Dr. phil. nat., 1912–1975, Dozent Frankfurt/Main 43, ao. Prof. 50

Ehrhardt, Albert, Dr. phil., 1904–1069, Dozent Rostock 37, Heidelberg 40, Posen 41, lecturer Münster/Westfalia 48, Honorary Prof. 57

Engel, Horst, Dr. phil. Prof., born 1901, Dozent Münster/Westfalia 37, ao. Prof. Technical College Danzig 42, Dozent Hamburg 48, o. Prof. 54

Erhard, Hubert, Dr. phil. Prof., 1883–1959, ao. Prof. Gießen 22, o. Prof. Fribourg (Switzerland) 28, Prof. Munich 37–45

Escherich, Karl, Dr. med. and phil. Prof., 1871–1951, o. Prof. Munich 14, Prof. emeritus 36, Rector Univ. Munich 33–35

Evenari, Michael (formerly Walter Schwarz), Dr. phil. Prof., 1904–1989, Dozent Technical College Darmstadt 31–33, dismissed and emigration to Palestine 33, lecturer Hebr. Univ. Jerusalem 34, senior lecturer 44, Prof. 50–74

Faber, Friedrich-Carl von, Dr. Prof., 1880–1954, o. Prof. Vienna 31, Munich 35, pensioned 48, Prof. emeritus 51

Feldberg, Wilhelm S., Dr. med., 1900–?, Dozent Physiology Berlin 30–33, dismissed 33, emigration to England, Australia, Nat. Inst. for Medical Research (London) 49

Feuerborn, Heinrich-Jacob, Dr. phil. Prof., 1883–1979, ao. Prof. Münster/Westfalia 27, head of dept. Inst. for Zoology Berlin 36–45, dismissed 46

Figdor, Wilhelm, Dr. phil. Prof., 1866–1938, ao. Prof. Vienna 1909

Firbas, Franz, Dr. rer. nat. Prof., 1902–1964, Dozent Frankfurt/Main 31, Göttingen 33, ao. Prof. 37, o. Prof. Agricultural College Hohenheim 39, ao. Prof. Göttingen 46, o. Prof. 58

Fischel, Werner, Dr. phil., 1900–1977, Dozent Leipzig 42, Munich 50, ao. Prof. 53, Prof. Leipzig 54

Fischer, Eugen, Dr. med., sc. h.c., and med. h.c. Prof., 1874–1967, o. Prof. Freiburg 18, Berlin 27, Rector 33–34, Prof. emeritus 42, Dir. KWI for Anthropology 27–42

Fischer, Ilse, Dr. phil., born 1905, Fel. KWI for Biology 34–38, Dozent Münster 42, ao. Prof. 49

Fisher, Ronald Aylmer, Prof., 1890–1962, Galton-Prof. Univ. College London 33, Prof. Cambridge 43

Fitting, Johannes, Dr. phil. Prof., 1877–1970, o. Prof. Bonn 12

Fraenkel, Gottfried Samuel, Dr. phil. Prof., born 1901, Dozent Frankfurt/Main 31–33, dismissal and emigration 33, Res. Assoc. Univ. Coll. London 33–35, lecturer Imperial College London 36–48, emigration to the U.S. 48, Prof. Univ. Illinois 48

Frank, Hans, 1900–1946, executed as criminal of war in the Nuremberg prison, Governor-General of Poland 39–45

Franz, Viktor, Dr. phil. Prof., 1883–1953, ao. Prof. Jena 19, o. Prof. 36

Freisleben, Rudolf, Dr. sc. nat., 1906–1943, Dozent Halle/Saale 36

Freund, Ludwig, Prof., 1878–1953, ao. Prof. Prague 30, dismissed 38, sent to Theresienstadt concentration camp 43, o. Prof. Halle 49

Friederichs, Karl, Dr. phil. Prof., 1878–1969, ao. Prof. Rostock 21, o. Prof. Posen 40, Prof. emeritus Göttingen 58

Friedrich-Freksa, Hans, Dr. rer. nat. Prof., 1906–1973, Fel. KWI for Biochemistry 37, head of dept. KWI for Biochemistry and Dozent Tübingen 1946, Dir. MPI for Virus Research Tübingen 54 and ao. Prof Univ. Tübingen

Frisch, Karl von, Dr. phil. Prof., 1886–1982, studied medicine and zoology in Vienna and Munich, Asst. of R. Hertwig (Munich) 10, Dozent Munich 12, ao. Prof. 19, o. Prof. Rostock 21, Breslau 23, Munich 25, Graz 46, Munich 50, Nobel prize for medicine 73 (with K. Lorenz and N. Tinbergen)

Führer, Wilhelm, Dr., born 1904, regional head of the NSDB Munich, since 40 senior government official in the REM, SS Captain

Gaffron, Hans, born 1902, Asst. KWI for Biochemistry Berlin 25–36, Guest Researcher KWI for Biology 36–37, emigration to the U.S. 37, Univ. of Chicago 39, Prof. 52–60, Florida State Univ. 60–72

Gams, Helmut, Dr. phil. Prof., 1893–1976, Dozent Innsbruck 29, dismissed 38, ao. Prof. 45

Gassner, Gustav, Prof., 1881–1955, o. Prof. Technical College Brunswick 17–33, dismissal and emigration to Turkey 33, return to Germany 39, head of a private plant-breeding firm, Prof. Technical College Brunswick 45

Gegenbaur, Carl, Dr. med. Prof., 1826–1903, o. Prof. of anatomy and zoology Jena 58, Heidelberg 73

Geitler, Lothar, Dr. phil. Prof., born 1899, Dozent Vienna 28, ao. Prof. 37, head of dept. KWI for Cultivated Plant Research near Vienna 43, o. Prof. and Dir. Botanical Garden Vienna 48

Gerlach, Joachim, Dr. med. Prof., born 1908, Dozent Würzburg 50, ao. Prof. 56

Gierer, Alfred, Dr. rer. nat. Prof., born 1929, Dozent Tübingen 58, head of Dept. for Molecular Biology MPI for Virus Research 60, ao. Prof. Tübingen 64

Giersberg, Hermann, Dr. phil. Prof., 1890–1981, ao. Prof. Breslau 28, o. Prof. Frankfurt 37

Gieseler, Wilhelm, Dr. phil. Prof., 1900–1976, ao. Prof. Tübingen 33, o. Prof. 38, Dir. Inst. for Anthropology Tübingen 34–45 and 55–69

Globke, Hans, 1898–1973, jurist in the administration of the Reich Interior Ministry 29–45, coauthor of the official commentary to the Nuremberg Race Laws of 35, chief civil servant in Federal Chancellery 49, Secretary of State under Adenauer 53

Gluecksohn-Waelsch, Salome, Dr. phil. Prof., born 1907, Res. Assoc. Columbia Univ. New York 36–55, Assoc. Prof. Albert Einstein College of Medicine 55, Full Prof. 58

Glum, Friedrich, Dr. Sc. pol. Dr. jur. Prof., 1891–1974, Secretary General of the KWG 20–37, ao. Prof. Berlin 30

Göbel, Karl Ritter von, Dr. Prof., 1855–1932, o. Prof. for Botany Rostock 83, Marburg 87, Munich 91, morphologist and physiologist

Goetsch, Wilhelm, Dr. phil., Dr. rer. nat. Prof., 1887–1960, o. Prof. Santiago (Chile) 29–31, Breslau 34–45, Honorary Prof. Graz 47

Goldschmidt, Richard-Benedikt, Dr. phil. nat. Prof., 1878–1958, ao. Prof. Munich 09–14, o. Prof. Tokyo 24–26, KWI for Biology 14–35, Second Dir. 19, dismissal and emigration 35, Prof. Univ. Calif. Berkeley 36–46

Gottschewski, Georg, Dr. phil. Prof., 1906–1975, Dozent Königsberg/Prussia 40, ao. Prof. Vienna 45, ao. Prof. Gießen 48, head of dept. MPI for Developmental Biology 56–75

Graetz, Erich, Dr., born 1903, Asst. Berlin 29, dismissal and emigration 35, Panama Univ.

Greiser, Arthur, 1897–1946, NSDAP 28, SS 30, SS General 42, President of the Senate of Danzig 34–39, District Leader and Governor of the Reichsgau Wartheland 39, responsible for the mass deportation and extermination of Poles and Jews, distinguished himself by being particularly cruel, sentenced to death by a Polish court and hanged in Posen 46

Greite, Walter, Dr., born 1907, head of the Section for Biology and Medicine of the Notgemeinschaft 35–37, government official in the Genetics Dept. of the Reich Health Office 37–45, SS Captain in the SS Main Office for Race and Settlement

Groß, Fabius, 1906–1950, Asst. KWI for Biology 30, dismissed 33, emigration to England, Marine Biol. Association 33, Univ. of Edinburgh 37

Groß, Walter, Dr. med. Prof., 1904–1945, suicide, chief of the Office for Population Policy and Racial Care (later Race Policy Bureau) of the NSDAP, Dozent Berlin 35, Honorary Prof. of Race Anthropology Berlin, NSDAP 25

Gruber, Friedrich, born 1900, head of dept. KWI for Breeding Research, Müncheberg 33

Grüneberg, Hans, Dr. phil. and med. Prof., 1907–1982, Asst. Inst. for Anatomy Freiburg, dismissal and emigration to England 33, Fel. University College London 33–46, reader in genetics 46–55, Prof. 56–74

Günther, Hans F. K., Dr. Prof., 1891–1968, o. Prof. Jena 30, Berlin 34, Freiburg 39, author of *Rassenkunde des Deutschen Volkes* (Race theory of the German people), published under the pen name L. Winter after 45

Guttenberg, Hermann von, Dr. phil. Prof., 1881–1969, o. Prof. Rostock 23, Prof. emeritus 57

Hackbarth, Joachim, 1906–1977, head of dept. KWI Müncheberg 37, branch East Prussia 41–45, head of MPI for Breeding Research Scharnhorst 46

Haeckel, Ernst, Dr. med. Prof., 1834–1919, Prof. of zoology Jena 62–09

Haecker, Valentin, Dr. phil. Prof., 1864–1927, Asst. of A. Weismann Freiburg 89, ao. Prof. 95, o. Prof. Technical College Stuttgart 1900

Hallervorden, Julius, Dr. med. Prof., 1882–1965, dept. KWI for Brain Research Berlin 38, after 45 MPI for Brain Research Gießen

Hamburger, Viktor, born 1900, Dozent Freiburg/Breisgau 30, Asst. and Dozent Freiburg, dismissal and emigration 33, asst. Prof. Washington Univ. (St. Louis) 35, Assoc. Prof. 38, Prof. 41

Hämmerling, Joachim, Dr. phil. Prof., 1901–1980, Dozent Berlin 31, ao. Prof. 42, head of dept. KWI. for Biology Langenargen 46–49

Hannig, Emil, Dr. phil. Prof., 1872–1955, o. Prof. Münster/Westfalia 22

Harder, Richard, Dr. phil. Prof., 1888–1973, o. Prof. Technical College Stuttgart 23, Göttingen 32

Harms, Jürgen-Wilhelm, Dr. phil. Prof., 1885–1956, o. Prof. Tübingen 25, Jena 35–49

Harnisch, Otto, Dr. phil. Prof., born 1901, Dozent Cologne 27, ao. Prof. 33, ao. Prof. Kiel 36, pensioned 42

Hartmann, Max, Dr. phil., 1876–1962, Dozent zoology Univ. Gießen 03, researcher and later head of Dept. of Protozoology at the Inst. for Infectious Diseases Berlin (under Robert Koch), Dozent Berlin 06, ao. Prof. 09, Honorary Prof. of zoology and general biology Berlin 21, head of dept. and later Dir. KWI for Biology Berlin 14 (48 MPI, since 52 Tübingen), Honorary Prof. Univ. Berlin 34, Tübingen 47

Hartwig, Hermann, Dr. phil. Prof., born 1910, Dozent Cologne 51, ao. Prof. 71

Hayes, William, Sc.D., Dir. Medical Research Council's Microbial Genetics Research Unit, Hammersmith Hosp. London and senior lecturer Medical School of London

Heberer, Gerhard, Dr. sc. nat. Prof., 1901–1973, Dozent Tübingen 32, ao. Prof. Jena 39–45, lecturer in Göttingen since 49, Guest Prof.

Freie Univ. Berlin 61–62, head Anthropological Research Section Univ. Göttingen

Heidermanns, Curt, Dr. phil. Prof., 1894–1972, Dozent Bonn 28, ao. Prof. 35, o. Prof. Greifswald 38, Cologne 55

Heidt, Karl, Dr., born 1908, Dozent Gießen 40

Heilbronn, Alfred, Dr. phil. Prof., 1885–1961, ao. Prof. Münster 19, dismissal and emigration to Turkey 33, o. Prof. Istanbul 33, o. Prof. Münster 56

Heinroth, Katharina, born Berger, Dr. phil., 1897–?, scientific Dir. Zoological Garden Berlin 45–56, lecturer Technical University Berlin 53

Heinroth, Oscar A., Dr. med., 1871–1945, Managerial Asst. of the Zoological Garden Berlin 1904, founder and head of the Aquarium 11–44

Heinze, Hans B., Dr. med. Prof., 1895–1983, Dir. Psychiatric State Institution Goerden (Brandenburg), expert consultant on euthanasia, head of the Research Section Goerden, ao. Prof. 42, prison sentence of seven years USSR, Dir. of Clinic for Juvenile Psychiatry at the state hospital Wunstdorf (FRG) 53–60

Heitz, Emil, Dr. phil. Prof., 1892–1965, ao. Prof. Hamburg 31, dismissal and emigration to Switzerland 37, ao. Prof. Basel 37, Guest Prof. Tübingen 52–54, Honorary Prof. Tübingen 55

Henke, Karl, Dr. phil. Prof., 1895–1956, Dozent Göttingen 29, o. Prof. 37

Henseler, Heinz, Dr. phil. Prof., 1885–1968, o. Prof. Technical College Munich 20, Prof. emeritus 48

Hentschel, Ernst, Dr. phil. Prof., 1876–1945, Dozent Hamburg 19, ao. Prof. 32

Hermann, Siegwart, Dr., 1886–1956, Dozent German Univ. Prague 33, dismissal and emigration 39, via Paris to the U.S., Dir. laboratory of the pharmaceutical firm Norgine

Herre, Wolfgang, Dr. sc. nat. Prof., born 1909, Dozent Halle 36, ao. Prof. 42, Kiel 45, o. Prof. 51

Hershey, Alfred D., Ph.D., born 1908, American molecular geneticist, Nobel prize for medicine 69 (with M. Delbrück and S. E. Luria)

Hertwig, Paula, Dr. phil. Prof., born 1898, ao. Prof. Berlin 27, Prof. Halle 46, o. Prof. 48

Hertwig, Richard, Dr. med. Prof., 1850–1937, o. Prof. Königsberg 81, Bonn 83, Munich 85–25

Hertz, Mathilde, Dr. phil., 1891–1975, Asst. KWI for Biology 20–36, Dozent Berlin 30, dismissed 33, emigration to England 36, Fel. Cambridge Univ. 35

Hertzsch, Walter, Dr. agr., 1901–1975, KWI for Breeding Research, branch East Prussia 35, head of dept. MPI for Breeding Research Cologne 55

Himmler, Heinrich, Diploma in Agriculture, 1900–1945, suicide, Reichsführer-SS

Hippius, Rudolf, Dr., Dozent Posen 41, Prof. German Univ. Prague 42, Ministry for the Occupied Eastern Territories 42, Co-Dir. Inst. for European Ethnology and Peoples' Psychology of the Heydrich Foundation

Hirmer, Max, Dr. Prof., 1893–1981, ao. Prof. Munich, forced to retire 36

Hirt, August, Dr. med. Prof., 1898–1945, suicide, o. Prof. Greifswald 36, Frankfurt, Strasbourg 41, supported by SS Research and Teaching Foundation Das Ahnenerbe

Hochstetter, Ferdinand, Dr. Prof., 1861–1954, o. Prof. of Anatomy Vienna 08, Prof. emeritus 32

Hoerlein, Heinrich, Dr. phil. Hon. Prof., 1882–1954, Honorary Prof. Medical Academy Düsseldorf 21, Dir. IG Farben (Bayer), Wuppertal-Elberfeld till 45, acquitted in the Nuremberg physicians trial 1948

Hoffmann, Curt, Dr. phil. Prof., 1898–1959, Dozent Kiel 29, ao. Prof. 40, o. Prof. 56

Hoffmann, Walther, Dr. phil. nat. Prof., born 1910, head of dept. KWI for Breeding Research 36–42, KWI for Fiber Research 42–46, Prof. Halle 49–58

Höfler, Karl, Dr. phil. Prof., 1893–1973, ao. Prof. Vienna 30, o. Prof. 41

Hofmann, Elise, Dr. phil., 1889–1955, Dozent Vienna 35

Holst, Erich von, Dr. phil. Prof., 1908–1962, Dozent Göttingen 39, o. Prof. Heidelberg 46, head of dept. and Deputy Dir. MPI for Marine Biology Wilhelmshaven 46–54, Dir. MPI for Behavioral Physiology Seewiesen

Holtfreter, Johannes, Dr. phil. Prof., born 1901, Asst. KWI for Biology 28, Dozent Munich 33, dismissal and emigration 39, Cambridge Univ. 38–39, McGill Univ. (Canada) 42–46, Prof. Rochester 46

Holtz, Friedrich, Dr. med. and phil. Prof., 1898–1967, ao. Prof. Berlin 33, head of clinical-chemical section of the Dept. for Surgery of the Charité, Dir. Central Institute for Cancer Research Posen about 43, Prof. Halle 46–57

Huber, Bruno, Dr. phil. Prof., 1899–1969, ao. Prof. Darmstadt 32, o. Prof. Dresden 34, Munich 46

Husfeld, Bernhard, Dr. phil. Prof., 1900–1970, head of dept. KWI for Breeding Research Müncheberg 28–42, Dir. KWI for Vine Breeding Research Müncheberg 43–47, Dir. Research Inst. for Vine Breeding Geilweilerhof (FRG)

Huxley, Julian, Dr. Prof., 1887–1975, Prof. of zoology King's College London 25, President Inst. Animal Behaviour 35–42, First Director General of UNESCO 46–48

Ichii, Shiro, 1892–1958, physician in Japanese Army in Word War II

Ilse, Dora, Dr., 1898–?, guest at KWI for Biology 31–32, at Inst. for Zoology Munich 33, Zürich 34, Munich 34–35, emigration to England 36

Isaacs, Alick, M.D., 1921–1967, British virologist, discovered Interferone together with J. Lindenmann at the National Inst. for Medical Research London 1956

Jacob, François, Dr. med. Prof., born 1920, Officer of the French Free Forces 40–45, Asst. Inst. Pasteur (dept. of A. Lwoff) 50–56, head of dept. 56–64, Prof. for cell genetics at the Collège de France 64, Nobel prize for medicine 65 (with A. Lwoff and J. Monod) for discoveries of the genetic control of the synthesis of enzymes and viruses

Jacobs, Werner, Dr. phil., 1901–1972, Dozent u. Asst. Munich 30, ao. Prof. 37

Japha, Arnold, Dr. med. Dr. phil., 1877–1943, suicide, ao. Prof. Halle 21, dismissed 36

Jaretzky, Robert, Prof., born 1900, ao. Prof. Technical College Brunswick 31, o. Prof. 33

Jelinek, K. Dr. Habilitation, high school teacher

Jellinek, Walter, Dr. jur. Prof., 1885–1955, o. Prof. for Public Law Heidelberg 28, dismissed 35, o. Prof. Heidelberg 45

Jollos, Viktor, Dr. phil. Prof., 1887–1941, Dozent Berlin 21, Prof.

Cairo 26–29, ao. Prof. Berlin 30, KWI for Biology, dismissal and emigration 33, England, U.S.

Joseph, Heinrich, Dr. med. Prof., 1875–1941, suicide, ao. Prof. Vienna, Prof. emeritus 34

Jost, Ludwig, Dr. phil. Prof., 1865–1947, o. Prof. Heidelberg 19, Prof. emeritus 34

Just, Günther, Dr. phil. Prof., 1892–1950, Dozent Greifswald 23, ao. Prof. 28, Dir. Inst. for Genetics 33, head Genetics Research Inst. of the Reich Health Office Berlin Dahlem 37, ao. Prof. Würzburg 42, o. Prof. Tübingen 48

Kahlau, Gerhard, Dr. med. Prof., born 1908, Dozent Frankfurt 40, ao. Prof. 48

Kahmann, Hermann, Dr. phil., 1906–1990, Asst. and Dozent Munich 40

Kalmus, Hans, Dr. rer. nat. and Dr. med. Prof., born 1906, Asst. German Univ. Prague 31, Dozent 35, dismissal and emigration 38, Yugoslavia, England, Univ. London 39, lecturer 46, reader 54, prof. 66

Kaplan, Reinhard, Prof., born 1912, head of dept. KWI for Breeding Research 45, ao. Prof. Frankfurt 55, o. Prof. Frankfurt 63

Kappert, Hans, Dr. phil. Prof., 1890–1976, o. Prof. Berlin 31, Prof. emeritus 57

Kappus, Adolf, Dr. med. Prof., 1900–?, Dozent Göttingen 30, ao. Prof. 37

Karlson, Peter, Dr. rer. nat. Prof., born 1918, Dozent Tübingen 53, head of dept. MPI for Biochemistry Munich 56, o. Prof. Marburg 64

Katz, Sir Bernard, Dr. med. Prof., born 1911, Dr. med. Leipzig 35, emigration 35, Univ. College London 35–39, Australia 39–45, Univ. College London 46, Nobel prize for medicine 70

Kaudewitz, Fritz, Dr. rer. nat. Prof., born 1921, Dozent Tübingen 56, ao. Prof. 62, o. Prof. Munich 63

Kaufmann, Carl, Dr. Prof., 1900–1980, gynecologist, Dozent Berlin 32, ao. Prof. 36, Senior Physician at the Charité women's clinic

Kausche, Gustav Adolf, Dr., government official, head of the Laboratory for Experimental Virus Research at Biological Reich Inst. for Agriculture and Forestry Berlin

Keitel, Wilhelm, 1882–1946, General Field Marshal and 38–45 Chief of the Army High Command; subordinate only to Hitler, he passed on even those commands ordering mass murder, for example the decree of December 7, 1941, which permitted murder of persons endangering German security; sentenced to death in the Nuremberg war criminals trial and executed 46

Kiesselbach, Anton, Dr. med. and phil. Prof., 1907–1984, Dozent Greifswald 39, ao. Prof. Regensburg 55, ao. Prof. and Dir. Inst. Medical Academy Düsseldorf 55

Kisser, Josef, Dr. phil. Prof., 1899–1984, Dozent Vienna 27, ao. Prof. 34, dismissed 38, o. Prof. 46

Kliewe, Heinrich, Dr. med. Prof., 1892–1969, ao. Prof. Gießen 31 (hygiene and bacteriology), senior medical official, head of the medical bureau for infectious diseases of the State of Hessen

Knake, Else, Dr. Prof. born 1901, KWI for Biochemistry Berlin 43, head of Inst. for Tissue Culture Research of the German Research College Berlin, since 53 independent dept. of Tissue Culture Research of the MPI for Comparative Genetics and Hereditary Pathology Berlin

Knapp, Edgar, Dr. phil., 1906–1978, Dozent Berlin 36, o. Prof. Strasbourg 41, head branch Rosenhof of the MPI for Breeding Research 52, Honorary retired Prof. Heidelberg 53

Knoll, Fritz, Dr. phil. Prof., 1883–1981, o. Prof. Vienna 33, retired 48

Koegel, Anton Otto, Dr. med. Prof., 1889–1957, ao. Prof. Munich 27

Koehler, Otto, Dr. rer. nat. Prof., 1889–1974, ao. Prof. Munich 23, o. Prof. Königsberg 25, o. Prof. Freiburg 46

Koehler, Wilhelm, Dr. 1905–1943, Dozent Munich 39

Koernicke, Max, Dr. phil. Prof., 1874–1955, o. Prof. Bonn 08, Prof. emeritus 39

Koller, Gottfried, Dr. phil. Prof., 1902–1959, Dozent Kiel 30, o. Prof. Shanghai 34, German Univ. Prague 41, Saarbrücken 49

Koltsov, Nikolai K., Dr. Prof., 1872–1940, Prof. Univ. Moscow 18–30, Dir. of the first Soviet Inst. for Experimental Biology 17–39

Kornmüller, Alois, Dr. Prof., head of dept. KWI for Brain Research Berlin and MPI for Brain Research Göttingen

Kosswig, Curt Karl Ferdinand, Dr. phil. Prof., 1903–1982, ao. Prof. Technical College Brunswick 33, emigration to Turkey 37, o. Prof. Istanbul 37, o. Prof. Hamburg 55

Kramer, Gustav, Dr. phil., 1910–1959, Asst. KWI for Medical Research Heidelberg, German-Italian Institute for Marine Biology Rovigno (Italy) 33, Zoological Station Naples 37–41, Habilitation Heidelberg 45, head of dept. MPI for Marine Biology Wilhelmshaven 48

Krause, Gerhard, Dr. phil. Prof., born 1906, Dozent Berlin 43, Tübingen 58, o. Prof. Würzburg 58

Krause, Kurt, Dr. phil. Prof., 1883–?, Prof. 24, curator and Prof. at the Botanical Museum Univ. Berlin

Kretschmer, Ernst, Dr. med. Prof., 1888–1964, o. Prof. of psychiatry Marburg 26–46, o. Prof. and Dir. Univ. Neurological Clinic Tübingen 46

Krieck, Ernst, Dr. phil.h.c. Prof., 1882–1947, Pedagogical Academy Frankfurt 28, o. Prof. 33, Heidelberg 34, philosophy, pedagogy

Krieg, Hans, Dr. phil. Dr. med. Prof., 1888–1970, ao. Prof. Munich 27

Kronacher, Karl, Dr. med. vet. Prof., 1871–1938, o. Prof. Berlin 30

Kröning, Friedrich, Dr. phil. Prof., 1897–?, Dozent Göttingen 28, ao. Prof. 35

Krüger, Friedrich, Dr. phil. Prof., born 1902, Dozent Münster 36, ao. Prof. 42, Hamburg 56

Krüger, Paul, Dr. phil. Prof., 1886–1964, o. Prof. Vienna 29, Heidelberg 35, Prof. emeritus 58

Kubach, Fritz, Historian of Science, head of the Science Section of NSDStB

Kubart, Bruno, Dr. phil. Prof., 1882–1959, Dozent Graz 12, o. Prof. 20–36, 38–45

Kuckuck, Hermann, Dr. agr. Prof., 1903–1992, head of dept. KWI for Breeding Research Müncheberg 28, dismissed 36, Dozent Halle 45, o. Prof. 46, Berlin 48, Hanover 54, Prof. emeritus 69

Kuhl, Willi, Dr. phil. Prof., 1892–1972, Dozent Frankfurt 28, ao. Prof. 39, ao. Prof. Frankfurt 57

Kühn, Alfred, Dr. phil. and Dr. rer. nat. Prof., 1885–1968, Dozent 10, ao. Prof. Freiburg 14, Berlin 18, o. Prof. Göttingen 20, Second Dir. KWI for Biology Berlin 37, Tübingen 44, Dir. MPI for Biology Tübingen 51–58, o. Prof. Tübingen 45, Prof. emeritus 51

Kuhn, Otto, Dr. rer. nat. Prof., 1896–1978, Dozent Göttingen 28, ao. Prof. Göttingen 34, o. Prof. Cologne 35–45 and since 50

Kuhn, Richard, Dr. phil. Prof., 1900–1967, Dir. KWI for Medical

Research Heidelberg and o. Prof. Univ. Heidelberg 29, Nobel prize
for chemistry 39
Kühnelt, Wilhelm, Dr. phil. Prof., born 1905, Dozent Vienna 34, ao.
Prof. Vienna 41, ao. Prof. Graz 50, o. Prof. Vienna 53
Küster, Ernst, Dr. phil. Prof., 1874–1953, o. Prof. Gießen 20

Laibach, Friedrich, Dr. phil. Prof., 1885–1967, ao. Prof. Frankfurt 28,
o. Prof. 34, Prof. emeritus 48, head private Biological Research Inst.
Limburg
Lang, Anton, Dr. rer. nat. Prof., born 1913, Fel. KWI for Biology
Berlin 39–49, Res. Assoc. McGill Univ. 49, Vis. Prof. Agr. and Med-
ical College Texas 50, Res. Fel. Cal. Tech. 50–52, Assist. and Assoc.
Prof. Calif. Los Angeles 50–55, Prof. 55–59, Prof. Cal. Tech. (bi-
ology) 59–65, developmental biology of plants
Langelüddeke, Albrecht, Dr. med. Prof., 1889–?, psychiatrist, Dozent
Hamburg 30, ao. Prof. 35, State Senior Medical Official, Dir. State
Psychiatry Marburg
Lehmann, Ernst, Dr. Prof., 1880–1957, Dozent Kiel 09, ao. Prof. Tü-
bingen 13, o. Prof. Tübingen 22, pensioned 49, Prof. emeritus 52
Lehmann, Julius F., 1864–1935, founder of the national-volkish pub-
lisher J. F. Lehmann, Chairman of the Pan-Germanic League and
Co-founder of the German Völkische Schutz-und-Trutzbund 1919
Lehmensick, Rudolf, Dr. phil. and med. Prof., born 1899, Dozent
Bonn 37, Asst. Prof. 56, ao. Prof.
Leick, Erich, Dr. phil. Prof., 1882–1956, o. Prof. Greifswald 28, Dir.
Biological Research Institute Greifswald 30, Dir. Inst. for Plant
Ecology Greifswald 34
Leider, Kurt, Dr. phil. habil., born 1902, Asst. at the Philosophical
Seminar Königsberg, Posen 40–44, head of the Philosophical Acad-
emy Lübeck after 45
Leiner, Michael, Dr. phil. Prof., 1893–1972, Dozent Berlin 40, ao.
Prof. Mainz 46
Lengerken, Hanns von, Dr. phil. Prof., 1889–1966, ao. Prof. Berlin
20, o. Prof. 35, o. Prof. Halle 49
Lettré, Hans, Dr. phil. Habilitation, born 1908, Dozent Göttingen 38,
Berlin 41, chief chemist of the Cancer Dept. at the Rudolf Virchow
Hospital
Levi-Montalcini, Rita, Dr. Prof., born 1909, Assoc. Prof. Biology
Washington Univ. (St. Louis) 51–58, Full Prof. 58–77, Dir. Inst. for

Cellbiology Rome 77, Nobel prize for physiology 86 (with S. Cohen)

Levit, Solomon G., 1894–1937, Soviet physician and human geneticist, head of Medical-Biological Institute in Moscow 27, persecuted and executed under Stalin

Lipmann, Fritz A., Dr. med. and Dr. phil. Prof., 1899–1986, Asst. KWI for Biology Berlin 27–29, KWI for Medical Research 29–31, emigration 32, Denmark 32–39, emigration to the U.S. 39, Prof. of biological chemistry Harvard Univ., Rockefeller Univ. New York 57. Nobel prize for medicine 53 (with H. Krebs)

Loewi, Otto, Dr. Prof., 1873–1961, Prof. Pharmacology Graz 09, dismissed, imprisoned 38, emigration, Oxford 39–40, Prof. New York Univ., Nobel prize for medicine 36 (with Sir Henry Dale) for discoveries in the field of chemical transfer of information by transmitters at synapses

Löffler, Lothar, Dr. med., Prof., 1901–1983, Dozent Kiel 31, o. Prof. Königsberg 34, Vienna 42, chief of Race Policy Bureau of the NSDAP Königsberg and Vienna 33–45, Prof. emeritus Technical University Hanover 45

Lorbeer, Gerhard, Prof., born 1899, Asst. Univ. Freiburg/Breisgau 36, Dozent Prague 43, ao. Prof. Prague 44

Lorenz, Konrad, Dr. med. and phil. Prof., 1903–1989, Dozent Vienna 37, o. Prof. of psychology Königsberg 41, head of Division for Behavioral Physiology at the MPI for Marine Biology Wilhelmshaven 50, Honorary Prof. Univ. Münster 53, Deputy Dir. MPI for Behavioral Physiology Seewiesen 55, Dir. 61–73, Honorary Prof. Univ. Munich 57, Vienna and Salzburg 74, Nobel prize for medicine 73 (with K.v. Frisch and N. Tinbergen)

Lorenz, Werner, 1891–1974, member NSDAP 29, SS 31, SS General 36, chief of Volksdeutsche Mittelstelle, an office of the SS, 37, sentenced to twenty years of prison for war crimes 48, but released in 1955

Löwenstein, Otto E., Dr. phil. Prof., born 1906, zoologist, Asst. Univ. Munich 31–33, dismissal and emigration 33, Re. Scholar Birmingham Univ. 33–37, lecturer Univ. College Exeter and Glasgow Univ. 37–52, Mason Prof. Birmingham Univ. 52–74

Lüdicke, Manfred, Dr. phil. Prof., born 1911, Dozent Rostock 39, ao. Prof. Heidelberg 50

Lüdtke, Heinz, Dr. rer. nat. Prof., born 1908, Dozent Königsberg 40, ao. Prof. Freiburg 47

Ludwig, Wilhelm, Dr. phil. Prof., 1901–1959, Dozent Halle 30, ao. Prof. 38, Mainz 46

Luria, Salvador E., M.D. Prof., born 1912, Univ. Turin, emigration, U.S. 40, Assoc. Prof. for bacteriology Indiana Univ. 43–50, Prof. 50–59, MIT 59, Nobel prize for medicine 69 (with A. Hershey and M. Delbrück)

Lwoff, André, Sc.D., born 1902, Inst. Pasteur 21, Nobel prize for medicine 65 (with F. Jacob and J. Monod) for discoveries of the genetic control of the synthesis of enzymes and viruses

Lynen, Feodor, Dr. phil. Prof., 1911–1979, Dozent Munich 42, o. Prof. of biochemistry 53, Dir. MPI for Cell Chemistry 54, Nobel prize for physiology 64

Lysenko, Trofim D., 1898–1976, agronomist, worked at All Union Institute for Genetics and Breeding Odessa 29, scientific leader of a research station of the AN SSSR near Moscow 38, since 66 head of laboratory, President of the Lenin All Union Academy of the Agricultural Sciences 38–56 and 61–62, Dir. Inst. for Genetics of the AN SSSR 40–65

Mägdefrau, Karl, Dr. phil. nat. Prof., born 1907, Dozent Erlangen 36, ao. Prof. Strasbourg 42, ao. Prof. Munich 51, o. Prof. 56

Mainx, Felix, Dr. med. Dr. rer. nat. Prof., born 1900, Dozent Prague 29, ao. Prof. 36, dismissed 38, Dozent Vienna 47, ao. Prof. 49, o. Prof. 56

Mangold, Otto, Dr. rer. nat. Prof., 1891–1962, head of dept. KWI for Biology Berlin 23, Dozent Berlin 24, ao. Prof. 29, o. Prof. Erlangen 33, Freiburg 37, Prof. emeritus 53, head of dept. Heiligenberg-Inst. Heiligenberg/Baden 46

Marchet, Arthur, Dr. phil. Prof., 1892–1980, Dozent Vienna 23, o. Prof. 40, retired 45

Marcus, Ernst, Dr. phil. Prof., 1893–1968, ao. Prof. Berlin 31, dissmised 35, emigration 36, Prof. Univ. São Paulo (Brazil) 36–63

Markgraf, Friedrich Karl, Dr. phil. Prof., born 1897, Dozent Berlin 27, ao. Prof. 34, Munich 48

Marquardt, Hans, Dr. phil. nat. Dr. phil., born 1910, Dozent Freiburg/Breisgau 40, ao. Prof. 46, personal o. Prof. 54, o. Prof. 63

Martius, Heinrich, Dr. med. Prof., 1885–?, gynecology, o. Prof. Göttingen 26

Mattes, Otto, Dr. phil. Prof., born 1897, Dozent Marburg 29, ao. Prof. 35

Matthes, Ernst, Dr. phil. Dr. h.c. Prof., 1889–1958, ao. Prof. Breslau 24, o. Prof. Greifswald 27–37, Prof. emeritus 52, dismissal and emigration 36, Prof. contratado Coimbra (Portugal) 36

May, Eduard, Dr. phil. nat. Prof., 1905–1956, Dozent Munich 42, o. Prof. Freie Univ. Berlin 51

Meitner, Lise, Dr. Prof., 1878–1968, Austrian physicist, head of Physics Dept. KWI for Chemistry Berlin 1917, fled from Germany and emigrated to Sweden 1938, Prof. Technical College Stockholm 47

Meixner, Josef, Dr. phil. Prof., 1889–1946, ao. Prof. Graz 32, o. Prof. 39

Melchers, Georg, Dr. phil. Prof., born 1906, Asst. KWI for Biology Berlin 34, head of dept. Research Division for Virology KWIs for Biochemistry and Biology 41, Dir. KWI for Biology Tübingen 47, Honorary Prof. Tübingen 47

Mengele, Josef, Dr. phil. and med., 1911–1985?, Asst. of Prof. Verschuer in Frankfurt, SS camp doctor in Auschwitz 43–45, carried out selections for the gas chambers and conducted deadly medical experiments with prisoners, did research for Verschuer in Auschwitz

Mentzel, Rudolf, Dr. Prof., 1900–?, President of the DFG 37, chief of Science Dept. in the REM 39, ao. Prof. Technical College Berlin 34, o. Prof. 35, after 45 three years in American POW camps

Mertens, Robert, Dr. phil. Prof., 1894–1975, Dozent Leipzig 32, ao. Prof. 39, Honorary Prof. Frankfurt/Main 53

Merton, Hugo, Prof., 1879–1940, ao. Prof. Heidelberg 30, emigration to Scotland 36

Metzner, Paul, Dr., born 1893, ao. Prof. Tübingen 29, o. Prof. Greifswald 30

Meusel, Hermann, Dr. sc. nat. Prof., born 1909, Dozent Halle 39, ao. Prof. 46, Prof. 48

Mevius, Walter, Dr. phil. Prof., 1893–1975, ao. Prof. Münster/Westfalia 30, o. Prof. Berlin 32, Münster/Westfalia 35, Rector 37, o. Prof. Hamburg 45

Meyer, Konrad, Dr. phil. Prof., 1901–1973, Dozent Göttingen 30, o. Prof. Jena 34, Berlin 34–45, Technical College Hanover 56, chief of General Biology and Agricultural Science Section in the RFR 37–45, chief of the Planning Dept. of the Reich Security Main Office of the SS, author of the General Plan East

Meyer-Abich, Adolf, Dr. phil. Prof., 1893–1971, ao. Prof. Hamburg 30, Dir. German-Dominican Tropics Research Inst. 39, ao. Prof. Hamburg 46, o. Prof. 58

Meyerhof, Otto, Dr. med. Prof., 1884–1951, Prof. Kiel 13, KWI for Biology 24, Dir. at the KWI for Medical Research Heidelberg 30, dismissed 38, emigration to France 38, U.S. 40, Prof. Univ. of Pennsylvania 40–51, Nobel prize 22

Michaelis, Peter, Dr. phil., 1900–1975, head of dept. KWI for Breeding Research Müncheberg 35, Voldagsen 45, MPI for Breeding Research Cologne 55

Mitscherlich, Alexander, Dr. med. Prof., 1908–1982, medical scientist and psychologist, Dozent Heidelberg 46, ao. Prof. 52, Dir. Psychosomatic Clinic Heidelberg, Prof. Frankfurt 67, Dir. Sigmund Freud Inst. Frankfurt 60–76

Mitschurin, Ivan W., 1855–1935, Russian botanist and plant breeder, advocated Lamarckian theories of heredity

Möbius, Karl August, Dr. phil. Prof., 1825–1908, o. Prof. Kiel 1868, Berlin 1887

Monod, Jacques, Sc.D. Prof., 1910–1976, Asst. (zoology) Univ. Paris 34–45, head of laboratory Inst. Pasteur 45–53, head of dept. 54, Dir. Inst. Pasteur 71, Prof. Univ. Paris 59, Collège de France 67, Nobel prize for medicine 65 (with F. Jacob and A. Lwoff) for discoveries of the genetic control of the synthesis of enzymes and viruses

Moritz, Otto, Dr. phil. Prof., born 1904, Dozent Kiel 31, ao. Prof. 37, o. Prof. 59

Mothes, Kurt, Dr. phil. Prof., 1900–1983, Dozent Halle 28, o. Prof. Königsberg 35–45, Halle 49, President German Academy der Naturforscher Leopoldina, Halle 54–74

Möwus, Franz, Dr. phil., 1908–1959, Asst. KWI for Biology Berlin 37, for Medical Research Heidelberg 38, Dozent Heidelberg 42

Mrugowsky, Joachim, Dr. sc. nat. Dr. med., 1905–1947, chief of Institute Hygiene of the Waffen-SS 37, Dozent Berlin 39, SS Brigadier General, sentenced to death by the American Military Court 1947

Muckermann, Hermann, Dr. Prof., 1877–1962, head of dept. KWI for Anthropology 27–33, dismissed, o. Prof. Berlin 48

Müller, Hans Aurel, Dr. med. Dozent, born 1910, Dozent Marburg 52, gynecology and obstretrics

Müller, Wilhelm, Dr. phil. Prof., 1880–?, applied mathematics and mechanics, ao. Prof. Technical College Hanover 28, German Technical College Prague 28, o. Prof. Technical College Aachen 34, Munich 39

Munk, Klaus, Dozent born 1922, Dozent Munich 61, Heidelberg 62, Dir. Inst. for Virology Heidelberg

Nachmansohn, David, Dr. med. Prof., 1899–1972 (74)?, Dr. med. Univ. Berlin 26, Fel. KWI for Biology Berlin 26–30, asst. physician and head of chem. lab. Rudolf Virchow Hospital, emigration to France 33, Sorbonne Paris 33–39, since 39 in U.S., Yale Univ. 39–42, Prof. Columbia Univ. 42

Nachtsheim, Hans, Dr. phil. Prof., 1890–1979, ao. Prof. Berlin 23, head of dept. KWI for Anthropology, Berlin 41, o. Prof. Humboldt Univ. Berlin 46, Dir. MPI for Comparative Heredity and Pathology 46–60, o. Prof. Freie Univ. Berlin 49

Needham, Joseph, Ph.D. Prof., born 1900, Prof. Univ. Cambridge

Neuberg, Carl, Dr. phil. Prof., 1877–1956, o. Prof. of Biochemistry Berlin 19–34, dismissed, Acting Dir. KWI for Biochemistry Berlin 13, Dir. 25, compulsory resignation 35, emigration to Palestine 38, Prof. Hebrew Univ. (Jerusalem) 39, Res. Prof. New York Univ. and Brooklyn Polytechnic Inst. 40

Noack, Kurt, Dr. phil. Prof., 1888–1963, Dozent Strasbourg 18, Freiburg 19, ao. Prof. Bonn 21, o. Prof. Erlangen 22, o. Prof. Berlin 31, Prof. emeritus 56

Nürnberg, Ursula, Dr. rer. nat. Prof., born 1920, Prof. Humboldt Univ. Berlin 45

Nyiszli, Miklós, Dr. med., 1901–1956, prisoner and Asst. of J. Mengele in Auschwitz.

Oehlkers, Friedrich, Dr. phil. Prof., 1890–1971, Dozent Tübingen 22, ao. Prof. 25, o. Prof. Technical College Darmstadt 28, Freiburg/ Breisgau 32

Orth, Reinhard, Dr. phil. nat. Prof., born 1907, Dozent Heidelberg 37, pl. ao. Prof. Munich 42–45, compulsory retirement 45

Osenberg, Werner, Dr.-Ing. Prof., 1900–1974, Asst. Technical College

Dresden 25–38, o. Prof. Technical College Hanover 38, chief of Four Year Plan at the Technical College Hanover 41–45

Overbeck, Fritz, Dr. phil. Prof., 1898–?, Dozent Frankfurt/Main 26, ao. Prof. 34, o. Prof. Bonn 39

Panschin, Igor, born 1914, Fel. Genetics Dept. KWI for Brain Research 44–45

Pascher, Adolf, Dr. phil. Prof., 1881–1945, o. Prof. Prague 27

Pätau, Klaus, Dr. Prof., 1908–1975, KWI for Biology 35, head Research Division for Biology Berlin 45, emigration to U.S. 1948

Pavlov, Ivan P., Dr. med. Prof., 1849–1936, o. Prof. of Physiology St. Petersburg 96–24, Nobel prize for medicine 04

Pax, Ferdinand Albert, Dr. phil. Prof., 1885–?, ao. Prof. Breslau 15, Prof. emeritus 51, Dir. Inst. for Marine Research Bremerhaven 48–50

Pekarek, Josef, Dr. phil. Prof., 1899–?, Dozent Graz, ao. Prof. 41

Penners, Andreas, Dr. phil. Prof., 1890–1951, ao. Prof. Würzburg 27, ao. Prof. and head Zoological Inst. Vienna 37, dismissed 38, ao. Prof. Würzburg 46

Pernkopf, Eduard, Dr. med. Prof., 1888–1955, o. Prof. Vienna 33

Perron, Oskar, Dr. phil. Prof., 1880–1975, o. Prof. Heidelberg 14, Munich 22, Prof. emeritus 51

Peterfi, Tibor, Dr. med. Prof., 1883–1953, o. Prof. Bratislava (Ungarn) 19, Guest Researcher KWI for Biology Berlin 22, emigration to England 34, Univ. Copenhagen 36–39, Univ. Istanbul 39–46, Prof. 44, o. Prof. Univ. Budapest 44–48

Pfankuch, E., Dr. government official, collaborator at Laboratory for Experimental Virology at the Biological Reich Inst. for Agriculture and Forestry Berlin

Philip, Ursula, Dr. phil., born 1908, res. KWI for Biology Berlin 31–34, emigration 34, res. Univ. College London 34

Pichler, Alexander, Dr. med. Prof., born 1906, Dozent Vienna 39, pl. ao. Prof. 40

Pirschle, Karl, Dozent, 1900–1945, Dozent Munich, Berlin, KWI for Biology Berlin 40

Pirson, André, Dr. phil. Prof., born 1910, ao. Prof. Marburg/Lahn 44, o. Prof. 51, Göttingen 58

Pisek, Arthur, Dr. Prof., 1894–1975, Dozent Innsbruck 30, o. Prof. 33

Planck, Max, Dr. Prof., 1858–1947, o. Prof. Berlin 92, Nobel prize for physics 18, President KWG 28–37

Ploetz, Alfred, Dr. med., 1860–1940, founder of German racial hygiene theory

Plötner, Dr. med., camp doctor in Dachau

Pohl, Franz, Prof., 1896–?, Dozent Prague 31, ao. Prof. Prague 41

Pohl, Oswald, 1892–1951, NSDAP 26, SS General and General of the Waffen-SS 42, head of the Economic Planning and Administration Main Bureau of the SS 42, sentenced to death by American Military Court 47 and hanged as a war criminal 51

Pringsheim, Ernst-Georg, Dr. phil. Prof., 1881–1970, ao. Prof. Berlin 21, Prague 23, dismissed 38, emigration 39, Botanical School Cambridge 39–53, Honorary Prof. Univ. Göttingen 53

Propach, Hermann, born 1906, collaborator, KWI for Breeding Research since 33

Przibram, Hans, Dr. Prof., 1874–1944, ao. Prof. Vienna 30, dismissed 38, sent to Theresienstadt concentration camp, died of starvation

Radu, Georg, Dr., born 1910, Fel. Genetics Dept. KWI for Brain Research 42–45

Rajewsky, Boris, Dr. phil. nat. Dr. h. c. Prof., 1893–1974, o. Prof. Frankfurt/Main 34, Dir. KWI for Biophysics 36

Rapoport, I.A., Soviet geneticist, after 48 transferred to "practical" work in a pharmaceutical institute

Rascher, Sigmund, Dr. med., ?–1945, army physician, SS member, mass murderer who conducted experiments on prisoners in Dachau concentration camp, executed by the SS because his wife abducted babies

Rau, Werner, Dr. rer. nat. Prof., born 1927, Dozent Munich 62, ao. Prof. 69, Botany

Rauch, Konrad von, Dr., born 1905, botanist, SS-Obersturmführer, collaborator of Ernst Schäfer in SS Research and Teaching Foundation Das Ahnenerbe

Rauh, Werner, Dr. rer. nat., born 1913, Dozent Heidelberg 40, ao. Prof. 46, ao. Prof. 56, o. Prof.

Reichensperger, August Carl Alexander, Dr. phil. Prof., 1878–1962, o. Prof. Bonn 28–48

Reinmuth, Ernst, Dr. phil. Prof., born 1901, Dozent Rostock 36, ao. Prof. 44, Dir. 48, Prof. 51

Reiter, Hans, Dr. med. Hon. Prof., 1881–?, ao. Prof. Rostock 20, Senior Medical Official 26, Honorary Prof. Rostock 28, President Reich Health Office Berlin 33, Honorary Prof. 34

Remane, Adolf, Prof., 1898–1976, ao. Prof. Kiel 30, o. Prof. Halle 34, Kiel 36

Renner, Otto, Dr. phil. Prof., 1883–1960, o. Prof. Jena 20, Munich 48, Prof. emeritus 52

Rensch, Bernhard, Dr. phil. Prof., 1900–1990, head of dept. Zoological Museum Berlin 25, ao. Prof. Münster/Westfalia 43, o. Prof. 47

Riehl, Nikolaus, Dr. phil. Prof., born 1901, Dir. Scientific Main Lab. Auer Company Berlin 39, head German scientific group in the USSR 45–55, ao. Prof. Technical College Munich 57

Ries, Erich, Dr. phil. Prof., 1908–1944, Dozent Cologne 33, Leipzig 34, o. Prof. Münster 41

Rietschel, Peter, Dr. rer. nat. Prof., born 1903, Dozent Frankfurt/Main 39, ao. Prof. 50

Risse, Otto, Dr. med. Prof., 1895–?, ao. Prof. Berlin 34, head of dept. Inst. for Radiation Research of the Univ.

Roberg, Max, Dr. phil. Prof., born 1900, Dozent Breslau 37–45

Roggenbau, Christel Heinrich, Dr. med. Prof., 1896–?, Dozent Berlin 35, ao. Prof. 40, Senior Physician at the psychiatry of the Charité clinic

Römer, Theodor, Dr. phil. Prof., 1883–1951, o. Prof. Halle 19

Rompe, Robert, Dr. phil. Prof., born 1905, o. Prof. of experimental physics Humboldt Univ. Berlin 46

Rose, Gerhard, Dr. med. Aviation-Medical Research Institute of the German Air Force, chief of Dept. for Tropical Medicine at the Robert Koch Inst., consultant hygienicist to the chief of the Medical Dept. of the Air Force, sentenced to life in prison in the Nuremberg physicians' trial 47, later released

Rosenberg, Alfred, 1883–1946, executed, ideologist, Minister for the Occupied Eastern Territories 42–45, sentenced in Nuremberg trials

Rössle, Robert, Dr. med. Prof., born 1876, o. Prof. Jena 11–22, Basel 22–29, Berlin 29, Dir. Pathological Inst. Charité clinic

Rotmann, Eckhard, Dr. phil. nat. Prof., 1907–1950, Dozent Freiburg/Breisgau 36, ao. Prof. Cologne 42, o. Prof. 48

Roux, Wilhelm, Dr. med. Prof., 1850–1924, ao. Prof. Breslau 1986, founder and head of the Inst. for Entwicklungsmechanik (Experimental Embryology) 88, o. Prof. Innsbruck 89, founder of Archiv für Entwicklungsmechanik 94, o. Prof. Halle 95–21

Rüdin, Ernst, Dr. med. Prof., 1874–1952, o. Prof. Basle 25, Honorary Prof. and Dir. KWI for Psychiatry Munich 28, o. Prof. 33

Rudorf, Wilhelm, Dr. sc. nat. Prof., 1891–1969, o. Prof. Leipzig 34, o. Prof. Berlin and Dir. KWI for Breeding Research Müncheberg 36, Voldagsen 45, Cologne-Vogelsang 55, Honorary Prof.

Ruff, Siegfried, Dr. med., Dir. Aviation Medical Inst. of the German Research Inst. for Aviation, took part in deadly low pressure experiments with prisoners of the Dachau concentration camp, acquitted in the Nuremberg physicians trial, employed by the U.S. Army, later Dir. Inst. for Aviation Medicine near Bonn

Ruge, Ulrich, Dr. phil. Prof., born 1912, Dozent Greifswald 40, ao. Prof. Kiel 47, o. Prof. Technical College Hanover 49

Ruhenstroth-Bauer, Gerhard, Dr. med. rer. nat. Prof., born 1913, Dozent Tübingen 51, Munich 57, ao. Prof. 58

Ruhland, Wilhelm, Dr. phil. Dr. rer. nat. h.c., 1878–1960, o. Prof. Tübingen 18, Honorary Prof. Erlangen 47

Ruska, Helmut, Dr. med. Habilitation, 1908–1973, biophysics and electron microscopy, Habilitation, Berlin 43, Prof. 48

Rust, Bernhard, 1883–1945, suicide, high school teacher, NSDAP member since 22, REM Minister 34–45, President of the RFR 40

Ruttner, Franz, Dr. phil. Prof., 1882–1961, head Biological Station Lunz 23, Dozent Vienna 25, ao. Prof. 27, Prof. emeritus 57

Saller, Karl, Dr. phil. and Dr. med. Prof., 1902–1969, Dozent Göttingen, dismissed 35, o. Prof. Munich 48

Sauckel, Fritz, 1894–1946, District Leader in Thüringen 27, Minister-President of Thuringia 32, Governor in Thüringen 33–45, Plenipotentiary for the Use of Labor 42–45, in that capacity responsible for deportation of millions of people to the occupied territories of Eastern Europe and for the extermination of tens of thousands of Jewish workers in Poland, sentenced to death as war criminal 46

Sauerbruch, Ferdinand, Dr. med. Prof., 1865–1951, surgeon, o. Prof. Zürich 10, Munich 18, Berlin 28, head of Medical Section of the RFR 37–45

Schäfer, Ernst, Dr. phil. Prof., 1910–1992, zoologist, Tibet researcher, SS member since 33, SS Lieutenant Colonel, Dozent Munich 42, Dir. Reich Inst. Sven Hedin Munich, after 45: Prof. Univ. Central de Venezuela, Honorary member German Ornitholog. Society and Royal Asiatic Society (London), member Academy of Natural Sciences (U.S.)

Schäfer, Konrad, Dr. med., Physician at the Research Inst. for Aviation Medicine, took part in deadly experiments with prisoners of the Dachau concentration camp, acquitted in the Nuremberg physicians trial, employed by the U.S. Army.

Schäfer, Werner, Dr. med. vet. Prof., born 1912, Dozent Gießen 52, ao. Prof. 58, Dir. MPI for Virology Tübingen 56

Scharnke, Johannes, Dr., 1907–1941, Asst. and Dozent Zoological Inst. Munich 38

Schattenberg, Dr., member of the SS, responsible for the "Plant-Breeding Commando" of the Auschwitz concentration camp

Schaxel, Julius, Dr. phil. Prof., 1887–1943, ao. Prof. Jena 16, dismissal and emigration to the USSR 33, died under obscure circumstances

Scheibe, Arnold, Dr. rer. tech. Prof., born 1901, Dozent Gießen 37, o. Prof. Technical College Munich 41, German Dir. of the German-Bulgarian Institute for Agricultural Research, Sofia, head of dept. MPI for Breeding Research Neuhof bei Gießen 48, o. Prof. Gießen 51, Göttingen 55

Schemm, Hans, died 1935, primary school teacher, founder of the NSLB 29, Bavarian Minister of Education 33

Scheuring, Ludwig, Prof., 1888–1970, ao. Prof. Munich 24

Schick, Rudolf, Dr. Prof., born 1905, head of dept. KWI for Breeding Research Müncheberg 28, dismissed 36, Prof. Rostock 51, National Prize GDR 51

Schiemann, Elisabeth, Dr. phil. Prof., 1881–1972, ao. Prof. Berlin 31, dismissed 40, head of dept. KWI for Cultivated Plant Research 43–49, head of Research Division for History of Cultivated Plants of the MPG 49–56

Schiller, Josef, Dr. Prof., 1877–1960, ao. Prof. Vienna 27

Schlottke, Egon, Prof., born 1901, Dozent Vienna 30, Technical College Danzig 40, ao. Prof. Strasbourg 42–45

Schmidt, Eberhard, Dr. jur. Prof., 1891–?, o. Prof. Breslau 21, Kiel 26, Hamburg 29, Leipzig 35, Göttingen 45, Heidelberg 48

Schmidt, Jonas, Dr. phil., 1885–1958, o. Prof. Göttingen 21, Berlin 35–45, Dir. KWI for Animal Breeding Research near Rostock 42, o. Prof. Agricultural College Hohenheim 46

Schmidt, Martin, Dr. phil. Prof., born 1905, KWI for Breeding Research 34–45, Dozent Posen 43, lecturer Berlin 49, Dir. Central Research Inst. for Plant Breeding Müncheberg 50, Prof. Humboldt Univ. Berlin 51

Schmidt, Wilhelm J., Dr. phil. Dr. med. h.c. Prof., 1884–1974, o. Prof. Gießen 26, Prof. emeritus 53

Schmidt-Ott, Friedrich, 1860–1956, jurist, Prussian Minister of Education 17, founder of the Notgemeinschaft 20, President of Notgemeinschaft till 34

Schmitz, Heinrich, Dr. phil. nat. Prof., born 1904, Dozent Heidelberg 33, o. Prof. Freiburg/Breisgau 36, ao. Prof. Hamburg 55

Schmucker, Theodor, Dr. phil. Prof., 1894–1970, ao. Prof. Göttingen 33, o. Prof. 37

Schnarf, Karl, Dr. phil. Prof., 1879–1947, ao. Prof. Vienna 31

Schneider, Hermann, Dr. phil. Prof., 1886–1961, o. Prof. Tübingen 21, Rector Univ. Tübingen 45

Schramm, Gerhard, Dr. phil. Prof., 1910–1969, head of dept. KWI for Biochemistry Berlin 41, Dozent Berlin 44, ao. Prof. Tübingen 53, Dir. MPI for Virology Tübingen 56

Schratz, Eduard, 1901–1977, Dozent Münster 33, ao. Prof. 39

Schraub, Alfred, Dr. rer. nat. Prof., born 1909, KWI for Biophysics Frankfurt, Dozent Frankfurt/Main 59, o. Prof. Gießen 62

Schreiber, Hans, Dr.-Ing. Prof., born 1902, Dozent Berlin 34, ao. Prof. 41, Asst. Univ. Inst. for Radiation Research

Schrödinger, Erwin, Dr. Prof., 1887–1961, Austrian physicist, o. Prof. Berlin 27, emigration to Oxford 33, Graz 36, emigration to Dublin 38, Vienna 56, Nobel prize for physics 33 (with P. Dirac)

Schubert, Gerhard, Dr. med. Prof., born 1907, Dozent Göttingen 43, ao. Prof. 49, o. Prof. Hamburg 50

Schulze, Karl Ludwig, born 1911, DFG Scholar Göttingen 39, Berlin 40

Schumacher, Walter, Dr. phil. Prof., 1901–1976, Dozent Bonn 33, o. Prof. 41

Schumann, Erich, Dr. phil. Dr. rer. nat. Prof., born 1898, o. Prof. 33, Dir. II. Physics Inst. Berlin till 45, chief Research Division Army Ordnance Dept. 34, head of dept. for scientific research and technology in the REM, head of Science Section in the Reich War Ministry and Ministerial Manager 38–45, Dir. without Inst. Berlin 51

Schütrumpf, Rudolf, Dr. phil. nat. Prof., 1909–1986, Habilitation Cologne 69, ao. Prof. Cologne 70, collaborator of Eduard May in SS Research and Teaching Foundation Das Ahnenerbe

Schütte, not clear whether identical with Schütte, Ernst, Dr. med. and rer. nat. Prof., born 1908, Habilitation Leipzig 45, Dozent Frankfurt/Main 46, o. Prof. Freie Univ. Berlin 51

Schwartz, Oskar, Dr. phil., 1901–1945. Dozent Hamburg 39

Schwartz, Wilhelm, Dr. phil. Prof., 1896–?, Dozent Technical College Karlsruhe 38, ao. Prof. Technical College Brunswick 48

Schwartz, Prof., Biological Reich Inst., Potato Beetles Inst. Kruft

Schwarz, Walter, see Evenari, Michael

Schwarze, Paul, Dr. phil., head of dept. KWI for Breeding Research 37, Dr. phil. habil.

Schwemmle, Julius, Dr. rer. nat. Prof., 1894–1979, Dozent Tübingen 25, Berlin 29, o. Prof. Erlangen 30

Seel, Hans K. E., 1898–1961, Dozent Halle 28, Hamburg 30, head Pharmacology Dept. KWI for Physical Chemistry Berlin 33, head Research Inst. Clinical Pharmacology Berlin 35, Dir. Pharmacology Dept. Central Institute for Cancer Research Posen 43, Dozent Humboldt Univ. Berlin 52, Prof. 54

Seidel, Friedrich, Dr. phil. Prof., 1897–?, Dozent Königsberg 26, ao. Prof. 30, o. Prof. Berlin 37–45, Marburg 54

Seiler, Jakob, Dr. Prof., 1886–1970, Dozent Munich 23, ao. Prof. 24, o. Prof. ETH Zürich 37

Sengbusch, Reinhold von, Dr. 1898–1985, head of dept. KWI for Breeding Research 28–37, Dir. Dept. for Cultivated Plant Research at the MPI for Breeding Research Hamburg

Serebrovski, Alexandr S., 1892–1948, Soviet geneticist, head of Genetics Dept. of Inst. for Experimental Biology Moscow

Seybold, August, Dr. phil. Prof., 1901–1965, Dozent Cologne 29, o. Prof. Heidelberg 34

Sievers, Wolfram, SS Colonel Managing Director of SS Research and Teaching Foundation Das Ahnenerbe, head of the Institute for Practical Research in Military Science of the SS, executed 47

Söding, Hans, Dr. rer. nat. Prof., 1898–?, Dozent Technical College Dresden 28, ao. Prof. 34, ao. Prof. Münster 41, WissR. Hamburg 47, head of dept. State Inst. 55

Spatz, Hugo, Dr. med. Prof., 1888–1969, Dozent Munich 23, Dir. KWI for Brain Research 37–45, Dir. MPI for Brain Research 48–57

Speer, Albert, Dipl. Ing., 1905–1981, Reich Minister for Armament and War Production 42–45, member of NSDAP 32, member SA 31, sentenced to twenty years in prison 46

Spek, Josef, Dr. phil. nat. Prof., 1895–1964, Dozent Heidelberg 20, ao. Prof. 25, o. Prof. Rostock 47

Spemann, Hans, Dr. Prof., 1869–1941, o. Prof. Rostock 08–14, Second Dir. KWI for Biology Berlin 14, o. Prof. Freiburg 19, Prof. emeritus 37

Spengler, Oswald, 1880–1936, high school teacher (mathematics) and author of *Der Untergang des Abendlandes* (The decline of the West, 1920)

Sperling, Karl, Dr. rer. nat. Prof., born 1941, o. Prof. Freie Univ. Berlin 76, Dir. Inst. for Human Genetics

Stammer, Hans-Jürgen, Dr. phil. Prof., 1899–1968, Dozent Breslau 31, ao. Prof. 37, o. Prof. 39

Stark, Johannes, Dr. rer. nat. Prof., 1874–1951, o. Prof. Aachen 19, Würzburg 20–22, President of the Physical Technical Reich Inst. 33–39, President of the Notgemeinschaft 34–36, NSDAP member 30, together with P. Lenard founder and leading advocate of "German Physics," Nobel prize for physics 19

Stavenhagen, Kurt, Dr. Prof., Prof. Posen 41

Stein, Emmy, Dr. phil., 1879–1954, Asst. Inst. for Genetics and Breeding Research Berlin 23–39, researcher KWI for Biology 40, (48 MPI)

Steinböck, Otto, Dr. phil. Prof., 1893–1969, ao. Prof. Innsbruck 30, o. Prof. 37

Steiner, Maximilian, Dr. phil. Prof., 1904–1988, Dozent Technical College Stuttgart 36, ao. Prof. Göttingen 40, ao. Prof. Bonn 48, o. Prof. 51

Steiniger, Fritz, Dr. phil. Habilitation Prof., born 1908, Dozent Greifswald 37, ao. Prof. 45, Dir. State Office for Hygienic and Bacteriological Investigations

Steinitz, Walter, Dr. phil. and med., 1882–?, Dozent Breslau 30, dismissed 35, emigration to Palestine 35, physician in Ramot Hashavim

Stent, Gunther, Ph.D. Prof., born 1924 (Berlin), Belgium 38, emigration to U.S. 40, Prof. of bacteriology and virology Univ. Calif. 59, Prof. of molecular biology 63

Stepp, Wilhelm, Dr. med. Prof., 1882–1963, o. Prof. Jena 24, Breslau 26, Munich 34, internal medicine

Stern, Curt, Dr. phil., 1902–1981, Asst. KWI for Biology Berlin 24–33 and Dozent Berlin 28–33 Fel. Rockefeller Foundation (U.S.) 32–33, emigration to U.S. 33, Univ. Rochester 33–47, Assist. Prof. 35, Assoc. Prof. 37, Prof. 41, Univ. Calif. Berkeley 47

Stickl, Otto, Dr. med. Prof., 1897–1951, Dozent Greifswald 28, o. Prof. 34, Tübingen 36, Dir. Inst. for Hygiene Univ. Tübingen, Rector Univ. Tübingen till 45, imprisoned by the Americans

Stieve, Hermann, Dr. med. and phil. Prof., 1886–1952, o. Prof. Halle 21, Berlin 35, Dir. Univ. Inst. for Anatomy

Stocker, Otto, Dr. phil. nat. Prof., 1888–1979, o. Prof. Technical College Darmstadt 34, Prof. emeritus 57

Stokar, Neuforn von, Walter , Dr. phil. Prof., 1901–?, ao. Prof. Cologne 39, Dir. Univ. Inst. for Prehistory

Stoppel, Rose, Dr. phil. Prof., 1874–1970, ao. Prof. Hamburg 28

Storch, Otto, Dr. phil. Prof., 1886–1951, o. Prof. Graz 29, dismissed 38, o. Prof. Vienna 45

Stosch, Hans Adolf von, Dr. phil., born 1908, Dozent Königsberg/Prussia 40, Technical College Darmstadt 48

Straub, Josef, Dr. Prof., 1911–1987, KWI for Biology Berlin 40, Dozent Berlin 42–47, Tübingen 48–49, o. Prof. Cologne 50

Streit, Hanns, 1896–?, Curator of the Reich University Posen 41–45, District Chief of NSDB, Managing Dir. Central Institute for Cancer Research Posen 43–45, head of ethnic issues *(Volkstumsfragen)* in the District Administration Wartheland of the NSDAP

Stresemann, Erwin, Dr. phil. Prof., 1889–1972, Prof. Berlin 30, Prof. Humboldt Univ. Berlin 46–61

Strouhal, Hans, Dr. phil. Prof., 1897–1969, Dozent Vienna 33, dismissed 38, ao. Prof. Vienna 46

Strugger, Siegfried, Dr. phil. Prof., 1906–1961, Dozent Greifswald 33, Jena 35–39, Dir. Botanical Inst. Technical College Hanover 39–48, o. Prof. Münster/Westfalia 48

Stubbe, Anna-Elise, born 1907, Fel. Genetics Dept. KWI for Brain Research Berlin-Buch 40–45

Stubbe, Hans, Dr. Prof., 1902–1989, head of dept. KWI for Breeding Research Müncheberg 28, dismissed 36, Fel. KWI for Biology 36, Dir. KWI for Cultivated Plant Research Vienna 43–45, later Gatersleben (GDR), o. Prof. Halle 46

Studnitz, Gotthilft von, Dr. phil. Prof., born 1908, Dozent Kiel 35, ao. Prof. Halle/Saale 41, o. Prof. 42–46, Dir. Natural History Museum Lübeck 51

Suomalainen, Esko, Ph.D. Prof., born 1910, Prof. of genetics Univ. Helsinki 48

Svedberg, Theodor, Dr. phil. and med. Prof., 1884–1971, Prof. of chemistry Uppsala Univ. 12, Nobel prize for chemistry 26, Dir. G. Werner Inst. for Nuclear Chemistry

Telschow, Ernst, Dr. phil., 1889–1988, General Director KWG 37, managing member of the Verwaltungsrat of the MPG 48

Thienemann, August, Dr. phil. Dr. agr. Prof., 1882–1960, o. Prof. Kiel 24, Dir. Hydrobiological Station of the KWG Plön 17

Timoféeff-Ressovsky, Elena A., died 73, geneticist at KWI for Brain Research Berlin 25–45, return to USSR to her husband N. V. Timoféeff-Ressovsky in scientific internment camp 47, research in radiobiology as scientific colleague of her husband till 73

Timoféeff-Ressovsky, Nikolai Vladimirovic, 1900–1981, Asst. Moscow 21, Asst. KWI for Brain Research 25, head of dept. 29, head of independent Genetics Dept. at KWI for Brain Research 36–45, arrested in USSR 45 and imprisoned and interned in labor camp 45–47, research in scientific internment camp in Sungul 47–55, head of Dept. for Biophysics and Radiobiology in Laboratory of the Academy of Sciences Sverdlovsk 55–64

Tinbergen, Nikolaas, Dr. phil. Prof., born 1907, Dozent Leiden 36–40, lecturer 40–47, o. Prof. 47–49, Oxford Univ. 60; Nobel prize for medicine 73 (with K. von Frisch and K. Lorenz)

Tischler, Georg, Dr. phil. Dr. med. h.c. Prof., 1878–1955, o. Prof. Kiel 22

Tiselius, Arne W. K., Dr. phil. Prof., 1902–1971, Prof of biochemistry Uppsala Univ. 38, Nobel prize for chemistry 48

Tobler, Friedrich, Dr. phil. Prof., 1879–1957, o. Prof. Technical College Dresden 24

Todt, Fritz, 1891–1942, Reich Minister for Armament and Munitions 40–42. NSDAP 22, SA General

Troll, Wilhelm, Dr. phil. Prof., 1897–1978, o. Prof. Halle/Saale 32, o. Prof. Mainz after 45

Tschermak-Seyssenegg, Erich von, Dr. phil. Prof., 1871–1962, o. Prof. of Botany Vienna 09

Turba, Friedrich, Dr. rer. nat. Prof., 1917–1965, o. Prof. Würzburg 58, physiological chemistry

Twort, Frederick W., Dr. Prof., 1877–1950, Prof. of bacteriology Univ. London 19–40

Ubisch, Gerta von, Dr. rer. nat. Prof., 1882–1965, ao. Prof. Heidelberg 29, dismissed 35, emigration, head of dept. Inst. Butantan São Paolo (Brazil) 35–38, return to Heidelberg 52

Ubisch, Leopold von, Dr. jur. et phil. Prof., 1886–1965, o. Prof. Münster/Westfalia 27, dismissed 35, emigration to Norway, return to Germany 56, Honorary Prof. Hamburg 56

Uexküll, Jakob von, Dr. med. h.c. and phil. h.c. Prof., 1864–1944, Honorary Prof. Hamburg 25

Ullrich, Hermann, Dr. phil. Prof., born 1900, Dozent Leipzig 29, Prof. 37, Berlin 38–43, head of Dept. of Plant Physiology at the KWI for Breeding Research Müncheberg 37–43, ao. Prof. Technical College Stuttgart 43, o. Prof. 50, Bonn 53

Ulrich, Werner, Dr. phil. Prof., born 1900, Dozent Berlin 29, ao. Prof. 38, o. Prof. 48

Vavilov, Nikolai Ivanovic, Dr. Prof., 1887–1943, Prof. of botany 1917, head of Dept. of Applied Botany and Selection of the All Union Institute Moscow 1920, Dir. of the Institute 30–40, arrested as opponent to Lysenkoism 40 and died in detention in a prison in Saratov

Verschuer, Otmar von, Dr. med. Prof., 1896–1969, Dozent Tübingen 27, ao. Prof. Berlin 33, o. Prof. Frankfurt/Main 35, Dir. KWI for Anthropology 42–45, o. Prof. Münster/Westfalia 51

Vielmetter, Walter, Dr. rer. nat. Prof., o. Prof. of genetics and microbiology Cologne 67

Virchow, Rudolf L., Dr. med. Prof., 1821–1902, o. Prof. Würzburg 49–56, Berlin 56

Vogel, Christian, Dr. rer. nat. Prof., born 1933, Dozent Kiel 66, Prof. 70, o. Prof. Göttingen 72, anthropology, comparative behavioral sciences

Vögler, Albert, 1877–1945, suicide, Member of Parliament representing the NSDAP since 33, General Director of the United Steel Company, President of the KWG 41–45

Vogt, Oskar, Dr. med. Prof., 1870–1959, Dir. KWI for Brain Research

19–37, forced to leave inst. because of political pressure, head of private inst. for Brain Research in Neustadt (Black Forest) 37

Volk, Otto Heinrich, Dr. phil. nat. Prof., born 1903, Dozent Würzburg 38, ao. Prof. 49, ao. Prof. 57

Wacker, Otto, Dr. phil., 1899–?, Minister of Cultural Affairs and Education in the state of Baden 35, Vice-President of the RFR 37–39, chief of Science Section in the REM till 39, SS member

Wagler, Erich, Dr. Prof., 1884–?, ao. Prof. Munich 32–45

Wagner, Gerhard, Dr. med., 1888–1939, chief Office for People's Health in the Reich leadership of the NSDAP and Chief Physician of the Reich (Reichsärzteführer)

Wahrman, Jacob, Dr. Dept. of Genetics Hebrew Univ. (Jerusalem), Chairman of the Genetics Society of Israel about 1960

Walter, Heinrich, Dr. phil. Prof., 1898–?, Dozent Heidelberg 24, ao. Prof. 27, Technical College Stuttgart (Agricultural College Hohenheim) 32, o. Prof. 39, Posen 41, o. Prof. Hohenheim after 45

Warburg, Otto Heinrich, Dr. phil. and med. Prof., 1883–1970, Dir. KWI for Cell Physiology 31–53, MPI for Cell Physiology 53, Nobel prize for medicine 31

Weber, Friedrich, Dr. phil. Prof., 1886–1960, Dozent Graz 18, o. Prof. 36, Prof. emeritus 57

Weber, Hermann, Dr. rer. nat. Prof., 1899–1956, ao. Prof. and Dir. Zoological Institute Technical College Danzig 30, o. Prof. Freiburg 35, Münster 36, Vienna 39, Strasbourg 41–44, Guest Prof. Tübingen 48

Weber, Ulrich, Dr. Prof., 1898–1954, ao. Prof. Würzburg 32, ao. Prof. Marburg 48

Weidel, Wolfhard, Dr. rer. nat. and med. Prof., 1916–1964, KWI for Biochemistry, Dozent Tübingen 53, ao. Prof. 57, Dir. MPI for Biology 56

Weinert, Hans, Dr. rer. nat. Prof., 1887–1967, ao. Prof. Berlin 32, o. Prof. and Dir. Inst. for Anthropology Kiel 35–55

Weismann, August F., Dr. med. Prof., 1834–1914, ao. Prof. of zoology Freiburg 66, o. Prof. 74–12

Weiss, Paul, Dr. phil. Prof., born 1898, developmental biologist, Biological Research Inst. Vienna 22–29, Fel. KWI for Biology Berlin 29–31, emigration, Yale Univ. 31–33, Asst. and Assoc. Prof. Univ. Chicago 33–42, Prof. 42–54, Rockefeller Univ. 54–64

Went, Friedrich A. C., Dr. Prof., 1863–1935, Dutch plant physiologist

Westergaard, Mogens, born 1912, Prof. of genetics Univ. Copenhagen

Wettstein, Fritz von, Prof., 1895–1945, Dozent Berlin 23, o. Prof. Göttingen 25, Munich 31, First Dir. KWI for Biology Berlin and o. Prof. Univ. Berlin 34

Wettstein, Otto von, Dr. phil. Prof., 1892–1967, Dozent Vienna 41, pensioned 45, tit. ao. Prof. 54

Wettstein, Wolfgang von, Dr.-Ing. Prof., 1898–1984, head of dept. KWI for Breeding Research Müncheberg 28, Dir. Inst. for Cellulose Breeding Technical College Karlsruhe 42–45, head of Dept. for Forestry and Forest Plant Breeding Mariabrunn 47

Wetzel, Karl, Dr. rer. nat. Prof., 1893–1945, Dozent 27, o. Prof. 36

Wetzel, Robert, Dr. med. Prof., 1898–1962, ao. Prof. Würzburg 32, o. Prof. Tübingen 36, forced to retire 45

Weygand, Friedrich, Dr. rer. nat., born 1911, Dozent Heidelberg 41, b. ao. Prof. Strasbourg 43, ao. Prof. Heidelberg 48, ao. (with civil servant status) Prof. Tübingen 53, o. Prof. Technical Univ. Berlin 55, Technical College Munich 58

Widder, Felix, Dr. phil. Prof., 1892–1974, Dozent Graz 26, ao. Prof. 36, o. Prof. 50

Wideröe, Rolf, Dr.-Ing. Prof., born 1902, developed first European Betatron in Hamburg 44–45, Dozent Zürich 53, Prof. 62

Wieland, Heinrich, Dr. phil. Prof., 1877–1957, o. Prof. Technical College Munich 17, Freiburg 31, Univ. Munich 25

Wieland, Theodor, Prof., born 1913, o. Prof. Frankfurt 51

Willstätter, Richard, Dr. phil. Prof., 1872–1942, Dozent Chemie Munich 96, ao. Prof. 02–12, o. Prof. Zürich 12–15, Munich 15–25, resigned 25 to protest against anti-Semitism at the university, emigration to Switzerland 39, Nobel prize for chemistry 15 (chemistry of plant pigments, including Chlorophyll)

Wimmer, Dr. med., collaborator of A. Hirt

Winkler, Hans, Dr. phil. Prof., 1877–1945, o. Prof. Hamburg 12

Winkler, Hubert, Dr. phil. Prof., 1875–?, ao. Prof. Breslau 21, ao. Prof. 39

Witsch, Hans von, Dr. phil. Prof., born 1909, Dozent Göttingen 39, ao. Prof. 48, o. Prof. Technical College Munich 56

Wittmann, Heinz-Günther, Dr. Prof., 1927–1990, Res. Assoc. Berkeley 56, MPI for Biology Tübingen 57, Dir. MPI for Molecular Genetics Berlin 64

Wohl, Kurt, Dr. phil. Prof., 1896–1962, Dozent Berlin 29, Research at IG Farben 36–38, emigrated to England 39, U.S. 42, Prof. Chem. Engr. Univ. Delaware (Newark) 47

Wolf, Ernst, Dr., born 1902, Dozent Heidelberg 30, dismissed 33, Harvard Univ. 31–47, American Optical Company 47–55

Wolf, P. M. head of Dept. of Radiology of Auer Company

Wollman, Elie, Sc.D., born 1917, French molecular biologist, Inst. Pasteur 42, Vice-Dir. 66

Wollman, Elisabeth, French molecular biologist, murdered in Auschwitz 1943

Wollman, Eugène, French molecular biologist, murdered in Auschwitz 1943

Wunder, Wilhelm, Dr. phil. Prof., 1898–?, ao. Prof. Breslau 30, ao. Prof. 44, Dozent Erlangen 47, ao. Prof. 58

Wüst, Walther, Dr. phil. Prof., 1901–?, o. Prof. Munich 35, Rector Univ. Munich, President of SS Research and Teaching Foundation Das Ahnenerbe, SS Lieutenant Colonel

Zachau, Hans Georg, Dr. rer. nat. Prof., born 1930, molecular biologist Munich

Zarapkin, S. R., Soviet geneticist, Fel. Genetics Dept. KWI for Brain Research Berlin-Buch 25–45, arrested in USSR 45 and interned 45–47, in scientific internment camp Sungul 47, transferred to middle Asia around 52, died soon afterward

Zierold, Kurt, Dr. jur., Ministerialrat, 1899–1989, head of dept. Niedersächsisches Ministry of Education 45, Vice-President Notgemeinschaft 49–52, Secretary General of DFG 52

Zimmer, Carl, Dr. phil. Prof., 1873–1950, o. Prof. Berlin 24, Prof. emeritus 37

Zimmer, Karl Günter, Dr. phil. Prof., 1911–?, Genetics Dept. KWI for Brain Research Berlin-Buch 37–45, research in Soviet scientific internment camp in Sungul 45–55, Dozent Hamburg and Stockholm 56, ao. Prof. Heidelberg 57, o. Prof. 58, Dir. Inst. for Radiation Biology, Reactor Station Karlsruhe 57

Zimmermann, Walter, Dr. rer. nat. Prof., 1892–1980, ao. Prof. Tübingen 29, regional commissioner for nature and landscape protection Südwürtt.-Hohenz. 46, o. Prof. Tübingen

Zur Straßen, Otto, Dr. phil. Prof. 1869–1961, Dozent Leipzig 1896, ao. Prof. 01, o. Prof. Frankfurt/Main 14

Biologists in the Study

The following are the names of the 445 biologists who form the statistical basis of the present study. In 1933 or by 1945, at the latest, they had obtained *Habilitation* at a German or Austrian university or pursued scientific work at a KWI during the same period.

O. Abel, J. Abromeit, W. Ahrens, F. Alverdes, K. Andersen,
H. André, W. E. Ankel, O. Appel, K. Arens, A. Arndt, M. Auerbach,
H. Autrum, F. Bachmann, A. Barthelmeß, R. Bauch, H. Bauer,
E. Baur, R. Beatus, I. Beleites, W. Benecke, E. Bergdolt, G. Bergold,
W. Berndt, A. Berr, E. Bersa, L. Bertalanffy, R. Beutler, R. Biebl,
A. Bluhm, F. Boas, D. Bodenstein, C. R. Boettger, A. Borgert,
H. J. Born, H. Borriss, P. Branscheidt, L. Brauner, G. Bredemann,
H. Breider, E. Bresslau, A. Brewig, F. Brieger, F. Brock, L. Brüel,
C. Brunner, P. Buchner, H. Budde, W. Buddenbrock-Hettersdorf,
J. Buder, H. Buhr, F. Bukatsch, E. Bünning, H. Burgeff,
H. Cammerloher, E. Cartellieri, A. Catsch, P. Claussen, C. J. Cori,
A. T. Czaja, V. Czurda, R. Danneel, E. G. Daumann, P. Deegener,
R. Demoll, V. Denk, E. Diebschlag, L. Diels, L. Döderlein,
W. Döpp, H. Döring, H. Driesch, G. Duncker, B. Dürken,
F. Duspiva, K. Eberhardt, T. Eckart, F. Eckert, F. Eggers, B. Eggert,
K. Egle, H. Engel, H. Erhard, A. Erhardt, I. Esdorn, P. Esser,
F. C. v. Faber, H. J. Feuerborn, P. Filzer, F. Firbas, W. Fischel,
I. Fischer, O. Fischnich, J. Fitting, A. Fleischmann, H. Fortner,
G. S. Fraenkel, V. Franz, R. Freisleben, L. Freund, K. Friederichs,
H. Friedrich-Freksa, K. von Frisch, G. Funk, H. Gaffron, H. Gams,
T. Gante, G. Gassner, B. Geinitz, L. Geitler, M. Gersch, F. Gessner,

J. Gicklhorn, H. Giersberg, E. Gilg, R. Gistl, H. Glück, W. Goetsch,
R. B. Goldschmidt, G. Gottschewski, H. Gradmann, P. Graebner,
K. Grell, J. G. Grimpe, F. Groß, F. Gruber, K. Günther,
H. von Guttenberg, J. Hackbarth, K. von Haffner, V. Hamburger,
J. Hämmerling, E. Hannig, R. Harder, J. W. Harms, O. Harnisch,
M. Hartmann, O. Hartmann, J. Hauer, R. Heberdey, G. Heberer,
C. Heidermanns, K. Heidt, A. Heilbronn, E. Heitz, F. Hempelmann,
K. Henke, H. Henseler, E. Hentschel, C. Herbst, S. Hermann,
W. Herre, K. Herter, P. Hertwig, M. Hertz, W. Hertzsch, T. Herzog,
R. Hesse, R. Heymons, M. Hirmer, C. Hoffmann, H. Hoffmann,
R. W. Hoffmann, W. Hoffmann, K. Höfler, E. Hofmann,
E. von Holst, J. Holtfreter, B. Huber, B. Husfeld, E. Irmscher,
W. Jacobs, E. Janchen, O. Janson, A. Japha, R. Jaretzky, V. Jollos,
H. Joseph, L. Jost, G. Just, H. Kalmus, R. Kaplan, H. Kappert,
H. H. Karny, A. Kießelbach, J. Kisser, B. Klatt, H. Klebahn,
G. Klein, W. Klein, A. Kloiber, E. Knapp, F. Knoll, A. Koch,
A. O. Koegel, W. Koehler, M. Koernicke, O. Köhler, R. Kolkwitz,
G. Koller, C. Kosswig, G. H. L. Krause, K. Krause, H. Krieg,
F. Kröning, C. Knonacher, F. Krüger, P. Krüger, B. Kubart,
H. Kuckuck, W. Kuhl, A. Kühn, E. Kuhn, O. Kuhn, W. Kühnelt,
E. Küster, L. Lämmermayr, F. Laibach, W. Lamprecht, S. Lange,
E. Lehmann, R. Lehmensick, K. Lehnhofer, E. Leick, M. Leiner,
H. G. Lengerken, W. Lenkeit, Friedrich Lenz, W. Limpricht,
W. Lindenbein, T. List, H. Lohmann, H. Lohwag, G. Lorbeer,
K. Lorenz, M. Lüdicke, H. Lüdtke, W. Ludwig, H. Lüers,
K. Mägdefrau, F. Mainx, O. Mangold, E. G. Marcus,
W. von Marinelli, F. K. Markgraf, H. Marquardt, O. Mattes,
E. Matthes, J. Meixner, G. Melchers, E. Menner, E. Merker,
R. Mertens, H. Merton, P. Metzner, H. Meusel, W. Mevius,
A. Meyer-Abich, C. Mez, P. Michaelis, H. Molisch, C. Montfort,
O. Moritz, G. Mosebach, K. Mothes, F. Möwus, A. Mühldorf,
E. Murr, H. Nachtsheim, E. Neresheimer, W. Neuhaus, K. Noack,
M. Nordhausen, F. Oehlkers, F. Oltmanns, R. Orth, F. Overbeck,
K. Paech, I. Panschin, A. Pascher, K. Pätau, F. A. Pax, J. Pekarek,
A. Penners, O. Pesta, T. Peterfi, O. Pflugfeder, U. Philip, H. Piepho,
R. Pilger, K. Pirschle, A. Pirson, A. Pisek, L. Plate, F. Pohl,
H. Precht, E. G. Pringsheim, H. Propach, H. Przibram, G. Radu,
B. Rajewsky, W. Rauth, J. Reibisch, E. Reichenow,
A. C. A. Reichensperger, E. Reinmuth, E. Reisinger, A. Remane,

O. Renner, B. Rensch, O. Richter, W. Riede, E. Ries, P. Rietschel,
K. Rippel, M. Roberg, K. von Rosenstiel, W. Rößler, E. Rotmann,
W. Rudorf, U. Ruge, W. Ruhland, F. Ruttner, W. Sandt, R. Schaede,
R. Scharfetter, J. Scharnke, J. Schaxel, G. Schellenberg, W. Scherz,
L. Scheuring, R. Schick, E. Schiemann, V. Schiffner, J. Schiller,
W. Schleip, C. Schliper, E. Schlottke, G. Schmid, H. Schmidt,
M. Schmidt, O. C. Schmidt, W. J. Schmidt, H. Schmitz,
T. Schmucker, K. Schnarf, K. C. Schneider, G. Schramm, E. Schratz,
A. Schraub, G. Schubert, K. L. Schulze, P. Schulze, W. Schumacher,
P. N. Schürhoff, B. Schussnig, F. Schwanitz, O. Schwartz,
V. Schwartz, W. Schwartz, W. Schwarz, P. Schwarze,
P. Schwemmle, F. Seidel, R. Seifert, J. Seiler, A. Seybold, H. Sierp,
S. V. Simon, H. Söding, S. Spek, H. Spemann, A. Sperlich, E. Spohr,
H. J. Stammer, O. Steche, E. Stein, O. Steinböck, F. Steinecke,
M. Steiner, F. Steiniger, W. Steinitz, W. Stempell, C. Stern,
A. Steuer, O. Stocker, H. A. Stolte, R. Stoppel, O. Storch,
H. A. von Stosch, O. zur Strassen, J. Straub, E. Stresemann,
H. Strouhal, S. Strugger, A. E. Stubbe, H. Stubbe, G. von Studnitz,
K. Suessenguth, F. Süffert, W. Szeliga-Mierzeyewski,
A. Thienemann, N. W. Timoféeff-Ressovsky, G. Tischler, E. Trojan,
W. Troll, G. von Ubisch, L. von Ubisch, J. von Uexküll,
E. Uhlmann, H. Ullrich, W. Ulrich, J. Versluys, O. H. Volk, F. Voß,
E. Wagler, A. Wagner, B. Wahl, H. Walter, F. Weber, H. Weber,
U. Weber, F. Werner, F. von Wettstein, O. von Wettstein,
W. von Wettstein, A. Wetzel, K. Wetzel, H. Weyland, F. Widder,
Hans Winkler, Hubert Winkler, H. von Witsch, T. Wohlfahrt,
E. Wolf, R. Woltereck, H. D. Wulff, W. Wunder, H. Wurmbach,
H. Zickler, H. Ziegenspeck, C. Zimmer, K. G. Zimmer, Karl
Zimmermann, Klaus Zimmermann, W. Zimmermann.

Abbreviations

APS	American Philosophical Society, Philadelphia
Arch. AdW.	Zentrales Archiv der Akademie der Wissenschaften, Berlin
Arch. MPG	Archive for the History of the Max Planck Society, Berlin
BAK	Bundesarchiv, Koblenz
BDC	Berlin Document Center
DFG	Deutsche Forschungsgemeinschaft—German Research Association
DFR	Deutscher Forschungsrat—German Research Council
KWG	Kaiser Wilhelm Gesellschaft—Kaiser Wilhelm Society
KWI	Kaiser Wilhelm Institute
MPG	Max Planck Gesellschaft—Max Planck Society
MPI	Max Planck Institute
Ms. SPSL	Documents of the Society for the Protection of Science and Learning, Bodleian Library, Oxford University
NA	National Archives, Washington D.C.
NSDAP	Nationalsozialistische Deutsche Arbeiterpartei—National Socialist German Workers' Party (Nazi party)
NSDB	Nationalsozialistischer Dozentenbund—National Socialist League of University Lecturers
NSLB	Nationalsozialistischer Lehrerbund—National Socialist Teachers' League
OKL	Oberkommando der Luftwaffe—Air Force High Command
OKM	Oberkommando der Marine—Navy High Command
OKW	Oberkommando der Wehrmacht—Army High Command
RAC	Rockefeller Archive Center, Tarrytown, New York
REM	Reichsministerium für Wissenschaft, Erziehung, und Volksbildung—Reich Ministry for Science, Education, and People's Education
RFR	Reichsforschungsrat—Reich Research Council

RfW	Reichsamt für den Wirtschaftsausbau
RM	Reichsmark
RSHA	Reichssicherheitshauptamt of the SS
RuSHA	Rasse- und Siedlungshauptamt of the SS
SCI	Science Citation Index
SD	Sicherheitsdienst des Reichsführers SS
TH	Technische Hochschule—Technical College
TU	Technical University, Berlin
UA	University archives at various German and Austrian institutions
uk	"Unabkömmlich"—exemption from military service

Notes

Introduction

1. Mentzel was president of the German Research Association from 1937 to 1942 and from 1939 on ministerial director in the Reich Education Ministry. Schumann headed the research division of the Army Ordnance Office from 1934 on, headed the Science and Technology Section in the Office for Science in the Reich Education Ministry, and in 1938, as ministerial leader, was put in charge of the Science Section of the Reich War Ministry (Zierold 1968, p. 192).

2. As the most glaring example of this, Zierold (1968) mentions Hitler's order in 1940 that all research which did not lead to any results by the end of the year was to be stopped (p. 271).

3. For this reason Speer arranged in 1942 for the reorganization of the council and in 1943 for the creation of a planning office—like the one that had been set up in 1942 in the economic sector for the distribution of raw materials—also in the area of science (Zierold 1968, pp. 248ff.). In a 1942 memorandum on the Reich Research Council—the author, though anonymous, is presumed to be Professor Osenberg—we read: "Since research, especially basic research, had been severely neglected in Germany during the Third Reich, the government decided . . . in 1942 to put research work on a new basis and to intensify it in every way" (BAK, R 73/29).

4. Änne Bäumer (1990) does not address the expulsion of Jewish scientists, the content of biological research, and the impact of National Socialist university and research policy on the field of biology.

5. *FIAT-Reviews of German Science, 1939–1946,* esp. vols. 21 and 22 (1949) and vols. 52–55 (1948) (see Chapter 3). K. Macrakis (1989, pp. 165–213) emphasizes the continuation of basic biological research under National Socialism at the Kaiser Wilhelm Institutes; an appraisal of genetic and botan-

ical research in Berlin, including the Nazi period, can be found in Plarre 1987 and Schnarrenberger 1987.

6. An attempt was made to include in the list the names of all biologists at these Kaiser Wilhelm Institutes with a Ph.D. but without *Habilitation* if they had received DFG support or fellowships, were dismissed in 1933 or later, and acquired *Habilitation* by 1961 at the latest (based on information in *Kürschners Deutsches Gelehrten Lexikon*).

7. KWI for Biology in Berlin-Dahlem (founded 1912), Hydrobiological station in Plön (taken over by the Kaiser Wilhelm Society in 1917), the genetics department (founded 1925) of the KWI for Brain Research in Berlin-Buch, the KWI for Breeding Research in Müncheberg (founded 1927) with its branches, and the KWI for Biophysics in Frankfurt (founded 1937).

1. The Expulsion and Emigration of Scientists, 1933–1939

1. The determination of who was a Jew, a half-Jew, and so on, was made independent of the religion of the person concerned or that of his or her parents. For the National Socialists, who made the concept of race the basis of their state, it was only logical to be guided not by religion but by presumed racial membership. To discover that membership they had to go back as far as grandparents, since there were hardly any Jewish-Christian baptisms before 1800.

2. Of the twenty-eight members of the senate elected or appointed in May 1933, fourteen were chosen by the Central Committee of the KWG. Three of them, Franz von Mendelssohn, Alfred Merton, and Paul Schottländer, were Jews. The formerly Social Democratic members were not reelected; fourteen members of the senate were appointed by the ministries.

3. Arch. MPG, Abt. I, Rep. 1a.

4. M. Planck in *Die Naturwissenschaften* 22 (1934): 344.

5. M. Planck in *Die Naturwissenschaften* 25 (1937): 376.

6. She was dismissed for being non-Aryan because her grandmother had been Jewish before she was baptized in 1839 (Interior Ministry to F. Glum, March 3, 1934, Arch. MPG, Abt. I, Rep. 1A/543).

7. Du Bois-Reymond to F. Glum, March 30, 1934, Arch. MPG, Abt. I, Rep. 1A/543.

8. Knoll to State Commissioner Plattner, April 27, 1939, archival material with Doris Baumann, Vienna.

9. Adjunct of the Biological Experimental Institute (name indecipherable) to the curator of scientific institutes *(Hochschulen)*, Vienna, May 4, 1940, archival material with Doris Baumann, Vienna.

10. Ms. SPSL 538/2.

11. Przibram to R. Goldschmidt, January 9, 1941, R. Goldschmidt Papers, Bancroft Library.

12. RAC, RF, RG 1.1/200, box 163, folders 2004 and 2005.

13. R. Goldschmidt to the American Friends Service Committee, February 24, 1941, APS, Ms. coll. 5, C. Stern, Goldschmidt file #4.

14. Ibid.

15. Gerta von Ubisch, "Lebenserinnerungen," no date, ca. 1955, Heidelberg Hs. 4029 (university library Heidelberg).

16. C. Stern to M. Hartmann, May 16, 1933, APS, Ms. coll. 5, C. Stern, M. Hartmann file.

17. Zentrales Staatsarchiv Potsdam, Best. REM No. 1393, fol. 26.

18. *Albert Einstein / Max Born: Briefwechsel 1916–1955,* new edition (Munich: Nymphenburger Verlagshandlung, 1991), p. 153.

19. Cited after Goldstein 1987, p. 153.

20. This group, as well, does not include the dismissed university assistants without *Habilitation* and the dismissed students of biology.

21. The specific circumstances of the dismissal were told to me by Dr. P. W. Wohlers in a letter of February 17, 1993.

22. Archive of the Research Foundation for Jewish Immigration, Technische Universität Berlin, Zentrum für Antisemitismusforschung, archive, C. Kosswig; compare Erichsen 1991.

23. Nachmansohn 1979 acknowledges the scientific accomplishments of German Jewish biochemists. Among them were many Nobel laureates, for example, Otto Meyerhof, director at the KWI for Medical Research in Heidelberg, and Otto Loewi, professor at the University of Graz.

24. Compare the correspondence between R. Goldschmidt and colleagues in R. Goldschmidt Papers, Bancroft Library, and Goldschmidt 1956a and 1940.

25. Arch. MPG, Abt. I, Rep. 1A/1538.

26. Professor W. Rau, Munich, 1973, archive of the Research Foundation for Jewish Immigration, TU Berlin, Zentrum für Antisemitismusforschung, archive, L. Brauner.

27. More detailed information can be found, for example, in Hamburger 1988; Oppenheimer and Willier 1974; Horder, Witkowski, and Wylie 1985; Nakamura and Toivonen 1978; Sander 1990.

28. On the current state of research on the "organizer" and inductive molecules, see Gilbert and Saxén 1993. Scientists now believe that it is not a single "master molecule" which is responsible for the formation of Spemann's organizer, but that we are dealing with a process brought on by several maternal and zygotic factors. The gene of one of the proteins responsible for the development of the dorsal lip of the blastopore into the mesoderm with its

inductive capacities was cloned in 1992. The identity of the diffusible factors is still unknown.

29. Richard Hesse in an assessment for the DFG, September 18, 1933, DFG file Holtfreter.

30. Hamburger analyzed the contributions of the four authors and came to the conclusion that only Holtfreter's experiments (130 manifestly strong cases) gave evidence of the inductive capacity of dead tissues; the experiments of the coauthors showed only 4 weak cases (1988, pp. 96–99).

31. Hamburger attributed this to Spemann's organismic notions, which led him, when it came to choosing experiments, to prefer those in which the embryo or the tissue remained as intact as possible (Hamburger 1980, p. 308). However, as described above, this did not mean that Spemann decided to forgo an experimental causal analysis of the properties of the organizer (self-differentiation, induction, regulation).

32. The experiments on which this paper were based were carried out between 1930 and 1932.

33. V. Hamburger, "My First Encounters with the Developing Nervous System," unpublished essay, February 1991. Unless otherwise noted, the following information about Hamburger's research comes from this essay and from his letters to me of April 15 and June 5, 1991.

34. Hamburger to author, April 15, 1991.

35. On the advantages of chick embryos as research subjects for neuroembryology, for example the fact that their nerve centers—owing to their size and the exact timetable they keep in their development—are much more suitable for experimental analysis, see R. Levi-Montalcini 1987, p. 1155.

36. See also the autobiography of R. Levi-Montalcini (1988).

37. Hamburger 1990, p. 127. Levi-Montalcini has described the story of the nerve growth factor as "a long sequence of unanticipated events" (1987, p. 1161).

38. Personal communication, August 15, 1991.

39. V. Hamburger to author, June 5, 1991, and UA Cologne, personal file O. Kuhn.

40. Personal communication, October 21, 1991.

41. Ibid.

42. Fürst von Fürstenberg had made his castle available for this institute.

43. One of these students, Martin Schnetter, a professor in Freiburg, "remembered" Hamburger and Holtfreter (Gluecksohn-Walesch is not mentioned) as follows in a speech on the Spemann era at the Zoological Institute of the University of Freiburg: "Hamburger, later assistant and assistant professor at the institute, worked on the development of nerveless extremities and later went to the U.S. . . . Holtfreter, one of the most important collaborators of Spemann and Mangold, who is still today working in Rochester

(USA) . . ." (*Berichte der Naturforschenden Gesellschaft Freiburg* 58 [1968]: 99).

44. Personal communication from Professor Ursula Nürnberg, Alt-Ruppin, April 3, 1990. Nürnberg was a student at the Institute for Heredity in Berlin until 1945, she no longer remembered the name of the woman party comrade whose dissertation Schiemann rejected.

45. Personal communication from Hermann Kuckuck, June 14, 1989 (compare Kuckuck 1988).

46. An alternative hypothesis, according to which the dance of the bees was merely intended to incite the bees to swarm, while the actual information about the food source was passed on via olfactory substances, was a source of controversy for many years (see, for example, Manning 1979, pp. 95ff.). J. L. Gould (1976), however, gave unequivocal experimental evidence of the communicative meaning of the dance.

47. Information from letter from E. Bergdolt to the REM, May 15, 1936, BDC, file Frisch.

48. Political evaluation of Dr. Karl von Frisch by W. Führer, Gau leadership Munich of the NSDAP, October 19, 1937, ibid.

49. See n. 47. Otto Löwenstein worked as assistant from 1931 to 1932 and from 1932 to 1933 as a free researcher at the institute, before emigrating to England (Birmingham) in 1933. Curt Stern, assistant at the KWI for Biology, emigrated in 1933 because of persecution relating to his Jewish descent.

50. *Das Reich* 5 (1941). Bergdolt considered this article the result of a campaign started by "Wettstein and his friends" in an effort to influence the decision concerning Frisch via the Forschungsdienst, the RFR, and Göring (Bergdolt to W. Führer, February 6, 1941, BDC, file Frisch). The weekly *Das Reich* began publication June 26, 1940; its editorial writer was Joseph Goebbels. Subsequent quotations and documents discussed in the text in recounting the decision on Frisch, are from the BDC file Frischi.

51. F.-C. von Faber to W. Gerlach, November 6, 1940, quoted after Beyerchen 1977, p. 166.

52. UA Munich, personal file Ruth Beutler, E-II-N.

53. Ibid., Klaus Clusius file, August 19, 1946.

54. I am indebted to Dr. P. W. Wohlers for pointing out the importance of Gassner, personal communication, February 17, 1993.

55. F. Oehlkers to O. Renner, November 12, 1954, UA Tübingen, 275/3.

56. F. Oehlkers to E. Bünning, November 12, 1954, UA Tübingen, 275/3.

57. H. Grüneberg to H. Nachtsheim, September 6, 1954, H. Grüneberg Papers, Wellcome Institute for the History of Medicine.

58. Ibid.

59. H. Nachtsheim to H. Grüneberg, April 23, 1955, Grüneberg Papers.

60. The best-known example of the extent of such indemnification is the

case of Gerstenmaier. Eugen Gerstenmaier, president of the Bundestag (Lower House of Parliament), received in 1965 a retroactive payment of DM 280,000 as compensation for the professorship that was blocked by the Nazis. His application was processed within nine months (Herbst and Goschler 1989, p. 240). As the examples discussed below indicate, this case is utterly different from the experience of most of the persecuted.

61. Ministry of Culture of Baden-Württemberg to G. v. Ubisch, February 17, 1953, UA Heidelberg, personal file G. v. Ubisch.

62. The widow of G. P. A. Merton to the rector of the University of Heidelberg, July 24, 1950, UA Heidelberg, personal file H. Merton.

63. Declaratory decree by the State Office for Compensation in Karlsruhe to Mrs. Gertrud Merton, February 13, 1953, UA Heidelberg, personal file H. Merton.

64. Restitution *(Wiedergutmachung)* decision by the Ministry of Culture of Baden-Württemberg against the State of Baden-Württemberg, August 17, 1956, UA Heidelberg, personal file H. Merton.

65. Unless otherwise indicated, the dates come from the university archive Heidelberg, personal file G. v. Ubisch, and from Ubisch's unpublished "Memoirs," "Lebenserinnerungen," no date, ca. 1955, Heidelberg Hs. 4029.

66. According to Ubisch, this attitude to genetics could be considered the typical attitude of most professors of biology ("Lebenserinnerungen").

67. Ubisch describes in her memoirs the reaction of the physicist Philipp Lenard to her planned *Habilitation:* "It is a sad decision you have made. A woman's purpose is something very different. Are you unmarried? If you had sent me the announcement of your engagement, I would have been very happy, but this . . ." At the faculty vote, he was the only one to vote against her *Habilitation.* Lenard was one of the founders of "Aryan physics," and he had already called attention to himself in the twenties through his anti-Semitic and antiforeign attitude.

68. G. v. Ubisch, July 5, 1956, UA Heidelberg, personal file G. v. Ubisch.

69. "Expert Opinion by Professor Jost on the *Habilitation*'s dissertation of v. Ubisch," December 8, 1922, ibid.

70. Ibid.

71. Ms. SPSL 206/3, 332–334.

72. Statement by professor Walter Hoffmann, March 23, 1954, UA Heidelberg, personal file G. v. Ubisch. Hoffmann stated that the NSDStB had initially spread the assertion that Ubisch had advocated liberalist, unwelcome views in her lectures, and that he subsequently put pressure on the students to boycott the lectures.

73. G. Berger, University of Edinburgh, to the Academic Assistance Council, September 13, 1934 (Ms. SPSL, 206/3).

74. Letter of Ubisch to the Württemberg Ministry of Culture, undated,

1953 (UA Heidelberg, personal file G. v. Ubisch). In "Lebenserinnerungen" Ubisch in this context talks about the example of Reinhard Heydrich: "No non-German will understand that the widow of the 'general protector' of Bohemia, Heydrich, who outside of Germany is called the 'mass butcher of Bohemia' and who has thousands upon thousands of Czechs and Poles on his conscience—that his widow, retroactive from 1950 (or even earlier), had been granted the widow's pension of high officer, presumably because there are no laws about nonpayment of a pension to the widow of a mass murderer in a high government position."

75. Letter of Ubisch, April 4, 1956 (UA Heidelberg, personal file G. v. Ubisch).

76. Ibid.

2. NSDAP Membership, Careers, and Research Funding

1. BAK, R21 appendix, nos. 10,000ff.

2. This number refers to all physicians who were registered with the Reich Physician Chamber between 1936 and 1945.

3. In the SA, especially, there were a number of resignations in the first years of Nazi rule, which were presumably connected with the unclear political role of the organization during that time.

4. These were almost exclusively geneticists who were simultaneously botanists or zoologists.

5. REM, WA no. 2160/42, re Reichhabilitationsordnung, BAK, R 21/10807.

6. Ibid.

7. The Reichdozentenführer (head of the NSDB), 1938, BAK, NSD 33/1.

8. Ibid.

9. Decree of the REM on May 14, 1938, BAK, R 21/11193.

10. BAK, NS 15/298.

11. The evaluations mentioned here are in the archives of Humboldt University.

12. Ibid., ZDI/148.

13. Ibid., Z/B2/1861/A:2.

14. Ibid., ZDI/742.

15. Ibid., Z/B2 1885/A:4.

16. For example, according to Eleonore Bünning, a lot of political pressure was put on her husband, Erwin Bünning—who was not a member of the party—and his family, as a result of which they had to leave Jena in 1935 (personal communication, September 13, 1989).

17. UA Vienna, personal file Paul Krüger.

18. UA Munich, personal file Friedrich-Carl von Faber, E-II-N.

19. Dean A. Marchet to the REM, September 14, 1944, UA Vienna, personal file O. Abel.

20. According to information furnished by the UA Graz.

21. Ibid.

22. Thus the rector of the University of Tübingen, for example, refused to appoint Karl Henke to the chair (full professorship) of zoology in Tübingen in 1936 because of an evaluation from the NSDB, in which Henke was criticized for political indifference (BAK, R 21/10417, rector to the REM, October 16, 1936). Shortly afterward Henke was appointed to the chair in Göttingen.

23. BDC, file Hans Stubbe.

24. Prussian education minister to L. von Ubisch, December 4, 1934, UA Heidelberg, personal file G. von Ubisch.

25. G. von Ubisch to the Ministry of Culture of Baden Württemberg, undated, about February 1953, ibid.

26. Preußisches Geheimes Staatsarchiv (Berlin; hereafter Geh. Staatsarch.) Rep. 76/293.

27. BDC, file H.-J. Feuerborn.

28. Ibid.

29. Geh. Staatsarch. Rep. 76/293.

30. Ibid., W. Mevius (Münich professor) to K. Meyer (of the REM), October 20, 1935. According to Mevius, Feuerborn used the students as "combat troops" *(Kampftruppe).*

31. These and other conflicts with the party and the SS led Kosswig to emigrate to Turkey to escape imminent arrest (see Chapter 1, Section 6). Breider was not given the assistant position in Braunschweig. He became department head at the KWI for Breeding Research.

32. Geh. Staatsarch. Rep. 76/293, dean of the faculty of philosophy and natural sciences to the REM, December 7, 1935.

33. Ibid., Mevius to Meyer, October 20, 1935.

34. Ibid. Mevius also emphasized that he believed it was necessary to choose German biologists for international congresses with greater care. They were to help improve the hostile international attitude toward Germany, because "the Jewish press has been masterful at inciting the so-called intellectual circles against us."

35. Ibid.

36. According to volkish ideology, which originated in the nineteenth century, the Germans could attain a dominant position in Europe, which they were entitled to as the largest nation in Europe in terms of population size, by becoming conscious again of their Volkdom *(Volkstum).* The state was seen as the natural "organic community," resting on the German Volkdom. The Western European notion of the nation state as a historical and political

functional union was rejected. The volkish movement demanded the cleansing of the real Germandom from all "foreign elements" of other cultures, for example from Englightenment ideas, urbanization, and parliamentarianism. Characteristic of the volkish ideology was its anti-Western and anti-Semitic orientation. In its first years the National Socialist movement was substantially influenced by such volkish anti-Semitic groups as the Thule Society, which had emerged out of the Teutonic Order. See Bracher 1979, pp. 22ff.

37. E. Lehmann to W. Wüst, October 26, 1938, BDC, file E. Lehmann. It is not known how long Lehmann belonged to these organizations.

38. E. Lehmann to Rector Stickl, October 19, 1944, UA Tübingen, 126/373-1. The membership list of the German Association of Biologists from February 1932 contains in fact, with one exception, no names of Jewish professors and lecturers (as far as they are known to me). By contrast, a large part of the non-Jewish university professors and lecturers were members (*Der Biologe* 1 [1931/1932]: 147–158).

39. Memorandum from Lehmann, December 3, 1936, UA Tübingen, 126/373-5.

40. UA Tübingen, 136/130. At the request of the Lehmann Publishing House, H. F. K. Günther wrote, among other works, his *Rassenkunde des Deutschen Volkes* (Racial anthropology of the German Volk). This book, first published in 1922 and reprinted many times without any substantial changes, became the standard work of National Socialist racial doctrine.

41. BDC, NSDAP file E. Lehmann.

42. UA Tübingen, 201/PA, personal file Lehmann.

43. Lehmann to SS Lieutenant Colonel Prof. Wüst, October 26, 1938, BDC, file Lehmann.

44. Ibid., he himself thought the only remaining possibility was that a charge of "unsuitability of character" had led to the refusal to admit him.

45. Petition by Lehmann dated October 23, 1937, UA Tübingen, 201/PA, personal file Lehmann.

46. In his letter to Wüst, Lehmann wrote about a letter to Himmler dated April 26, 1937, and a petition to Hitler in June of 1938.

47. Memorandum by Lehmann dated December 13, 1936, UA Tübingen, 126/373-5; all quotes in this section have been taken from this memorandum.

48. Statement by the University Commission of the NSDAP from December 1938 on Lehmann's conduct in Munich in 1934, UA Tübingen, 126/373.

49. Wettstein during the preliminary proceedings against Lehmann in December of 1938, UA Tübingen, 126/373.

50. Statement by Lehmann, May 24, 1940, UA Tübingen, 126/373-18.

51. Ibid.

52. On "German physics" see, for example, Beyerchen 1977 and Richter 1980.

53. In addition, "German physics" had produced two theoretical approaches that diverged from general physics, Lenard's ether theory and Stark's atomic theory, neither of which could be sustained scientifically (Beyerchen 1977; Richter 1980).

54. As an illustration, consider a statement by Lenard in the *Völkische Beobachter* (May 13, 1933): "The foremost example of the damaging influence upon natural science from the Jewish side was presented by Mr. Einstein, with his 'theories' mathematically blundered together out of good, preexisting knowledge and his own arbitrary garnishes, theories which now are gradually decaying, which is the fate of all procreations alien to nature." Quoted from Beyerchen 1977, p. 101.

55. K. Beurlen to the DFG, February 23, 1937, BAK, R 73/15952. The president of the DFG, R. Mentzel, originally intended this publication to make the journals *Naturwissenschaften* and *Nature* obsolete. *Die Naturwissenschaften* an "unwelcome but hitherto necessary liberal scientific journal (was to be) crushed by legal means." In 1937, however, he no longer considered this goal attainable (Mentzel to F. Kubach, August 5, 1937, ibid.).

56. According to information from the Rosenberg Office, the following people belonged to the so-called Munich circle: Führer, Kubach, Bergdolt, Wüst, Dingler, Thüring, and, since 1941, May (Hauptamt Wissenschaft, file E. May, Arch. YIVO Institute, New York).

57. Lehmann received positive reviews for the DFG from Harder, Knoll, and Wettstein (R 73/12627); in 1936 and 1937 he received just under RM 6,000 from the DFG; and his publications were cited sixty-one times by others in the SCI 1945–1954.

58. In the scientific area, too, Lehmann saw himself as the victim of unjustified attacks and subjective criticism. In 1942 he spoke about the "polemic of the Renner school" against his publications (UA Tübingen, 126/373-7), and in 1944 he complained to the rector of the University of Tübingen: "v. Wettstein and his students have, not only in Dahlem but also in Tübingen, what are under the current conditions quite unusual possibilities of carrying out their work unimpeded" (UA Tübingen, 126/373-1).

59. Brücher had charged that pressure from Lehmann had been responsible for his not being allowed to publish, in his doctoral dissertation, results that could only be explained by cytoplasmic inheritance, and Lehmann subsequently considered suing him for libel (UA Tübingen, 126/373-10, November 1938).

60. G. v. Ubisch, "Lebenserinnerungen," no date, ca. 1955, Heidelberg Hs. 4029 (university library, Heidelberg).

61. UA Tübingen, 126/373. He was accused, among other things, of embezzling money and illegally opening mail.

62. BDC, file Lehmann, note of October 25, 1944, no name. Lehmann,

too, was of the opinion that Wetzel was behind the accusations (Lehmann to H. Schneider, rector of the University of Tübingen, July 9, 1945, UA Tübingen, 149/36). Wetzel had been professor of anatomy at the University of Tübingen since 1936. After 1945 he became *Professor zur Wiederverwendung* (waiting for a new position after de-Nazification), but had teaching assignments at the TH Stuttgart and the University of Tübingen.

63. UA Tübingen, 126/373-16. These activities included statements like the following: "Rosenberg has two wives and no children; what do you expect if a common sailor becomes Reichstatthalter; you can't possibly believe that Fritsch is a National Socialist."

64. UA Tübingen, 126/373-9.

65. UA Tübingen, 126/373-1.

66. Lehmann on December 22, 1938, UA Tübingen, 126/373-5.

67. Lehmann to the rector, September 12, 1945, Landesverwaltung für Kultus, Stuttgart and Military Government, UA Tübingen, 149/38.

68. Ibid.

69. Lehmann on point three of the accusation, undated, ca. 1940, UA Tübingen, 126/373-18.

70. Lehmann to the rector, September 12, 1945, Landesverwaltung für Kultus, Stuttgart and Military Government, UA Tübingen, 149/38.

71. *Der Biologe* 4 (1935): 98–99.

72. Lecture by Lehmann, 1937, UA Tübingen, 126/373-7a.

73. UA Tübingen, 126/373-2.

74. Ibid.

75. UA Tübingen, 126/373-11.

76. BAK, DFG file A. Kühn. Kühn inquired of the Notgemeinschaft on July 3, 1934, whether or not he should accept the offer. He noted that he had never before applied for support from the Rockefeller Foundation or any other foreign source, because "it goes without saying that I prefer to draw upon the funding sources of my fatherland for my research work." The answer has not been documented.

77. In 1972 all the files of the Notgemeinschaft and the DFG from 1920 to 1946 that were saved in the war were transferred to the BAK. They contain hardly any information about grants for biologists betweeen 1920 and 1932. For that reason I have evaluated the approved grant applications for the years 1930 to 1932 according to the "Lists of individual applications reviewed by the special committees (of the Notgemeinschaft)" (BAK, R 73/109–R 73/118). These files contain the lists nearly in their entirety.

78. According to K. Noack, Berlin, Mothes was not very popular with his Königsberg faculty because as a National Socialist he pushed himself into the limelight (letter of March 10, 1952, to Fritz Bopp, Munich). Mothes, according to Noack, was one of the biologists who made a political about-face in

1933; before that he had leaned to the left (letter of December 9, 1933, to L. Jost, Heidelberg). Both letters are part of K. Noack's posthumous correspondence, 1928–1952 (in the possession of E. Höxtermann, Berlin).

79. Ibid.; Mothes's good contact with Koch is also confirmed by Hermann Kuckuck, Hannover (personal communication, June 14, 1989).

80. Report of the RFR of August 19, 1944, BAK, R 26III/276.

81. F. Markgraf, to the REM, July 27, 1942, BAK, R 21/10985.

82. Max Planck in the preface to *25 Jahre Kaiser-Wilhelm-Gesellschaft* (Planck 1936).

83. Wettstein to the minister of education and culture, Munich, July 11, 1934. UA Munich, E-II-N personal file F. v. Wettstein.

84. Ibid.

85. A. Kühn to the DFG, September 3, 1940, BAK, R 73/11227.

86. See Chapter 4, Section 5.

87. Correspondence between M. Demerec and Timoféeff-Ressovsky from January to June 1936, APS, B/D394, Demerec Papers, Timoféeff-R #2.

88. Timoféeff-Ressovsky to Demerec, May 11, 1936, ibid.

89. Evaluation by Graue, head of NSDB Group Independent Research Institutes Berlin, June 13, 1938, UA Humboldt, Berlin.

90. Information from Dr. Scurla according to the minutes of the proceedings, Zentrales Staatsarchiv Potsdam, REM 3191. K. Macrakis (1989, pp. 224ff.) has evaluated this file with respect to the KWG.

91. Mentzel, according to the minutes of the proceedings, Zentrales Staatsarchiv Potsdam, REM 3191.

92. Ibid.

93. Scurla, according to the minutes of the proceedings, ibid.

94. Ibid.

95. Wettstein, according to the minutes of the proceedings, ibid.

96. District leader (name unreadable) to the district personnel office of the NSDAP, May 10, 1935, BDC, file A. Kühn.

97. SS Captain Walter Greite to Prof. Wüst, June 7, 1939. BDC, SS file Prof. Knoll.

98. "In humans, where natural selection is much less intense, especially among civilized peoples, the conditions for the preservation and spread of even strongly pathological mutations are even more favorable; this explains why the human population is also burdened by a number of dominant hereditary diseases. For the science of human genetics as well as racial hygiene, it would be of particular importance not only to determine the percentage of genetically diseased people, but also to gradually carry out an analysis of the geographic diffusion and concentration of heterozygotic genetic carriers. This would not only support supervision in terms of racial hygiene, it would also make it easier to clarify many difficult questions relating to the etiological

and genetic classification of certain genetic diseases" (Timoféeff-Ressovsky 1935).

99. *Neues Volk,* (December 1938), pp. 26–30. The Office of Race Policy was set up by the the NSDAP in 1934 to coordinate the training and propaganda in the areas of population policy and racial policy.

100. Timoféeff-Ressovsky to the Reich education minister concerning his research in 1937/1938, BAK, R 21/11065.

101. General Director Telschow wrote to the Reich education minister on December 8, 1938, that Boehm, a member of the board of trustees of the KWI for Brain Research, had adopted Timoféeff's-Ressovsky's position of not changing his citizenship and had represented it to the REM. BAK, R 21/11065, p. 58.

102. Because of his statements about the military strength of the Soviet Union, the REM asked the KWG in a letter (dated September 7, 1943) to comment on this and to investigate him. BAK, R 21/11065.

103. Telschow to the REM, October 1, 1943, ibid.

104. BDC, file Walther Arndt.

105. Ibid.

106. Meyer 1942, p. 255. He emphasized that, according to the führer's decree of October 7, 1939, the restructuring of the "German East" was under Himmler's command. We should recall that Meyer was the section chief in the RFR who decided most of the grant applications from biologists.

107. "The General Plan East," presented by SS Brigadier General Prof. Dr. Konrad Meyer, Berlin 1942, Institut für Zeitgeschichte, Munich, MA 1497. Meyer had presented a first, tougher version of his plan as early as July 15, 1941—a version that can no longer be located; see Heiber 1958. On the various versions of the General Plan East and the planned as well as actually realized expulsion and destruction of the Jewish, Polish, and Soviet populations of the East, see also Aly and Heim 1991, pp. 394–440.

108. The General Plan East was the reason why Meyer was charged by the military tribunal in Nuremberg after the war. However, he was acquitted of all main charges (Becker, Dahms, and Wegeler 1987, p. 421). He became full professor for town and country planning at the University of Hannover in 1956 and member of the Academy for Spatial Research and Town and Country Planning in 1957.

109. Institut für Zeitgeschichte, Munich, NO 2585, report on the meeting at the Ministry of the East on February 4, 1942.

110. H. Stubbe to the RFR, January 23, 1942, Arch. MPG, Abt. I, Rep. 1A/2564.

111. OKW, June 16, 1941, Stubbe Fond 152, Arch. AdW. Berlin.

112. In a 1940 meeting with Wettstein, Rudorf, and Ernst Telschow (General Director of the KWG), Backe emphasized that he considered the estab-

lishment of a central place for primitive strains of cultivated plants an urgent necessity and that he would promote this project on his own (file note by Telschow, January 31, 1940, Arch. MPG, Abt. I, Rep. 1A/2963).

113. Among the people who sent him congratulations on the occasion of this appointment were Albert Vögler, the president of the KWG, and Erich von Tschermak-Seysenegg from Vienna (BAK, NL 75 [Backe Papers], no. 10).

114. BAK, NL 75 (Backe Papers), no. 7: "Backe, die russische Getreidewirtschaft" (undated), pp. 171–178; Kempner 1987, pp. 282–293.

115. Sworn statement by the managing director of the Max Planck Society (Ernst Telschow) on January 3, 1949, in the conciliation proceedings against the posthumous papers of former Reich Minister Herbert Backe. BAK, NL 75 (Backe Papers), no. 9.

116. File note by Telschow, February 19, 1942, and March 9, 1942, Arch. AdW. Berlin, KWG Nr. 7, Bl. 4 and 4; compare Macrakis 1989, pp. 224ff.

117. File note by Telschow concerning a meeting in the office of State Secretary Backe with F. v. Wettstein and K. Meyer in the REM on February 9, 1942, Arch. AdW Berlin, KWG Nr. 7, Bl. 4, quoted by Macrakis 1989, p. 229.

118. Wettstein summarized the details of the plan presented at the meeting in a letter to Telschow (dated March 9, 1942), Arch. AdW. Berlin, KWG Nr. 7, Bl. 1–2.

119. Ibid.

120. Ibid. We can no longer determine today what Wettstein's intention was in making this remark.

121. File note by Telschow on February 10, 1942, ibid., Bl. 4.

122. Ibid.

123. One could argue that humanitarian motives were behind Rudorf's and Wettstein's intervention with regard to the Soviet institutes. They may have possibly intended to prevent the institutes from falling into the hands of unqualified biologists or the SS. In any case, they evidently believed that the areas in which these institutes were located would remain in German hands.

124. *Jahrbuch der Kaiser-Wilhelm-Gesellschaft, 1942*, p. 32.

125. "Urgent war-important projects of the RFR," BAK, R 26III/271.

126. DFG file Hackbarth, BAK, R 73/11415.

127. Personal Staff Reichsführer-SS to SS Lieutenant General Lorenz, June 6, 1943: "I'm sure you know that the Reichsführer-SS has been charged by the Führer with pushing ahead with all means at his disposal with the research, breeding, cultivation, and extraction concerning the rubber plant Koksaghyz. To this end all relevant institutions and offices (. . . 3. The Kaiser Wilhelm Institute . . .) have placed themselves under the leadership of the Reichsführer-SS." BDC, SS file Vogel.

128. Waffen-SS to Reichsführer-SS, February 12, 1943, BDC SS-HO 7139, p. 163. The name of the SS lieutenant colonel in charge was not given.

129. Ibid., p. 161.

130. Lorenz to Brandt, July 27, 1943, BDC, SS file Vogel.

131. Report of the working conference, BDC, SS-HO 7139a, pp. 28–31.

132. Ibid.

133. D. Czech, *Kalendarium der Ereignisse im Konzentrationslager Ausch-witz-Birkenau 1939–1945* (Hamburg: Rowohlt 1989), p. 209.

134. BDC, SS-HO 71396, p. 161.

135. After the war W. Rudorf continued as director of the KWI/MPI for Breeding Research. The institute was moved to Voldagsen in 1945 and to Cologne-Vogelsang in 1955.

136. Institut für Zeitgeschichte, MA 441/1.

137. BAK, R 73/14509.

138. These expectations for radioactive radiation were voiced, for example, in *Münchner Neueste Nachrichten,* July 27, 1939 (BAK, R 21/11065).

139. BAK, R 73/16017.

140. Timoféeff-Ressovsky to the DFG, March 7, 1942, BAK, R 73/16017.

141. Gerlach to Prof. Beuthe, March 3, 1945, BAK, R 26 III/515. Zimmer believed that the negative effects of small radiation doses that were considered harmless could be attributed to the below-normal health condition of the persons in question (Zimmer to Gerlach, December 19, 1944, D. Irving microfilm DJ 31, p. 350).

142. Zimmer to the OKH, February 23, 1943, D. Irving microfilm DJ 31, p. 306. Zimmer and Alexander Catsch carried out radiobiological research also in collaboration with the research institute of Reich Postal Minister Ohnesorge. Ibid., p. 342.

143. Gerlach to the Office of Wartime Economy of the RFR, February 26, 1945. D. Irving microfilm DJ 29, p. 1159.

144. Rajewsky to Gerlach, April 9, 1944, D. Irving microfilm DJ 31, p. 316.

145. Gerlach to Colonel Geist, February 22, 1945, BAK, R 26 III/515.

146. File note by Gerlach concerning a meeting on October 26, 1944, D. Irving microfilm DJ 29, p. 1144.

3. The Content and Result of Research at Universities

1. BAK, R 73/13117 and R 73/14103.

2. In only a few cases do we have reports about the research contracts of the Wehrmacht or the Reich Office for Economic Development.

3. Private communication, December 13, 1987.

4. BAK, R 73/12557.

5. Institute for Food Research to F. Bukatsch, February 26, 1943, BAK, R 26III/123.

6. Contracts by the OKM, BAK, R 26/279.

7. Contracts by the OKL, October 1944, BAK, R 26/279.

8. RFR to H. Brücher, January 1945, BAK, R 26III/407.

9. BAK, R 73/11556.

10. UA Tübingen, 180/4.

11. Rector of the University of Münster (Mevius) to the curator of the university, March 9, 1943, BAK, R 21/10800.

12. F. Boas of the TH Munich described the chair in Tübingen for this reason as the only "Wettstein-free" chair in Germany (BDC, file E. Lehmann).

13. These were genes passed on in accordance with Mendel's rules but capable of influencing several characteristics of the phenotype simultaneously.

14. With some characteristics Lehmann also posited an extragenomic cause, but he never relinquished his belief in the unifactorially dividing inhibiting factor.

15. This becomes clear in Stubbe 1940, for example. According to Oehlkers 1943, this hope was presumably the driving motive behind the experiments on chemical mutagenesis that were carried out again and again in spite of the many failures.

16. H. Stubbe to the DFG, January 9, 1941, BAK, R 73/15057.

17. R. Bauch to the DFG, July 24, 1943, BAK, R 73/10176.

18. BAK, R 73/10176.

19. R. Bauch to the DFG, January 10, 1944, ibid.

20. Here, and in other publications as well, Seybold discussed the methods and results of Richard Willstätter's work on pigments. Willstätter, appointed full professor in Munich in 1915, had resigned in 1925 in protest against anti-Semitism and had emigrated to Switzerland in 1939.

21. Mothes was, until 1967, director of the Institute for the Biochemistry of Plants in Halle (newly established in 1963), and until 1974 president of the Leopoldina. His important publications include "Über den Einfluß des Kinetins auf Stickstoffverteilung und Eiweißsynthese," Flora 147 (1959): 445–464; and (with Schütte), Die Biosynthese der Alkaloide (Berlin, 1969).

22. In his 1940 publication he confirmed the findings he had made with onion epidermis cells: when living plasma is dyed with acridine orange, it turns green, whereas dead plasma, similarly dyed, turns dull red. The dye substance itself does not interfere with the normal developmental cycle of the amoeba.

23. Note that Guttenberg was not actually supported by the DFG or the RFR.

24. FIAT-Reviews, vol. 532, p. 127. The botanist Theodor Butterfaß

(Frankfurt) is of the opinion that of Laibach's scientific achievements, "what has lasted, as far as I can tell, is above all his fortunate move of introducing *Arabidopsis thaliana* into genetic research, that was about 1938" (personal communication, June 27, 1990).

25. Seybold in his review of an application by Harder, April 26, 1937, BAK, R 73/11489. In his application for funding of vernalization research, Harder, who had previously worked mainly in the area of photosynthesis, highlighted its importance for the national food supply.

26. Lysenko distinguished four phases of plant development. In his view, passage through the first, light-dependent, phase was a precondition for passage through the following, photoperiodic, phase, in which the lengths of daylight were important. It was supposedly possible to accelerate the developmental course of plant through the influence of moisture, temperature, and darkening at the beginning of germination. Lysenko's notion of phases proved untenable (Kappert and Rudorf 1958, p. 240). Moreover, the spread, for purely political and ideological reasons, of his neo-Lamarckian ideas in the USSR (Lysenko claimed that the characteristics acquired by plants through environmental factors, for example temperature, are inherited) led to the breakup of Soviet genetics).

27. Morphologists prior to Darwin were searching for the ideal type or the archetypal form among the great variability of observed manifestations in the field of the various taxonomic categories of living beings or their organs or other structures. While no serious idealistic morphologists were found in zoology after 1850, in botany a school of this orientation has existed to this day, and one of its more recent exponents was Troll. Other morphological directions are phylogenetic morphology and evolutionary morphology (see, for example, Mayr 1982, pp. 366ff.).

28. BAK, R 73/118.

29. From the beginning of the century, racial hygienists and anthropologists drew parallels from domestication research between domestication-related characteristics of domesticated animals and alleged manifestations of degeneration in "civilized" man. For example, in 1940 Hans Nachtsheim invoked Eugen Fischer, who had stated as early as 1914 that "all characteristics that appear among humans as racial differences . . . appear as such also in breeds of domesticated animals; conversely, most peculiarities of domesticated animals . . . are found again in humans as racial characteristics" (Nachtsheim in Just 1939). These alleged parallels gave rise to calls for the need of intensified selection, that is, eugenic measures.

30. Verschuer in a DFG evaluation (1936) on Schmidt. In 1942 Verschuer became director of the KWI for Anthropology. He did not shrink from examining for his work the organs of twins that Mengele sent to him from Auschwitz (Müller-Hill 1988, pp. 70f.).

31. SS Lieutenant Colonel Vogel to Himmler, January 27, 1942, BAK, NS 19/2920.

32. Himmler to Henseler, March 31, 1942, BAK, NS 19/2920. Henseler's response to this offer is not documented. After the conquest of Poland, Hans Frank became governor general there and was responsible for the extermination of the Jews and the destruction of the Polish elite. He was executed as a war criminal in 1946.

33. SS Lieutenant Colonel Vogel to Himmler, January 27, 1942, BAK, NS 19/2920.

34. Report by F. Duspiva to the DFG in 1944, BAK, R 73/10796.

35. BAK, R 73/13911.

36. W. Goetsch: "Stand der Forschungsaufgaben am 1.10.44," BAK, R 73/11277.

37. BAK, R 73/13138.

38. A. Meyer-Abich, "Gedanken über die Organisation der wissenschaftlichen Forschung in den Kolonien," November 12, 1940, BAK, R 73/13138.

39. Ibid.

40. Meyer to Prof. A. Kappus, August 31, 1938, BAK, R 73/13138.

41. Report by R. Lehmensick on his research expedition from June 15 to September 22, 1938, BAK, R 73/12641.

42. H. Krieg to the DFG, 1937, BAK, R 73/12416.

43. Henseler to Himmler, February 21, 1942, BAK, NS 19/2920.

44. Himmler to Henseler, March 31, 1942, BAK, NS 19/2920.

45. DFG application by the KWG, September 7, 1940, BAK, R 73/11227.

46. "Man . . . must understand the fundamental necessity of the workings of nature and comprehend just how much his existence, too, is subject to these laws of the eternal struggle upwards" (Friederichs 1937, p. 86).

47. The zoologist Hermann Weber, too, said the following in a lecture at the first plenary meeting of the faculty of the Reich University of Strasburg on January 28, 1942: "The pair of terms 'organism and environment,' the topic of this evening, means nothing other in the language of biology than the phrase 'blood and soil' in the language of politics" (*Der Biologe* 11 [1942]: 57).

48. "Befürwortete Anträge der Fachausschüsse der Notgemeinschaft vom November 1933," BAK, R 73/12475.

49. Minutes by A. Kühn of July 26, 1934, on the meeting of the commission of the Notgemeinschaft, BAK, R 73/12475.

50. Ibid.

51. Ibid.

52. Kühn to the DFG, 1937, BAK, R 73/12434.

53. Arch. MPG, Abt. I, Rep. 1A/1538.

54. Hertwig indicated in some publications (for example *Zeitschrift für*

induktive Abstammungs- und Vererbungslehre 79 [1941]) that she was being supported by the DFG. However, relevant DFG files are missing.

55. In an evaluation by the president of the RGA (Reiter) for the DFG in 1937, Just was described as one of the few biologists who, rejecting one-sided mechanistic trains of thought, were trying to construct a "general biology" also with respect to terminology. As one example Reiter mentioned that Just was imparting real content from the biological side to the constitution problem as a holistic problem. In addition, he singled out Just's work in racial hygiene and selection processes during and after the school years for the practical management of the education system.

56. Here the term "mechanism" does not refer to the branch of physics that deals with masses, forces, and their effects, but means "natural cause" in the sense in which it was used by the philosophers of the eighteenth century, above all by Kant (Sander 1991).

57. Morgan and his collaborators Calvin Bridges, Hermann Muller, and Alfred Sturtevant founded in the United States the most important school of *Drosophila* genetics. Their research led to the hypothesis of the coupling and linear arrangement of the gene on the chromosomes.

58. An overview of the work concerning the organization of the insect egg can be found in Seidel, Bock, and Krause 1940.

59. R. Kuhn to the president of the RFR (Rust), October 30, 1942, BAK, R 73/12694.

60. According to E. Henning (1987), one reason for this exception was Hitler's fear of cancer, Warburg's area of concentration.

61. BAK, R 73/11786.

62. R. Kuhn to the president of the RFR (Rust), October 30, 1942, BAK, R 73/12694.

63. BAK, R 73/12388, notes of Dr. Breuer, undated, ca. 1937.

64. Ibid.

65. Like Gerhard Domagk and Richard Kuhn, winners, respectively, of the Nobel prize in medicine and chemistry in the same year, Butenandt was not allowed to accept this prize. On Hitler's orders German citizens were prohibited from accepting the Nobel prize after Carl von Ossietzky had been given the prize for peace in 1935 (Wistrich 1982, p. 261).

66. Butenandt to the DFG, March 5, 1940, BAK, R 73/11117. A detailed account of the work at the KWI for Biochemistry can be found in Karlson 1990.

67. Report by H. Friedrich-Freksa to the DFG, 1941, BAK, R 73/11117.

68. The first step toward the experimental genesis of tumors had been taken by the Japanese Yamagiwa and Ichikawa, when they produced genuine tumors in 1915 by dabbing tar onto the ears of rabbits.

69. BAK, R 73/12388, notes of Dr. Breuer, undated, ca. 1937.

70. Holtz, who since 1931 had been carrying out vitamin and hormone research at the Charité in collaboration with Sauerbruch, specifically on the problem of cancer, became head of this department in 1936, where he set up the "tumor farm" as the central place for the breeding of laboratory animals with implantation tumors.

71. R. Kuhn to the president of the RFR (Rust), October 30, 1942, BAK, R 73/12694.

72. Ibid.

73. Arch. MPG, Abt. I, Rep. 1A/2398.

74. Minutes of the meeting of the senate of the KWG on April 24, 1942, BAK, R 21/11062.

75. Arch. MPG, Abt. I, Rep. 1A/2398.

76. Minutes of the meeting of the senate of the KWG on April 24, 1942, BAK, R 21/11062.

77. E. v. Holst, "Medizinische und zoologische Physiologie, Bericht an die DFG 1936" (Medical and zoological physiology, report to the DFG, 1936) BAK, R 73/11781.

78. Report to the DFG, September 23, 1944, according to *FIAT-Reviews,* vol. 52, p. 154, not published.

79. There are hardly any publications by Kuhn himself. A few papers on hormone physiology from the Cologne Institute during this time were published by his collaborators, among them Hartwig and Rotmann 1940, and Hartwig 1940.

80. G. v. Studnitz, "Geheimbericht an das OKM 1942," in Autrum, "Sinne der Tiere," *FIAT-Reviews,* vol. 52, p. 176.

81. Unpublished papers by Autrum, Rose, and R. Schmidt from the Research Institute on Aviation Medicine, 1944; personal communication from H. Autrum, August 10, 1989; see also *FIAT-Reviews,* vol. 52, p. 176.

82. *Der Biologe 5* (1936): 221.

83. Alverdes did not submit any applications to the DFG between 1933 and 1945.

84. This essay was the first larger theoretical work by Lorenz and was dedicated to Uexküll.

85. "Vitalism" refers to the notion that higher immaterial principles guide the processes in the organism purposefully and appropriately.

86. K. Lorenz to Dr. Greite, DFG, March 13, 1937, BAK, R 73/12781. In this evaluation Lorenz praised Fischel as a good experimenter but said that "nonsense results when he philosophizes about the 'psyche' of squirrels as a species-changing factor."

87. A summary of Lorenz's theory of instinct and a discussion with other, especially American, theories of instinct can be found in Lorenz 1937c.

88. In Germany, Lorenz's theory of instinct has been criticized fundamen-

tally by Hanna-Maria Zippelius (1992). She shows that many statements cannot be experimentally substantiated. Gereon Wolters (1991) has criticized the content and claim of evolutionary epistemology from a philosophical perspective.

89. F. v. Wettstein to the DFG, December 14, 1937, BAK, R 73/12781. Noteworthy here is the mention of political attitude as a prerequisite for DFG grants. Whereas "Aryan" descent was a clear criterion of support after the *Gleichschaltung* of the Notgemeinschaft in 1934, a National Socialist attitude was in many cases not decisive, even for those who received larger grants (see Chapter 2). Presumably this criterion played a role among the Austrians during the time of the Schuschnigg government; for example, prior to 1938 the late member of the SS Josef Pekarek (Graz) was investigated by the DFG as to his political sentiments.

90. Geuter bases his account on information Hans Thomae obtained about the appointment process from the then chairman of the German Society for Psychology, O. Kroh (personal communication from H. Thomae, September 6, 1990).

91. According to Thomae, to prevent the appointment of Lorenz, various professors of psychology had turned to Kroh, who had been given assurance by the REM that he would be consulted on appointments in the field of psychology, after a number of appointments of people outside the field had occurred between 1933 and 1938 (personal communication from H. Thomae, September 6, 1990).

92. In *Der Biologe* 3 (1934): 193–202, Koehler raised the demand for the "integration of basic biological truths into the ideology that guides our political-volkish actions." He saw in Hitler the person who could save the German people from what was, according to Spengler (see below), its imminent downfall: "Adolf Hitler believes that the German people want to live. We biologists know: if a people only wills seriously enough for it to move from willing to correct and successful action, it will live."

93. On the dissolution of army and air force psychology, see Geuter 1984, pp. 390–404.

94. BDC, file K. Lorenz: "Vorschlag zur Ernennung von Lorenz zum o. Professor durch das REM" (proposal for the appointment of Lorenz as full professor). Neither this nor any other source provide information on what Lorenz's participation in the Office of Race Policy meant in specific terms. The Office of Race Policy of the NSDAP had been set up on May 1, 1934, with the goal of coordinating all training and propaganda work in the fields of population and race policy. It was headed by the race scientist Walter Groß; the district office chiefs included Karl Astel (president of the State Office for Racial Affairs in Thuringia) and Lothar Löffler (professor for racial hygiene in Königsberg).

95. Organs or patterns of behavior are homologous if, in spite of different forms, they can be traced back to a common basic building plan that has a genetic basis.

96. The zoologist Ernst Haeckel (1834–1919) had, together with August Weismann (1834–1914), helped establish Darwin's theory of evolution in Germany. By popularizing evolutionary biological hypotheses and founding a monistic worldview, Haeckel played a decisive role in spreading social Darwinist ideas in Germany. He called his evolutionary philosophy monism, because it was in his view the only explanatory principle of the world. In monism, nature, humankind, and society were part of an interconnected whole. The doctrine of evolution became with Haeckel a religion that he intended to make into the ideological and ethical basis of the state and the foundation of all education.

97. As other factors that contributed to raising the percentage of heterozygotes, and thus to the divergence between the genetic picture and the appearance, he named mutation-producing toxins, X rays, and radium radiation.

98. Private posthumous papers of G. Kramer, with Elisabeth Kramer, Weinheim. According to Romilde Kramer, Gustav Kramer's wife, her husband, at that time a zoologist at the Zoological Station in Naples, had always critized this thesis, but he discussed his reservations with very few friends, because he otherwise thought highly of Lorenz (personal communication, May 1, 1989).

99. On the concept of racial hygiene, see Chapter 4, Section 7. Prior to 1933, many of the goals of racial hygiene in Germany corresponded to those of eugenics in England (see Weiss 1990).

100. This was the opinion of the geneticist and racial hygienist Erwin Baur (e.g., 1932) and zoologist Otto Koehler (e.g., 1934), for example. After World War I, the prevention of the decline of the German people and state increasingly became a goal of racial hygiene in Germany (Weiss 1990).

101. "By contrast, what evidently requires a thorough discussion are those intellectual and ethical values that the understanding of the phylogenetic developmental processes creates by giving our personal and political endeavors certain lofty goals" (Lorenz 1940b). Haeckel already wanted to make his evolutionary philosophy into the explanatory principle not only of biology but also of humankind and society. His teacher Rudolf Virchow criticized Haeckel's theories as religious fantasies rather than scientific hypotheses (Gasman 1971).

102. The species-preserving value of behavior was, according to Lorenz, a fundamental issue in ethological research. "Alongside the self-evident questions of causal analysis, our approach must also contain the question about

the species-preserving utility of every form of behavior that is normal for an animal species" (Lorenz 1937a).

103. In his psychological profile of Lorenz, Norbert Bischof (1993) addresses the continued existence of fascist ideas in Lorenz and interprets them in terms of developmental psychology.

104. I am indebted to Heinz-Horst Deichmann and Gottfried Bitter for pointing out this biblical passage and for explaining the use of the Greek *semeioomai* in the New Testament.

105. Lorenz 1943a, p. 302, similarly pp. 99, 314; Lorenz 1940a, p. 71; Lorenz 1943b, pp. 120, 125.

106. Lorenz 1940b, p. 25.

107. Lorenz 1940a, p. 75, similarly pp, 57, 69; Lorenz 1943a, p. 400.

108. Lorenz 1940a, p. 69.

109. Ibid., p. 70.

110. Ibid., p. 66; similarly Lorenz 1943a, p. 301.

111. I thank Michael Hubenstorf for the reference to this source.

112. Information from Otto Koehler (1963, p. 392). According to Koehler, from whom (as from all other biographers) we hear nothing about Lorenz's activity with the foundation, Lorenz was during this time psychiatrist in the reserve hospital in Posen.

113. The biography by F. M. Wuketits (1990), too, contains no reference to this work. With respect to Lorenz's statements during National Socialism, Wuketits adopts Lorenz's own argument (in Lorenz and Kreuzer 1988, pp. 96f.) and cleanses it of the word "murder" (see text): by writing that Lorenz had been politically naive and did not see that the Nazis meant "elimination" *(Ausmerzen)* when they said "selection" *(Auslese)* (p. 106), he suggests that Lorenz himself did not use the term *Ausmerzen.*

114. "Inclusive fitness," a concept of sociobiology, is measured by the individual reproductive success plus the reproductive success of relatives, weighted according to the degree of kinship. Behavior that is damaging to the species can thus also be adaptive, as Christian Vogel and his Göttingen work group, for example, were able to demonstrate once again with the example of "infanticide carried out by males of the Indian monkey species *Presbytis entellus* who have newly acquired the position of harem chiefs" (Vogel 1988).

115. It is surely no coincidence that sociobiology, which has played the biggest role in filling the concept "adaptive" with new content, originated in the United States. The readiness to hear about the failures of humankind is, as Lorenz noted with regret in 1988, correspondingly low there: "My book *Der Abbau des Menschlichen* [The waning of humaneness] was a bestseller in Germany, in the United States nobody buys it because it is directed against the American Way of Life" (Brügge 1988).

116. Holst informed Lorenz of this statement by Frisch in a letter dated February 13, 1950, Arch. MPG, Abt. III, Rep. 29 349.

117. K. Lorenz to G. Kramer, August 29, 1950, private posthumous papers of G. Kramer with E. Kramer, Weinheim.

118. Ibid.

119. E. v. Holst to A. Kühn, November 13, 1950, Arch. MPG, Abt. III, Rep. 29/320.

120. Arch. MPG Abt. III, Rep. 29/744.

121. Ibid., Rep. 29/196, Holst to M. Hartmann, January 4, 1951.

122. Ibid., Rep. 29/349.

123. W. C. Allee to K. v. Frisch, July 7, 1949, University of Chicago Library, W. C. Allee Papers. I thank Gregg Mitman for showing me this correspondence.

124. Frisch to Allee, undated, ca. July 1949.

125. N. Tinbergen to Margaret M. Nice, June 23, 1945 (original in English), W. C. Allee Papers, box 20, folder 12. I am grateful to the University of Chicago Library for permission to quote this letter in full.

4. The Content and Result of Research at Kaiser Wilhelm Institutes

1. *Die Naturwissenschaften* 21 (1933): 438–443, 449; 22 (1934): 359–364, 370–371; 23 (1935): 431–436, 443–444; 24 (1936): 35–40, 44–45; 25 (1937): 392–401, 408–409; 26 (1938): 342–349, 357; 27 (1939): 343–352, 357–358, 361; 28 (1940): 493–500, 504–505, 508; 29 (1941): 439–446, 449–452; 30 (1942): 627–633, 636–637; 31 (1943): 533–539, 542–545; 38 (1951): 409–419; 39 (1952): 507–509.

2. Franz Möwus, who studied sex-determining substances in *Chlamydomonas* under Kuhn and who was later exposed as a fraud, was not involved in these experiments of Hartmann's.

3. This effect refers to the increased growth and higher yield that hybrid plants have to show in some cases in comparison with their homozygotic parents.

4. After using *Neurospora* as a research object, G. W. Beadle described the difference in the research approach as follows: "Through control of the constituents of the culture medium we could search for mutations in genes concerned with the synthesis of already known chemical substances of biological importance" (Beadle 1951, p. 224).

5. As early as 1933, Maria Schönefeldt and Heinz Wülker had carried out research on *Neurospora* in Munich under Wettstein. Schönefeldt (1935) studied the development of the fungus; Wülker (1935) carried out tetradanalyses to discover the percentage of prereduction and postreduction for some characteristics.

6. H. Döring to H. Stubbe, December 14, 1939, Stubbe Fond 40, Arch. AdW. Berlin. Döring stated that he had produced sterile mutants with iodine.

7. Stubbe to the DFG, March 14, 1936, BAK, R 73/10753.

8. Stubbe to O. Renner, November 13, 1942: "As for experiments with Neurospora, unfortunately we don't have anybody here who could devote himself to them"; Stubbe Fond 172, Arch. AdW. Berlin.

9. F. v. Wettstein to E. Telschow, December 13, 1940, Arch. MPG, Abt. I, Rep. 1A/2963.

10. Correspodence between Stubbe and Kausche 1938–1943, Stubbe Fond 106, Arch. AdW. Berlin.

11. Board of the Industriebank, February 13, 1941, to the KWG, Arch. MPG, Abt. I, Rep. 1A/2963.

12. Arch. MPG, Abt. I, Rep. 46.

13. Schramm was the only German scientist whose research in the early forties was later acknowledged by J. Watson in his book *The Double Helix* (New York: Atheneum, 1968).

14. "Yeast nucleic acid" was the term at the time for RNA, and "thymonucleic acid" the term for DNA. In early work on nucleic acid, RNA had been isolated from yeast and DNA from the thymus gland of calves. Until 1938, it was believed that DNA and RNA were tetranucleotides, that is, molecules composed of only four nucleotides. In 1938 it was shown that DNA was a large molecule with a molecular weight between 500,000 and 1 million (Gulland 1938).

15. Ibid. H. S. Loring (1938) determined the molecular weight of TMV nucleic acid to be 37,000, and inferred from this that the nucleic acid was composed of twenty-two tetranucleotide units. According to S. S. Cohen and W. M. Stanley (1942), the protein-free RNA from TMV had an average particle weight of 300,000.

16. G. Schramm to the DFG, October 1, 1944, BAK, R 73/14509.

17. Report by G. Kausche to the RFR, February 2, 1942, BAK, R 73/12063.

18. In a letter to E. Telschow, dated December 13, 1940, F. v. Wettstein stated that as far as support for the most urgent issues of biological research was concerned, the establishment of two KWIs for Virus Research and for Cultivated Plant–Primitive Plant Research seemed particularly important to him. Arch. MPG, Abt. I, Rep. 1A/2963.

19. Later Vavilov's theory was not tenable in this form. A summary of the results of modern research on the theory of gene centers can be found in Kuckuck 1962.

20. W. Rudorf to the president of the KWG (Bosch), January 11, 1940, Arch. MPG, Abt. 1, Rep. 1A/2963.

21. Report of Stubbe on his negotiations in Vienna in April of 1943, undated, Stubbe Fond 243, Arch. AdW. Berlin.

22. Report by Stubbe to the RFR on the expedition of 1941, Arch. MPG, Abt. I, Rep. 1A/2964.

23. E. Schäfer to the DFG, August 26, 1936, BAK, R 73/14198. This expedition was financed by, among others, IG Farben and the DFG.

24. File note by Telschow, November 1, 1942, Arch. MPG, Abt. I, Rep. 1A/2963.

25. Stubbe to Schäfer, February 18, 1944, Stubbe Fond 184/1, fols. 10ff., Arch. AdW. Berlin.

26. File note by Telschow on a meeting in the office of State Secretary Backe with F. v. Wettstein and K. Meyer in the REM on February 9, 1942, Arch. AdW. Berlin, KWG no. 7, fol. 4 (compare Chapter 2, Section 10.3).

27. Stubbe Fond 184/1, fols. 10ff., Arch. AdW. Berlin.

28. Stubbe to the Society for the Promotion of German Industry, April 29, 1944, Stubbe Fond 243, Arch. AdW. Berlin. The Fördergemeinschaft had supported the research of the institute with RM 60,000 in 1943.

29. Arch. MPG, Abt. I, Rep. 1A/2603. Reinhold von Sengbusch (until 1937 department head at the KWI for Breeding Research), in a retrospective (undated, after 1950), gave the following background to Darré's decision to stop payment of the institute's budget and to have the research supervised by a leader of the Farmers' League: "Darré was furious with Baur because prior to 1933 Baur had sharply criticized one of Darré's publications, and he now used the opportunity of his appointment as minister to revenge himself upon Baur." Papers of R. v. Sengbusch, Arch. MPG, Abt. III, Rep. 2, Ordn. 7.

30. The head of the Department of Physiology at the KWI for Brain Research, A. Kornmüller, wrote to psychiatrist Max de Crinis (on October 11, 1941) that at the time Lenin's brain had been worked on with histological techniques in Moscow by an assistant of the institute at the request of Vogt. The occasion for the letter was a request to secure the brain should Moscow be taken, because, among other reasons, Lenin was said to have suffered a syphilitic brain disease, which, in Kornmüller's opinion, would be of value at least propagandistically (BAK, R 21/11065).

31. The development of Russian genetics has been thoroughly analyzed by Mark Adams; see Mayr 1982.

32. Chetverikov was forced in 1929 under Stalin to give up his post as head of the Department of Genetics at the Institute for Experimental Biology in Moscow; he was no longer able to continue his genetic research (Mayr 1982, p. 447).

33. In a letter to F. Glum (secretary general of the KWG), dated August 29, 1936, Spatz quotes from a letter from Vogt: "The Americans had given Timoféeff an ultimatum. Because of this he wanted to accept the appointment in America. The Ministry of Culture found out about it and promised that

he could head a genetics department of the current size within the institute, and it held out the prospect of special funding for it." Arch. MPG, Abt. I, Rep. 1A/1582.

34. Correspondence between M. Demerec and N. Timoféeff-Ressovsky from January to June 1936, APS, B/D 394, Demerec Papers, Timoféeff-R #2.

35. BAK, R 73/15215.

36. Neutrons were produced by firing atomic nuclei of deuterium with great electric energy at beryllium, which then released the neutrons. The high-voltage apparatus of the Genetics Department allowed voltages of up to 600,000 volts. The first cyclotron in Germany was built in 1938 at the KWI for Medical Research and began test operation in 1944.

37. In 1933 Kühn and Timoféeff-Ressovsky began to test mutagentic effects of chemicals on insects (BAK, R 73/12575). There are no publications on this work. Timoféeff-Ressovsky wrote to the DFG in 1937 that he had tested the effect of fluorescent substances with negative results (BAK, R 73/15216).

38. On Delbrück's emigration and scientific importance as a pioneer of molecular biology, see Fischer 1985.

39. Dessauer introduced the "hit theory" *(Treffer-Theorie)* and along with it concepts of modern physics into biology in 1922. His theory, a formal mathematical representation of the biological effects of radiation, described a "single-hit result" as a linear dose-effect curve. Crowther modified this theory in 1924; he substituted a more concrete "target theory" for the hit theory and defined a hit as a physical event: a reaction, for example a mutation, is produced when ionization occurs in a defined spatial area, the target area. In this way the volume of the target area can be calculated, raising the question of whether this mathematical volume constitutes a biological structure. Neither the hit theory nor the target theory could be sustained in its original, rigid form, because many biological parameters had not been taken into consideration (see Auerbach 1976; Zimmer 1960).

40. Robert Olby (1974) has analyzed this three-man work in detail.

41. According to Fischer (1985), Delbrück believed that Jordan had not understood Bohr's idea, and that he was trying to save vitalism in terms of quantum theory. Incidentally, when Delbrück expressed his view in a letter to Jordan that no specific coupling forces between identical particles existed relative to distance, the latter responded, "I myself only half-believe in the resonance." Letter of November 13, 1940, California Institute of Technology, Archives, Delbrück Papers, folder 12.13.

42. Jordan states that he discussed this problem widely among friends at the Genetics Department in Berlin-Buch.

43. With respect to the significance of radiation genetics for the solution of the question concerning the nature of the gene, Delbrück states (1970, p. 1312): "It is true that our hope at that time to get at the chemical nature

of the gene by means of radiation genetics never materialized. The road to success effectively bypassed radiation genetics."

44. Report by Timoféeff-Ressovsky to the DFG, July 23, 1943, BAK, R 73/15217.

45. Timoféeff-Ressovsky to the DFG, 1942, BAK, R 73/15216.

46. In his report to the DFG in 1943, Timoféeff-Ressovsky indicated as the result that both kinds of radiation could produce chromosome mutations, that in X-ray radiation these mutations showed themselves to be independent of wavelength, and that they were produced proportionally to the radiation dose used, since the chromosome fractures were directly proportional to the radiation dose (BAK, R 73/15217).

47. Ibid.

48. Applications in 1939 and 1942, BAK, R 73/16017.

49. The historical term "thoriumisotopes" refers today to the isotopes of the thorium decay series, which include various radium, polonium, and lead isotopes. As for thorium's effect on humans, Gerlach stated that thorium X was harmless because of its rapid decay. He wrote, however: "Additional tests must clarify, among other things, how the radioactive daughter products of [thorium] X behave in the body, since they are in part more radiologically effective than [thorium] X itself."

50. Paul and Krimbas rely on statements by Robley D. Evans, who remains *the* authority in the field of radiation and health. A dose indication of radium injections in humans equivalent to 0.03 milligrams radium per test is found in P. M. Wolf and H. J. Born, "Über die Verteilung natürlich-radioaktiver Substanzen im Organismus nach parenteraler Zufuhr," *Strahlentherapie* 70 (1941).

51. For example, structural changes in leucocytes appeared in the physicist Rolf Wideröe after he had done work in nuclear physics, even though the radiation doses had been, according to contemporary notions, too small to cause biological effects. He thought it was possible that one had to reckon with stronger biological effects of radiation than had been hitherto assumed (Wideröe to W. Gerlach, December 4, 1944, NA RG 319, G-2, Gerlach File, box 6, folder 20).

52. Wolf and his collaborators considered the delayed excretion of the thoriumisotope to be favorable for its use, they said nothing about possible damage to organs. Later it turned out that thorotrast caused malignant tumors in the liver and bones, and it was withdrawn from diagnostic use.

53. Ms. SPSL 420/8.

54. Arch. MPG, Abt. I, Rep. 1A/2779.

55. The first experiments in the genetic mapping of *Drosophila* were carried out by Alfred Sturtevant, one of Morgan's collaborators.

56. "Aus der Tätigkeit der Berliner Gesellschaft für Eugenik," in *Doku-*

mente der eugenischen Bewegung, 1930. We have no information about such lectures by Nachtsheim during the Nazi period.

57. Personal communication by Jonathan Harwood, August 1993.

58. Gutachten der Dozentenschaft betr. n.b.a.o. Prof. Dr. Nachtsheim, November 17, 1936, Archiv Humboldt-Universität, ZGI/742. (Since the *Gleichschaltung* of the universities in 1933, the Reich education minister based his decisions about positions on expert opinions by the faculty body and the NSDAP concerning the scientific quality as well as the political attitude of the respective person.)

59. Nachtsheim to the DFG, May 14, 1937, BAK, R 73/13328.

60. Nachtsheim to Eugen Fischer, April 12, 1948. I am indebted to Diane Paul for this piece of information from Nachtsheim's papers at the Arch. MPG; unfortunately I myself was not given access to these papers.

61. Ibid.

62. Nachtsheim to the Prussian Ministry for Science, Art, and Education, May 17, 1934, Zentrales Staatsarchiv, Dienststelle Merseburg, Rep. 87B, no. 20283, fol. 383.

63. Nachtsheim referred to F. Sal y Rosas's work in the Peruvian journal *Rev. Neuro-Psichiatr.* 2 (1939): 373–390.

64. Raphael Falk, "Hans Nachtsheim: How to Be a Eugenicist in National Socialist Germany and Prevail," unpublished essay, Hebrew University, Jerusalem, 1993, p. 30.

65. Ibid.

66. Nachtsheim to the DFG, March 23, 1943, BAK, R 73/15342.

67. Nachtsheim to the DFG, March 15, 1944, BAK, R 73/15342.

68. According to Klee, the children's ward in Görden sent in at least two cases a total of about 100 eight- to ten-year olds to the neighboring extermination center in Brandenburg to be gassed.

69. Letter from H.-H. Knaape to B. Müller-Hill, December 5, 1988.

70. H. Nachtsheim, notes on his conversation with R. Havemann, May 5, 1946, Arch. MPG, Sign. N 47, folder 13, quoted in Weingart Kroll, and Bayertz 1988, pp. 572ff. Since I was not given access to the Nachtsheim papers in the archive of the MPG, I have had to rely on the secondary literature.

71. Ibid.

72. I would like to thank Georg Rieder for calling my attention to Stieve's publication and to the article by Hans Harald Bräutigam.

73. I would like to thank Diane Paul for showing me this letter from the Nachtsheim Papers in the Arch. MPG.

74. Falk, "Hans Nachtscheim: How to Be a Eugenicist in National Socialist Germany," pp. 27–28.

75. Nachtsheim gave the following reasons, among others: after he had

opposed the "new genetics" of Lysenko in two essays in the paper *Tages-spiegel*, he was supposedly accused of inciting the people. The printing of the new edition of his book *Vom Wildtier zum Haustier* was (he claimed) pro-hibited in the Soviet zone because the book did not do justice to the results of the "new genetics." An article on evolutionary factors by one of his col-laborators was not published in the Soviet zone because it paid no attention to Lysenko's views and mentioned Timoféeff-Ressovsky, who had been de-clared an enemy of the people by *Pravda* (Nachtsheim 1949).

76. H. Nachtsheim to F. Lenz, December 23, 1947, Arch. MPG, Sign. N 18, quoted in Weingart, Kroll, and Bayertz 1988, p. 583.

77. "We couldn't do anything else, we had been muzzled" (H. Nachtsheim, "Lamarck, Darwin und Lyssenko, 2: Teil des Vortrages im RIAS am August 2, 1950," in Nachlaß Ludwig, Heidelberg Hs. 3668.7, university library Hei-delberg).

78. Quoted in M. Nushdin, "Wen verteidigt Prof. Nachtsheim," in *Neue Welt*, date unknown, ca. 1948, in Nachlaß Ludwig, Heidelberg Hs. 3668.6.

79. The lists of oppressed, arrested, and murdered Soviet biologists given in Joravsky 1970 and Zacharov and Surikov 1991 reveal the extent of Sta-linist violence against biologists. On the contribution to National Socialist racial doctrine by anthropologists, geneticists, and medical researchers, see, for example, Müller-Hill 1988; Proctor 1988; Weindling 1989; Weinreich 1946.

80. G. Wagner to the Ministy of the Interior, 1937, no exact date, BAK, R 18/5585.

5. Scientific Research by the SS

1. In a letter to Ernst-Robert Grawitz (dated September 30, 1942) that dealt with experiments with homeopathic remedies, Himmler wrote: "A number of people trust us in the belief that they are finding in the SS an institution that has not yet given in to the chemical trusts but is also carrying out scientific research" (Heiber 1968, p. 146).

2. The question of how the son of an ultraconservative, Catholic director of a humanistic gymnasium could turn into one of the greatest mass murderers of world history has given rise to speculations about the historical and polit-ical resilience of humanism. On Himmler's father, see Andersch 1980.

3. Other examples can be found in Heiber 1968: pp. 97, 146–150, 202, 206–209, 220–223, 256, 259–261, 281–283, 295, 303, 306–307.

4. BAK, R 73/14198 (DFG file E. Schäfer).

5. Ibid.

6. Ibid.

7. BDC, SS file E. Schäfer.

8. E. Schäfer to H. Himmler, June 5, 1938, Zentrales Staatsarchiv (Bundesarchiv Abt.) Potsdam, ZM 1457 A 5.

9. "Goals and Plans of the Tibet Expedition of the Society 'Das Ahnenerbe' under the Leadership of SS-Obersturmführer Dr. Schäfer," undated (1938), BAK, R 73/12198.

10. Ibid.

11. Ibid.

12. E. Schäfer to the DFG, August 26, 1936, BAK, R 73/14198.

13. Kater 1966, p. 203.

14. BDC, Ahnenerbe file E. Schäfer.

15. Himmler to Schäfer, February 24, 1943, BDC, SS file E. Schäfer. An explanation for Himmler's reserve can be only speculative. Apparently Himmler saw the war against England as a necessary evil; the real target of his plans of conquest and annihilation was Eastern Europe. In several speeches to generals and SS leaders betweeen 1941 and 1944, he spoke with respect of the English, since they had built up a world empire (Smith and Peterson 1974, pp. 187–189). In a speech to SS leaders in 1943, he emphasized that the destruction of the British empire had never been a goal of German policy, for "in the final analysis it is, after all, a world empire of the white race." He even saw forces in England that had doubts that a war against Germany made sense (speech to generals in 1944).

16. BAK, R 26 III/278.

17. BAK, R 26 III/271.

18. A. Hirt to W. Sievers, October 14, 1943, BAK, NS 21/907.

19. BDC, Ahnenerbe file Brücher, therein Astel to Himmler, April 18, 1939.

20. Brücher, "Ernst Haeckel, ein Wegbereiter biologischen Staatsdenkens," *NS-Monatshefte* 6 (1935): 1088–1098; "Ernst Haeckel und die 'Welträtsel'-Psychose römischer Kirchenblätter," *NS-Monatshefte* 7 (1936): 261–265.

21. Brücher, "Ernst Haeckel, ein Wegbereiter," p. 1097. Brücher is citing a passage from Haeckel, "Die Lebenswunder," which reads: "The death penalty, too, works in a downright benevolent manner as an artificial process of selection."

22. Ibid., p. 1098.

23. BAK, R 26 III/175. The author is unknown, but it was probably Sievers, since the letter requests research funds for Brücher.

24. Report by Brücher on the SS collecting commando, 1943, BDC, Ahnenerbe file H. Brücher.

25. Ibid.

26. Brücher to Sievers, March 1, 1944, BDC, Ahnenerbe file H. Brücher.

27. Report by Brücher on the SS collecting commando, 1943, BDC, Ahnenerbe file H. Brücher.

28. Ibid.

29. Letter from unknown author to K. Meyer, September 30, 1943, BAK, R 26 III/175.

30. Ibid.

31. Letter from Ahnenerbe managing director to the apparatus commission of the DFG, dated December 9, 1943, BAK, R 26 III/231.

32. Letter from unknown author to K. Meyer, September 30, 1943, BAK, R 26 III/175.

33. Working report by Brücher, end of year 1944, BDC, Ahnenerbe file H. Brücher.

34. For that reason the president of the RFR informed Brücher, in January of 1945, about similar research by Erwin Bünning in Strasbourg on the oil-plant Perilla. BAK, R 26 III/282.

35. Michael Kater, *Das Ahnenerbe* (Heidelberg, 1966), p. 272.

36. Decree of the Reichführer-SS, January 29, 1942, BAK, NS 21/910.

37. Ibid.

38. Curriculum vitae of Eduard May, November 10, 1941, BAK, NS 21/910.

39. Ibid.

40. Ibid.

41. Curriculum vitae of E. May, undated, ca. 1948, Archiv der Freie Universität (FU) Berlin: Rektorat/Präsidialamt, Personalreferat, Personalakte Eduard May.

42. *Habilitation* certificate of E. May, March 13, 1942, and curriculum vitae, undated, ca. 1948, ibid.

43. Sievers, February 12, 1942, BAK, NS 21/910.

44. Sievers's diary, February 18, 1943, Institut für Zeitgeschichte, MA-1406/2.

45. BDC, Ahnenerbe file E. May.

46. Note by Sievers dated April 3, 1943, concerning a conversation with E. May, BAK, NS 21/33.

47. Sievers's diary, February 14, 1943, March 10, 1943, Institut für Zeitgeschichte, MA-1406/2.

48. E. May to R. Schütrumpf, March 29, 1943, BAK, NS 21/33.

49. Ibid.

50. Brandt (in Himmler's office) to Sievers, January 15, 1944, BDC, Ahnenerbe file E. May.

51. Himmler to Oswald Pohl, March 31, 1943, BAK, R 26 III/287.

52. Report by May on ongoing and planned work at the Entomological Institute, September 23, 1944, BDC, Ahnenerbe file E. May.

53. He was sent into retirement for political reasons in October of 1945, UA Graz, 540 from 45/46.

54. Report by May, March 31, 1944, BDC, Ahnenerbe file E. May.

55. Ibid.

56. Kater 1966 reported on a disagreement that existed among staff members of the Ahnenerbe with respect to the necessity of killing people, "in opposition to Rascher and Hirt, May and Plötner opposed the killing" (p. 291).

57. Minutes of March 17, 1943, meeting, BAK, NS 21/906.

58. University of Munich rectorate to the curator of the FU Berlin, October 9, 1956, archive of the FU Berlin: Rektorat/Präsidialamt, Personalreferat, Personalakte Eduard May.

59. Copy of the notification of the State Council, dated December 19, 1945, ibid.

60. Statement by the Hauptamt Wissenschaft (Main Office for Science) concerning Krieck, undated, BAK, NS 15/290.

61. "The 'Hauptamt Wissenschaft' on Holism," November 27, 1936, Institut für Zeitgeschichte, MA 141/6.

62. Hauptamt Wissenschaft on Krieck, undated, BAK, NS 15/290, 25f.

63. "Thoughts by W. Groß on the Collaboration by H. Weinert on a Larger Scale," January 18, 193(9?), Institut für Zeitgeschichte, MA 116/17.

64. In 1959 he delivered a talk at the Cold Spring Harbor Symposium on Genetics and Twentieth-Century Darwinism entitled "The Descent of Man and the Present Fossil Card." His book *Die Evolution der Organismen* was published in a third edition in 1968. After the war Heberer published a series of articles on the evolution of fossil and recent human forms (hominids). In 1965 he edited the book *Menschliche Abstammungslehre*, which contained essays by professors at German and Swiss universities. In 1968 he became editor of the journal *Archiv für Anthropologie*. In the SCI 1945–1954 he had seventeen citations (minus his own), of which three were book citations; in the SCI 1955–1965 he was cited thirty-five times (of which six were book citations).

65. "The Stabsführer [head] of the RuSHA [Main Office for Race and Settlement] to the Reichsführer-SS, re. G. Heberer, October 21, 1936": "From the perspective of blood and professional criteria nothing stands in the way of Dr. Heberer's admission into the SS. His work in the Main Office for Race and Settlement is welcome . . . A larger sphere of activity for Dr. H. at the university seems desirable. He can pursue questions relating to prehistorical racial anthropology and the Indo-Germanic peoples." BDC, SS file G. Heberer.

66. Sievers to Karl Wolff, February 28, 1936, BDC, SS file G. Heberer.

67. Sievers to K. Wolff, May 29, 1936, ibid.

68. Ibid.

69. Himmler to B. Rust, November 5, 1936, ibid.

70. BAK R 21/10417, pp. 80ff.

71. Letter of November 24, 1936, BAK, R 21/10417, p. 117.

72. Astel committed suicide in 1945 before he could be charged at the Nuremberg trials.

73. Dean of the Faculty of Mathematics and Natural Sciences, June 17, 1941, BAK, R 21/10304.

74. Heberer to Wilhelm Gieseler, September 25, 1941, ibid.

75. Dean of the Faculty of Mathematics and Natural Sciences to the REM, September 1, 1941. UA Tübingen, 13111.

76. BDC, G. Heberer file.

77. Heberer, written statement "Concerning the Fight against the Theory of Evolution," April 15, 1937, BDC, G. Heberer file.

78. Heberer to the curator of the Ahnenerbe, W. Wüst, February 21, 1941. BDC, G. Heberer file.

79. *Der Biologe*, 1936, p. 320.

80. These included lectures to the German Society for Cultural Morphology in Frankfurt (1942), the association Die Wittheit and the Reich Association for German Prehistory in Bremen (1942), the University of Greifswald (1943), and the Reich League for German Prehistory in Berlin (1943).

81. The term "long-headed races" refers to headlengths measured by anthropologists. According to Weinert (1938), the Nordic and the Mediterranean races were European long-headed races.

82. Thus prior to 1933 Eugen Fischer rejected the notion of a Dalo-Nordic and an Eastern Baltic race as scientifically untenable, in Baur, Fischer, and Lenz 1927.

83. According to the anthropologist Karl Saller, the Indo-Germanic peoples came to Europe from the southeast as peaceful immigrants. The group of the so-called band-ceramists originated in the Near East and wandered to Eastern Europe and eastern Germany. There they introduced agriculture, which was later taken over by the younger culture of the string-ceramists. The latter are presumably of southern Russian origin and moved in a northwestern direction (Saller 1964, pp. 467f.).

6. Research to Develop Biological Weapons

1. In this respect there were parallels to the German research on an atomic reactor and the atomic bomb; see Goudsmit 1947.

2. Chemical warfare research and the reasons why chemical weapons were not used during the war have been analyzed by R-D. Müller (1980) and Günther Gellermann (1986), among others.

3. This was the finding of the American military secret service in March of 1946, in NA, RG 319, G2, "P" file.

4. File note (secret command matter) for Colonel Münch, OKW, January

28, 1943, author unknown, possibly Kliewe, NA RG 319, G2, P-Project file, box 3.

5. Report of the ALSOS Mission, June 20, 1945, based on the statements by Kliewe and Hirsch, NA, RG 112, entry 295 A, folder 62.

6. File note for Colonel Münch, OKW, January 28, 1943.

7. NA, RG 319, G2, "P" file.

8. From the headquarters OKW, June 6, 1942, re.: use of pathogens of Texas fever and potato beetles, NA RG 319, P-Project file, box 3. An explanation or reason for Hitler's prohibition could not be found either in American secret service reports or in German documents concerning the planning for biological warfare.

9. File note by Kliewe on the "Blitzarbeiter" (presumably "Blitzableiter") meeting at the WFSt on July 21, 1943, ibid.

10. File note of July 22, 1943, ibid.

11. File note by Kliewe, September 23, 1943, NA, RG 319, G2, "P" file.

12. File note for Colonel Münch, OKW, January 28, 1943.

13. The American military intelligence service (*Scientific Intelligence Review* [March 1946]), NA RG 319, G2, "P" file.

14. Meeting of the working group Blitzableiter, March 30, 1943, NA RG 319, G-2, P-Project file, box 3.

15. According to a report by the American secret service, undated, ca. 1946, NA RG 330, CD 23-1-4.

16. Ibid.

17. Meeting of the working group Blitzableiter, July 22, 1943, NA RG 319, G-2, P-Project file, box 3.

18. Report by the American secret service, undated, ca. 1946.

19. Meeting of the working group Blitzableiter, July 22, 1943.

20. Minutes of Kliewe's meeting with Regierungsrat Bayer of the Science Office, October 19, 1943, NA RG 319, G-2, P-Project file, box 3.

21. NA RG 319, G-2, P-Project file, box 3.

22. "Biological warfare research in Germany was, in general, limited in extent and inadequate in experimental tests." Undated, ca. 1943, NA RG 330, CD 23-1-4. Quoted from "Biological Warfare in Germany, WW II." During his interrogation, Kliewe gave the following reasons for the limited extent of biological warfare research: "(1) The führer prohibited all offensive biological warfare research. Defensive methods, however, were supported. (2) The personnel was inadequate. (3) Kliewe was constantly called away from his work. (4) dissension existed in the Committee on Biological Warfare (Blitzarbeiten) as to the methods of study and procedure."

23. BAK, R 26 III/122.

24. Examination of witness K. Blome at the Nuremberg physicians trial, NA M 887, spool 22.

25. File note by Kliewe, February 23, 1944, NA RG 319, G-2, P-Project file, box 3.

26. Mentzel to Brandt, October 22, 1943, BAK, R 26 III/539a.

27. Ibid.

28. Grants of the RFR from July 26, 1943 (RM 200,000), and September 17, 1943 (RM 6,000), BAK, R 26 III/539a.

29. Grants from the RFR, January 9, 1945, BAK, R 26 III/203.

30. Sievers to Franz Mueller-Darss, January 9, 1945, BDC, SS file K. Blome.

31. BAK, R 73/11786.

32. BAK, R 21/11062.

33. Meeting of the working group Blitzableiter, September 24, 1943, September 25, 1943, NA RG 319, G-2, P-Project file, box 3.

34. File note by H. Kliewe, re.: conversation with Prof. Blome on February 23, 1944, NA RG 319, G-2, P-Project file, box 3.

35. According to Blome's testimony at the Nuremberg trials, Himmler's main interest was the plague.

36. "German Biological Warfare Activities in WW II," *Scientific Intelligence Review* 3 (March 29, 1946), NA RG 319, G2, "P" file.

37. BDC, SA file K. Blome.

38. In a letter to May, dated July 14, 1944, Sievers wrote: "At a meeting that took place on October 27, 1943, in the headquarters of the führer between the Reichsführer-SS, the plenipotentiary of the Reich Research Council, SA Group Leader Dr. Blome, and SS Colonel Sievers, the Reichführer-SS gave his approval to the research projects planned by Prof. Blome, and promised the support of the SS and its installations to promote war-important work. Consequently a close working agreement was made with the Institute for Practical Research in Military Science of the SS." BAK, NS 21/45.

39. Ibid.

40. Assignment of October 4, 1943, BAK, R 26 III/729.

41. BAK, NS 19/3016. Sievers wrote to Blome on September 19, 1944, that the tests concerning the "dropping of infected Malaria mosquitoes" had been concluded by May.

42. Report by May, September 23, 1944, BDC.

43. May to Blome, September 19, 1944, BAK, NS 19/3016.

44. Report by May, September 23, 1944, BDC.

45. NS 21/906.

46. May to Himmler, August 16, 1944, BAK, NS 21/45.

47. Ibid.

48. Sievers to Blome, October 18, 1944, ibid.

49. Contract between Ahnenerbe and Borchers A. G., December 4, 1944, ibid.

50. BAK, R 26 III/539a.

51. Undated notes, presumably 1943, BAK, R 26 III/278.

52. Testimony of Blome at the Nuremberg physicians trial, NA M887, R 22.

53. Notes of conference of Sievers, May, Blome, and Borstell, September 23, 1943. BAK, NS 21/910. The league was established a few years after the Nazi seizure of power. It concerned itself with carrying out large-scale actions to combat catastrophic forest pest infestations, and in the forties it was also charged with combating the larvae of the malaria mosquito on a large scale.

54. Blome to the REM, May 21, 1943, BAK, R 26 III/539a.

55. BAK, NS 21/910.

56. BDC, file E. May.

57. R.-D. Müller, "Die deutschen Gaskriegsvorbereitungen 1919–1945," *Militärgeschichtliche Mitteilungen* 1 (1980): 25–54.

58. Ibid.

59. *Der Prozeß gegen die Hauptkriegsverbrecher vor dem Internationalen Militärgerichtshof in Nürnberg,* vol. 21 (Nuremberg, 1949), p. 607.

60. The diaries of Sievers (Institut für Zeitgeschichte, MA 1406/1) show that Blome had experiments with neutron radiation carried out (entry of April 18, 1944; more detailed information on this is not available), and that he was contemplating using Rascher to carry out experiments with bacterial pathogens in Nesselstedt (entry of April 26, 1944). At the Central Institute for Cancer Research, the extract of a supposedly cancer-curing plant developed by SS Major Lützelburg together with nineteen collaborators was tested on humans (entry of June 1, 1944, and Kater 1966, p. 204).

61. According to Hansen (1990, p. 78), after the war Blome told American secret service agents that prior to his escape he had hidden his research papers in the Anatomical Institute in Posen. It is thus possible that they fell into the hands of the Soviet Army.

62. *Trials of War Criminals before the Nuremberg Military Tribunal,* vol. 2 (Washington, D.C.: National Archives), p. 235. On Blome's role in the killing of Poles suffering from tuberculosis, see Mitscherlich and Mielke 1978, pp. 232ff.

63. L. Hunt, "U.S. coverup of Nazi scientists." *Bulletin of the Atomic Scientists,* April 1985, pp. 16–24.

64. *Trials of War Criminals before the Nuremberg Military Tribunal,* vol. 2, p. 235.

65. Ishii and several of his colleagues were probably brought to the United States secretly to pass on their knowledge to American experts on biological warfare (Harris 1992, p. 158).

66. German scientists whom the U.S. military wanted to recruit for their knowledge despite their proximity to or involvement in war crimes or medical crimes were identified, following their interrogation by American officers,

with a paperclip on their personal files. In this way they were able to become American citizens in contravention of immigration laws (Bower 1988, p. 2).

67. Linda Hunt (1985) attributes this status to the incriminating interrogation reports. Tom Bower (1988) offers a different explanation: protests, especially from Jewish groups, against the immigration of Professor Walter Schreiber, who was accused of having arranged for deadly human experiments in concentration camps, led to an investigation of "Project 63" and, among other things, to the canceling of Blome's contract.

7. Aftereffects of National Socialism

1. Maria von Osietzki (1982, p. 368) adopts this line of argument in her analysis of the organization of science in West Germany after 1945. She notes that the universities, with the establishment of the DFG, were able to maintain their position, an outcome she attributes at least partially to the apolitical attitude of the universities and considers characteristic of the West German restoration of science organization.

2. German physicists had not succeeded in building an atomic reactor, let alone an atomic bomb. Thus Otto Hahn noted in his diary on August 6, 1945, after the German physicists interned at Farm Hall had gotten news of the Hiroshima bomb: "Gerlach is despondent, he evidently feels like a defeated general." "Tagebuch O. Hahn," in "Unterlagen des Berichts von Löhde," ZDF (Zweites Deutsches Fernsehen), "Der Kampf um die Atombombe," 1972; Deutsches Museum, Munich. In biology, crucial advances were made by the Americans both in physiological genetics and in molecular genetics, even though intensive research was carried out in Germany in these fields— in physiological genetics even earlier than in the United States.

3. Haevecker mentions, without citing any reasons, that Carl Neuburg was sent into retirement in 1934, and that Ernst Rabel emigrated from Germany in 1936. In regard to the National Socialist seizure of power, he writes: "A minor press feud over ideological questions . . . is soon resolved through understanding negotiations" (1951, p. 46).

4. From 1936 to 1939, H. Döring carried out mutation experiments with *Neurospora,* but this research was not continued after his death in 1940 (see Chapter 4 Section 1.

5. I have taken into consideration publications by the KWIs (MPIs) for Biology, Biochemistry, Virus Research, and Comparative Hereditary Biology and Pathology (after 1964 MPI for Molecular Genetics). Up to 1952 these publications are listed in *Die Naturwissenschaften,* after that in the yearbooks of the MPG. Review articles, dissertations, lectures, and books were not considered.

6. Of the twenty-one publications in 1963, five came from the MPI for

Biochemistry, five from the MPI for Biology (departments of Melchers and Weidel), six from the MPI for Virus Research, and five from the MPI for Comparative Hereditary Biology and Pathology.

7. H. Grüneberg wrote to H. Nachtsheim on November 12, 1947, that it was difficult in England to get anything—e.g., books—from the United States, and that it was getting, if anything, worse rather than better as far as the financial problems were concerned (H. Grüneberg Papers, Wellcome Institute for the History of Medicine, London). In the fifties, H. Hayes was making his culture dishes from old medicine bottles (Wollman 1972).

8. The biochemist H. G. Zachau, in a lecture at the Symposium of Science Organizations in Bonn on October 26, 1989, said: "I have been told that the support was generous, considering the circumstances, even during times when we were not well off at all."

9. Much the same holds true also for chemists; see Wallenfels 1953.

10. Personal communication, May 15, 1992. Wallenfels himself, during his time as assistant at the KWI for Medical Research in Heidelberg during the war, came into contact with what was at the time the latest genetic and biochemical literature, including the work of Oswald Avery in 1944 on the transformation of pneumococci, which showed for the first time that DNA (and not proteins) carried the genetic substance.

11. Personal communication, September 13, 1989.

12. Personal communication from H. Autrum, H. Kuckuck, G. Melchers, 1989; from A. Lang, 1990.

13. UA Heidelberg, Akten Nat.-Math. Fakultät, Vorgang 73a.

14. Ibid.

15. *Marburger Hochschultage 1946* (Frankfurt, 1947), p. 161.

16. Ibid., p. 50.

17. Up to 1965 other symposia on themes in genetics or molecular genetics were held: Nucleic Acids and Nucleoproteins (1947), Genes and Mutations (1951), Viruses (1953), Population Genetics (1955), Genetic Mechanisms: Structure and Function (1956), Exchange of Genetic Material (1958), Genetics and Twentieth Century Darwinism (1959), Cellular Regulatory Mechanisms (1961), Basic Mechanisms in Animal Virus Biology (1962), Synthesis and Structure of Macromolecules (1963), Human Genetics (1964).

18. M. Demerec in the preface to *CSH Symposia on Quantitative Biology* (1946).

19. M. Demerec to O. Mohr, February 11, 1947, APS, C. Stern coll.

20. The general secretary of the congress, Gert Bonnier, in agreement with most European geneticists, had approved the participation at the congress of German geneticists with a clean political record; H. Nachtsheim to H. Grüneberg, May 11, 1948, Grüneberg Papers.

21. The only German geneticist who mentioned the political past in ex-

pressing opposition to this location was G. Melchers; Caspari to Lerner, March 24, 1960, APS, B/L 563, M. Lerner.

22. The protests and comments are documented in an extensive correspondence of the "Permanent International Committee" of the International Congress of Geneticists, APS, B/L 563, M. Lerner.

23. Westergaard to Lerner, March 14, 1960, ibid.

24. Caspari to Lerner, March 24, 1960, ibid.

25. J. Wahrman, Department of Genetics of the Hebrew University, to E. Suomalainen, Institute for Genetics of the University of Helsiniki, March 8, 1960 (enclosure in the letter of Wahrmann to K. Sperling, June 9, 1986, Institute for Human Genetics of the Free University of Berlin).

26. Ibid. According to Professor A. Lang (Michigan), the reservations toward Germans were generally less pronounced among Jews who had had to emigrate from Germany than among those born in Israel (personal communication, September 17, 1990).

27. Mohr to Demerec, June 8, 1945, APS, C. Stern coll.

28. François Jacob describes in his autobiography (1988), which did not find a German publisher, his fight with the Free French Forces against Nazi Germany.

29. Personal communication, November 26, 1991.

30. Conversation between B. Müller-Hill and G. Melchers, in Müller-Hill 1984, pp. 164–165.

31. Muller to Delbrück, February 26, 1947, Delbrück Papers, California Institute of Technology, Archives; emphasis added.

32. On a visit to Germany in 1947, Delbrück had asked Melchers for a young biochemist "who had not been a Nazi." Melchers named Weidel (personal communication from G. Melchers, 1989).

33. Apart from the above-mentioned Eighth International Genetics Congress in Stockholm in 1948, Melchers participated in the Symposium on Microbial Genetics in March 1951 in Copenhagen (organized by the Rockefeller Foundation), and he gave a talk at the Ninth International Genetics Congress in Italy in 1953.

34. I will mention only briefly that this work led to important findings. One example was the discovery, made in the late fifties, that TMV-RNA was changed by nitrous acid, whereby cytidin was transformed into uridin, work which contributed to the deciphering of the genetic code.

35. The fact that the objects on which most of the molecular genetic insights had been made were not used at all at the time must be seen against the background of research practice in the forties: at that time scientists worked on the most modern object, the TMV.

36. Personal communication, November 11, 1989.

37. Personal communication, July 18, 1989.

38. According to W. Vielmetter, Schramm's political background did not do him any harm. As a scientist he was always held in high esteem (personal communication, June 28, 1989).

39. Personal communication, August 19, 1992.

40. *Journal of Biological Chemistry* 144 (1942): 589–598.

41. Personal communication, May 21, 1992.

42. "Der Nationalsozialistische Deutsche Dozentenbund Berlin an den Rektor der Universität," November 25, 1936, UA Humboldt University, 7DI/194.

43. Personal communication from G. Stent, October 2, 1991, and F. Jacob, November 26, 1991.

44. Delbrück to Fraser, August 28, 1947, Delbrück Papers, folder 4.5.

45. Delbrück oral history, p. 104; California Institute of Technology, Archives; see P. Fischer 1985, pp. 216f.

46. Letter dated November 13, 1956, Delbrück Papers, folder 4.5.

47. Personal communication from F. Jacob, November 26, 1991.

48. According to Judson (1979, p. 373), Delbrück for some time considered these experiments useless and maintained that the findings were the result of contamination.

49. Personal communication from C. Bresch, May 15, 1992.

50. *Report of the DFG, January 4, 1958–March 31, 1959,* (Bonn, 1959), p. 56.

Conclusion

1. Robert J. Havighurst, Rockefeller Foundation, October 6, 1947, RAC, collection RF, RG 1.1/717 box 3, folder 19.

2. In most cases Hartshorne 1937 and Ferber 1956 have been drawn on as sources. A critical discussion of these sources can be found in Fischer 1988.

3. Miklós Nyiszli, *Dr. Mengele Boncolóorvosa voltam az Auschwitz-I Krematórium ban* (A cimplas Ruzicskal György munkója, 1946).

4. The vernalization of plants with winter behavior, that is, their (nongenetic) transformation into plants with summer behavior, was not Lysenko's discovery but was first described by Gustav Gassner in 1918. As we now know, this process is based on the different methylization of DNA.

5. "The Situation in the Biological Sciences," meeting of the Lenin Academy of Agricultural Sciences of the USSR, July 31–August 7, 1948, stenographic report, Moscow 1948 (in German).

6. A critical discussion of Nolte's biography of Heidegger can be found in Thomas Sheehan, "A Normal Nazi," *New York Review,* January 14, 1993, pp. 30ff.

7. This statement is found, for example, in APS, Ms. Coll. 5 C. Stern, file Dunn.

8. This position found expression in a number of publications, for example Eickemeyer 1953 (p. 66), as well as in many letters of German scientists to American colleagues in the postwar period (see the Genetics Collection of the APS, Philadelphia, and the Goldschmidt Papers, Bancroft Library).

Epilogue

1. The letter is reprinted by permission of the Master and Fellows of Churchill College Cambridge and Ulla Frisch.

Sources

Archival Sources

Bundesarchiv Koblenz
R 18 Reichsministerium des Innern
R 21 Reichsministerium für Wissenschaft, Erziehung und Volksbildung
R 22 Reichsjustizministerium
R 26 Reichsforschungsrat
R 73 DFG
R 86 Reichsgesundheitsamt
NSD NS-Drucksachen, nos. 31–49 NSDB
NS 8 Kanzlei Rosenberg
NS 12 NS-Lehrerbund
NS 15 Amt Rosenberg
NS 19 Reichsführer-SS, Institut für wehrwissenschaftliche Zweckforschung
NS 21 Das Ahnenerbe

Institut für Zeitgeschichte, Munich
MA 116 Hauptamt Wissenschaft im Amt Rosenberg
MA 141 NSDAP-Akten
MA 294 Das Ahnenerbe
MA 295 and 297 Reichsführer-SS
MA 1406/1 and 2, Tagebuch (diary) Sievers, 1941–1945
MA 1497 "Der Generalplan Ost," presented by SS-Oberführer Professor Dr. Konrad Meyer, Berlin 1942

University Archives Berlin (Freie Universität and Humboldt-Universität), Cologne, Freiburg, Munich, Tübingen; Graz and Vienna
Faculty, institute, and personal files

Universitätsbibliothek Heidelberg, Handschriftenabteilung
    Papers of Prof. W. Ludwig
    "Lebenserinnerungen G. v. Ubisch" (Heid. Hs. 4029)

Berlin Document Center, Berlin
    NSDAP Zentralkartei
    NSDAP juristische Kartei
    Documents of the Reichsforschungsrat, Reichserziehungsministerium, SS,
       and SA

Preußisches Geheimes Staatsarchiv, Berlin
    Files of the Reichserziehungministerium

Archiv zur Geschichte der Max-Planck-Gesellschaft, Berlin
    Files of the General Administration of the Kaiser Wilhelm Society
    Files of Kaiser Wilhelm Institutes
    Papers of E. v. Holst, R. v. Sengbusch

Archives of the Zentrum für Antisemitismusforschung, TU West Berlin
    Archives of the Research Foundation for Jewish Immigration

Zentrales Staatsarchiv in Potsdam and Merseburg (today: Bundesarchiv)
    Reichserziehungsministerium
    Preußisches Ministerium für Wissenschaft, Kunst und Volksbildung

Dokumentationsarchiv des Österreichischen Widerstandes, Vienna
    Documents of different origin

Bodleian Library, University of Oxford
    Records of the "Society for the Protection of Science and Learning" (Ms.
      SPSL)

Wellcome Institute for the History of Medicine, London
    Hans Grüneberg Papers

Institut Pasteur, Service des Archives, Paris
    Fonds Jacques Monod
    Stagiaires I.P. 1934/1953

American Philosophical Society (APS), Philadelphia
    "Genetics Collection"

Bancroft Library, University of California, Berkeley
    Richard Goldschmidt Papers

California Institute of Technology, Archives, Pasadena
    Max Delbrück Papers and oral history

Columbia University, New York (Oral History Office)
L. C. Dunn and T. Dobzhansky Interviews

Hoover Institution Archives, Stanford University
Deutsche Kongresszentrale
ALSOS-Mission

Lilly Library, Indiana University, Bloomington
H. J. Muller and T. Sonneborn Papers

National Archives, Washington D.C.
M 887: "Nuremberg trials—the doctor's case," Rolls 22,23
RG 112: Intelligence Reports, Biological Warfare
RG 165: "Merck's file," "ALSOS-Mission"
RG 319: Record of the Army Staff
RG 330: Record of the Office of the Secretary of Defense

Library of Congress, manuscript division, Washington, D.C.
German captured materials, Reels 128 (Blome), 135 ("Ahnenerbe"), 136
(Kliewe)

Rockefeller Archive Center, Tarrytown
Collection Rockefeller Foundation, RG 1.1
Grants given to institutes and individuals in Germany
Grants given to German refugees in the U.S.
Post-WWII reports on Germany

YIVO-Institute, New York
Records of the Rosenberg Hauptamt Wissenschaft (in large parts identical
with microfilms at the Institut für Zeitgeschichte, Munich)

Interviews and Correspondence

Hansjochem Autrum, Munich; Doris Baumann, Vienna; Carsten Bresch, Freiburg; Eleonore Bünning, Tübingen; Prof. Theodor Butterfaß, Frankfurt; Seymour S. Cohen, Woods Hole, Mass.; Franz Duspiva, Heidelberg; Maria Ehrenberg, Würzburg; Wolf-Dietrich Eichler, Berlin; Liselotte Evenari, Jerusalem; Salome Glueksohn-Waelsch, New York; Viktor Hamburger, St. Louis, Missouri; Francois Jacob, Paris; Lothar Jaenicke, Cologne; Peter Karlson, Marburg; Dr. Romilde Kramer, Weinheim; Hermann Kuckuck, Hannover; Anton Lang, East Lansing, Michigan; Otto Lange, Würzburg; Hans Marquardt, Badenweiler; Roland Maly, Lucerne; Dr. Renate Mattick, Beverungen; Georg Melchers, Tübingen; Ursula Nürnberg, Alt-Ruppin; Werner Rauh, Heidelberg; Nikolaus Riehl, Baldham; Lutz Rosenkötter, Frankfurt; Werner Schäfer, Tübingen; Gunther Stent, Berkeley, Calif.;

Hildegard Strübing, Berlin; Jurij A. Treguboff, Frankfurt; Walter Vielmettter, Cologne; Kurt Wallenfels, Freiburg; Veronika Wieczorek, Berlin.

Books and Articles

Aach, H. G. 1957. "Serologische Untersuchungen an Mutanten des Tabak-mosaikvirus," *Zeitschrift für Naturforschung* 12b: 614–622.

———. 1958. "Spektrophotometrische Untersuchungen an Mutanten des Tabakmosaikvirus," *Zeitschrift für Naturforschung* 13b: 165–167.

Abir-Am, Pnina. 1992. "From multidisciplinary collaboration to transnational objectivity: International space as constitutive of molecular biology, 1930–1970," in E. Crawford, T. Shinn, and S. Sörlin, eds., *Denationalizing Science: The Context of International Scientific Practice.* Yearbook of the Sociology of Sciences 16. Dordrecht: Kluwer Academic Publishers.

Adams, Mark, ed. 1990. *The Wellborn Science. Eugenics in Germany, France, Brazil, and Russia.* New York: Oxford University Press.

Albrecht, Helmuth, and Armin Hermann. 1990. "Die Kaiser-Wilhelm-Gesellschaft im Dritten Reich (1933–1945)," in Rudolf Vierhaus, and Bernhard von Brocke, eds., *Forschung im Spannungsfeld von Politik und Gesellschaft: Geschichte und Struktur der Kaiser-Wilhelm-/Max-Planck-Gesellschaft.* Stuttgart: DVA.

Allen, Garland E. 1975. *Life Science in the Twentieth Century.* New York: Cambridge University Press.

———. 1985. "Thomas Hunt Morgan: Materialism and experimentalism in the development of modern genetics." *Trends in Genetics* 1: 186–190.

Altner, Guunter. 1988. "Lorenz zur Diskussion." *Natur* 11: 34.

Alverdes, Friedrich. 1935. *Die Totalität des Lebendigen.* Leipzig: Barth-Verlag.

———. 1937. "Das Lernvermögen der einzelligen Tiere." *Zeitschrift für Tierpsychologie* 1: 35–38.

———. 1938. "Zur Psychologie der niederen Tiere." *Zeitschrift für Tierpsychologie* 2: 258–264.

Aly, Götz, and Susanne Heim. 1991. *Vordenker der Vernichtung: Auschwitz und die deutschen Pläne für eine neue europäische Ordnung.* Hamburg: Hoffmann und Campe.

Anders, Fritz. 1991. "Contributions of the Gordon-Kosswig melanoma system to the present concept of neoplasia." *Pigment Cell Research* 3: 7–29.

Andersch, Alfred. 1980. *Der Vater eines Mörders.* Zurich: Diogenes.

Ash, Mitchell G., and Ulfried Geuter. 1985. "NSDAP-Mitgliedschaft und Universitätskarriere in der Psychologie," in C. F. Graumann, ed., *Psychologie im Nationalsozialismus.* Berlin: Springer.

Auerbach, Charlotte. 1976. *Mutation Research: Problems, Results, and Perspectives.* London: Chapman and Hill.

Auerbach, Charlotte, and J. M. Robson. 1946. "Chemical production of mutations." *Nature* 157: 302.

Autrum, Hansjochem. 1941. "Über Gehör- und Erschütterungssinn bei Locustiden." *Zeitschrift für vergleichende Physiologie* 28: 580–637.

Barthelmeß, Alfred. 1952. *Vererbungswissenschaft.* Freiburg: Alber.

Bauch, Robert. 1941. "Experimentell erzeugte Polyploidreihen bei der Hefe." *Die Naturwissenschaften* 29: 687–688.

———. 1942. "Experimentelle Auslösung von Gigas-Mutationen bei der Hefe durch carcinogene Kohlenwasserstoffe." *Die Naturwissenschaften* 30: 263–264.

Bauer, Hans. 1939. "Röntgenauslösung von Chromosomenmutationen bei Drosophila melangoster I." *Chromosoma* 1: 343–390.

Bäumer, Änne. 1990. *NS-Biologie.* Stuttgart: Hirzel, Wiss. Verlags.-Ges.

———. 1942. "Röntgenauslösung Drosphila II." *Chromosoma* 2: 407–458.

Baur, Erwin. 1932. "Der Untergang der Kulturvölker im Lichte der Biologie." *Volk und Rasse* 7: 65–79.

Baur, Erwin, Eugen Fischer, and Fritz Lenz. 1927. *Grundriß der menschlichen Erblehre und Rassenhygiene,* 3d ed. Munich.

Bautzmann, H. J. Holtfreter, H. Spemann, and O. Mangold. 1932. "Versuche zur Analyse der Induktionsmittel in der Embryonalentwicklung." *Die Naturwissenschaften* 20: 971–974.

Beadle, G. W. 1951. "Chemical genetics," in L. C. Dunn, ed., *Genetics in the Twentieth Century.* New York: MacMillan, pp. 221–240.

Beadle, G. W. and E. L. Tatum. 1941. "Genetic control of biochemical reactions in Neurospora." *PNAS* 27: 499–506.

Becker, Erich. 1942. "Über die Eigenschaften, Verbreitung und die genetisch-entwicklungsphysiologische Bedeutung der Pigmente der Ommatin- und Ommingruppe (Ommochrome) bei den Arthropoden." *Zeitschrift für induktive Abstammungs- und Vererbungslehre* 80: 157–219.

Becker, Heinrich. 1987. "Von der Nahrungssicherung zu Kolonialträumen: Die landwirtschaftlichen Institute im Dritten Reich," in Becker, Dahms, and Wegener 1987.

Becker, Henrich, Hans-Joachim Dahms, and Cornelia Wegener, eds. 1987 *Die Universität Göttingen unter dem Nationalsozialismus—Das verdrängte Kapitel ihrer 250-jährigen Geschichte.* Munich: K. G. Saur.

Bentwich, Norman. 1953. *The Rescue and Achievement of Refugee Scholars: The Story of Displaced Scholars and Scientists, 1933–1952.* The Hague: Nijhoff.

Bergdolt, Ernst. 1937. "Zur Frage der Rassenentstehung beim Menschen." *Zeitschrift für die gesamte Naturwissenschaft* 3: 109–113.

————. 1940. "Über Formwandlungen—zugleich eine Kritik von Artbildungstheorien." *Der Biologe* 9: 398–407.

Bericht der Deutschen Forschungsgemeinschaft 1952–1965.

Bericht der Notgemeinschaft der deutschen Wissenschaft 1949–1951.

Bernstein, Barton J. 1987. "Churchill's secret biological weapons." *Bulletin of the Atomic Sciences* 43: 46–50.

————. 1988. "America's biological warfare program in the Second World War." *Journal of Strategic Studies* 3: 292–317.

Bertalanffy, Ludwig von. 1941. "Die organismische Auffassung und ihre Auswirkungen." *Der Biologe,* 247–345.

————. 1944. "Bemerkungen zum Modell der biologischen Elementareinheiten." *Die Naturwissenschaften* 32: 26–32.

Beyerchen, Alan D. 1977. *Scientists under Hitler.* New Haven: Yale University Press.

Bhaskaran, Govindan, Stanley Friedman, and J. G. Rodriguez, eds. 1982. *Current Topics in Insect Endocrinology and Nutrition: A Tribute to Gottfried S. Fraenkel.* New York: Plenum Press.

Bischof, Norbert. 1991. *Gescheiter als alle die Laffen: Ein Psychogramm von Konrad Lorenz.* Hamburg: Rasch und Röhring.

Bodenstein, Dietrich. 1938. "Untersuchungen zum Metamorphoseproblem II. Entwicklungsrelationen in verschmolzenen Puppenteilen." *Archiv für Entwicklungsmechanik* 137 (1938): 636–660.

Boehm, Hermann. 1937. "Das Erbbiologische Forschungsinstitut in Alt-Rehse." *Ärzteblatt für Berlin* 42: 415–416.

Böhme, R. W. 1940. "Anbau und Züchtung von Kautschuk- und Guttaperchapflanzen in der gemäßigten Zone." *Zeitschrift für Pflanzenzüchtung* 23: 411–453.

Born, Hans Joachim, Anton Lang, and Gerhard Schramm. 1943. "Markierung von Tabakmosaikvirus mit Radiophosphor." *Archiv für die gesamte Virusforschung* 2: 461–479.

Born, Hans Joachim, Elena A. Timoféeff-Ressovsky, and P. M. Wolf. 1943. "Versuche über die Verteilung des Mangans im tierischen Körper mit 56 Mn als Indikator." *Die Naturwissenschaften* 31: 246–247.

Bower, Tom. 1988. *The Paperclip Conspiracy.* London: Paladin.

Bracher, Karl Dietrich. 1979. *Die deutsche Diktatur: Entstehung, Struktur, Folgen des Nationalsozialismus,* 6th ed., Frankfurt: Ullstein. *The German Dictatorship,* trans. Jean Steinberg, New York: Praeger Publishers, 1970.

Brachet, Jean. 1974. *Introduction to Molecular Embryology.* New York: Springer.

Bräutigam, Hans Harald. 1989. "Tod nach Kalender." *Die Zeit* 1: 20.

Broszat, Martin. 1961. *Nationalsozialistische Polenpolitik 1939–1945.* Stuttgart: DVA.

Browder, Leon W. 1986. *Developmental Synthesis*, vol. 2, *The Cellular Basis of Morphogenesis*. New York and London: Plenum Press.

Brücher, Heinz. 1938. "Die reziprok verschiedenen Art- und Rassenbastarde von Epibolium und ihre Ursachen I. Die Nichtbeteiligung von Hemmungsgenen." *Zeitschrift für induktive Abstammungs und Vererbungslehre* 75: 298–340.

―――. 1939. "Abschließende Stellungnahme." *Zeitschrift für induktive Abstammungs und Vererbungslehre* 76: 608.

Brügge, P. 1988. "Konrad Lorenz: Von der Gans aufs Ganze." *Der Spiegel* 45: 244–263.

Bullock, Alan. 1991. *Hitler and Stalin. Parallel Lives*. London: Harper Collins.

Bünning, Erwin. 1942. "Untersuchungen über den physologischen Mechanismus der endogenen Tagesrhythmik bei Pflanzen." *Zeitschrift für Botanik* 37: 433–486.

―――. 1944. "Otto Renner 60 Jahre." *Flora* 37: vii–x.

―――. 1948. *Theoretische Grundfragen der Physiologie*. Jena: G. Fischer.

Burleigh, Michael. 1988. *Germany Turns Eastwards. A Study of Ostforschung in the Third Reich*. Cambridge: Cambridge University Press.

Burleigh, Michael, and Wolfgang Wippermann. 1991. *The Racial State: Germany, 1933–1945*. Cambridge: Cambridge University Press.

Buselmeier, Karin, Dietrich Harth, and Christian Jansen, eds. 1985. *Auch eine Geschichte der Universität Heidelberg*. Mannheim: Quadrat.

Butenandt, Adolf. 1977. "The historical development of modern virus research in Germany, especially in the Kaiser-Wilhelm-/Max-Planck-Society, 1936–1954." *Medical Microbiology and Immunology* 164: 3–14.

Butenandt, Adolf, Wolhard Weidel, and Erich Becker. 1940. "Kynurenin als Augenpigmentbildung auslösendes Agens bei Insekten." *Die Naturwissenschaften* 28: 63–64.

Butenandt, Adolf, Hans Friedrich-Freksa, St. Hertwig, and G. Scheibe. 1942. "Beiträge zur Feinstruktur des Tabakmosaikvirus." *Zeitschrift für physiologische Chemie* 274: 276.

Cairns, J. G., G. S. Stent, and J. D. Watson, eds. 1966. *Phage and the Origins of Molecular Biology*. New York: Cold Spring Harbour Symposia on Quantitative Biology. Expanded ed., 1992.

Campbell, Donald. 1975. "Reintroducing Konrad Lorenz to psychology," in Richard I. Evans, *Konrad Lorenz: The Man and His Ideas*. New York: Harcourt, Brace, Jovanovich, pp. 88–118.

Caspersson, Tobjörn. 1936. "Über die Verteilung von Nukleinsäure und Eiweiß in den Chromosomen." *Die Naturwissenschaften* 24: 108.

―――. 1940. "Über die Rolle der Desoxyribonucleinsäure bei der Zellteilung." *Chromosoma* 1: 147–156.

Catsch, Alexander, Karl Günther Zimmer, and O. Peter. 1947. "Strahlen-biologische Untersuchungen mit schnellen Neutronen." *Zeitschrift für Naturforschung* 2b: 1–5.

Cavallo, Giorgio, and Gerhard Schramm. 1954. "Über die Reinigung eines temperierten Phagen." *Zeitschrift für Naturforschung* 9b: 579.

Chemie der Genetik, 9. 1959. Colloqium of the Society for Physiological Chemistry, April 17–19, 1958 in Mosbach. Berlin: Göttingen.

Clausen, Richard. 1964. *Stand und Rückstand der Forschung in Deutschland.* Wiesbaden: Steiner.

Cohen, Seymour S., and Wendell M. Stanley. 1942. "The molecular shape and size of the nucleic acid of tobacco mosaic virus." *Journal of Biological Chemistry* 144: 589–598.

Czech, Danuta. 1989. *Kalendarium der Ereignisse im Konzentrationslager Auschwitz-Birkenau 1939–1945.* Hamburg: Rowohlt.

Danneel, Rolf, and H. Paul. 1940. "Zur Physiologie der Kälteschwärzung beim Russenkaninchen IV." *Biologisches Zentralblatt* 60: 79–85.

Delbrück, Max. 1947. "Über Bakteriophagen." *Die Naturwissenschaften* 34: 301–305.

———. 1970. "A physicist's renewed look at biology: Twenty years later." *Science* 168: 1312–1315.

Dokumentationsarchiv des österreichischen Widerstands und Dokumentationsstelle für neuere österreichische Literatur, ed. 1977. *Österreicher im Exil 1934–1945.* Vienna.

Duggan, Stephan, and Betty Drury. 1948. *The Rescue of Science and Learning: The Story of the Emergency Committee in Aid of Displaced Foreign Scholars.* New York, Macmillan.

Dunn, Leslie C., ed. 1951. *Genetics in the Twentieth Century.* New York: MacMillan.

Duspiva, Franz. 1939. "Beiträge zur Histophysiologie des Insektendarmes I." *Protoplasma* 32: 211–250.

Eickemeyer, Helmut, ed. 1953. *Abschlußbericht des Deutschen Forschungsrates über seine Tätigkeit von seiner Gründung am 9.3.1949 bis zum 15.8.1951.* Munich: Oldenbourg.

Erichsen, Regine. 1991. "Die Emigration deutschsprachiger Naturwissenschaftler von 1933 bis 1945 in die Türkei in ihrem sozial- und wissenschaftshistorischen Zusammenhang," in H. A. Strauß, ed. *Die Emigration der Wissenschaft nach 1933.* Munich: K. G. Saur.

Evenari, Michael. 1987. *Und die Wüste trage Frucht: Ein Lebensbericht.* Gerlingen: Bleicher.

Evolution und Hominisation—Festschrift zum 60. Geburtstag von Gerhard Heberer. 1962. (Edited by Gottfried Kurth.) Stuttgart: G. Fischer.

Ferber, C. v. 1956. *Die Entwicklung des Lehrkörpers der deutschen Univer-*

sitäten und Hochschulen 1864–1954. Göttingen: Vandenhoeck and Ruprecht.

FIAT-Reviews of German Science, 1939–1946.

Fischel, Werner. 1934. "Abstammungslehre und Tierpsychologie." *Archiv für Geschichte der Medizin* 27: 511–515.

Fischer, Ilse, and Georg Gottschewski. 1939. "Gewebekultur bei Drosophila." *Die Naturwissenschaften* 27: 391–392.

Fischer, Klaus. 1988. "Der quantitative Beitrag der nach 1933 emigrierten Naturwissenschaftler zur deutschsprachigen physikalischen Forschung." *Berichte zur Wissenschaftsgeschichte* 11: 83–104.

Fischer, Peter. 1985. *Licht und Leben: Ein Bericht über Max Delbrück, den Wegbereiter der Molekularbiologie.* Konstanz: Universitäts Verlag.

Fleming, Donald. 1969. "Émigré Physicists and the Biological Revolution," in Fleming and Bailyn 1969.

Fleming, Donald, and Bernard Bailyn, eds. 1969. *The Intellectual Migration.* Cambridge, Mass.: Harvard University Press.

Flitner, Andreas, ed. 1965. *Deutsches Geistesleben und Nationalsozialismus: Eine Vortragsreihe an der Universität Tübingen mit einem Nachwort von Hermann Diem.* Tübingen: Wunderlich.

Forman, Paul. 1973. "Scientific internationalism and the Weimar physicists: The ideology and its manipulation in Germany after World War I." *Isis* 64: 151–180.

Fraenkel, Gottfried S. 1934. "Pupation of flies initiated by a hormone." *Nature* 133: 834.

———. 1935. "A hormone causing pupation in the blowfly Calliphora erythrocephala." *PNAS*, ser. b, no. 807, 118: 1–12.

Freisleben, Rudolf. 1940. "Die phylogenetische Bedeutung asiatischer Gersten." *Der Züchter* 12: 157–172.

Freisleben, Rudolf, and A. Lein. 1942. "Über die Auffindung einer mehltauresistenten Mutante nach Röntgenbestrahlung einer anfälligen reinen Linie von Sommergerste." *Die Naturwissenschaften* 20: 608.

Friederichs, Karl. 1934. "Vom Wesen der Ökologie." *Die Naturwissenschaften* 27: 277–285.

———. 1937. *Ökologie als Wissenschaft von der Natur.* Leipzig: J. A. Barth.

Friedrich-Freksa, Hans. 1940. "Bei der Chromosomenkonjugation wirksame Kräfte und ihre Bedeutung für die identische Verdopplung von Nucleoproteinen." *Die Naturwissenschaften* 28: 376–379.

———. 1940. "Grenzdosis, Kombinations- und Spätwirkung von Methylcholanthren und Benzpyren an der Haut von Mäusen, ein Beitrag zum Wirkungsmechanismus von kanzerogenen Kohlenwasserstoffen." *Biologisches Zentralblatt* 60: 498–524.

———. 1948. "Genabhängigkeit biochemischer Reaktionen bei Neurospora

(Nach Arbeitn von G. W. Beadle, E. L. Tatum, N. H. Horowitz und Mi-
tarbeitern)." *Zeitschrift für Naturforschung* 3b: 63–66.

Friedrich-Freksa, Hans, Georg Melchers, and Gerhard Schramm. 1946.
"Biologischer, chemischer und serologischer Vergleich zweier Parallelmu-
tanten pythopathogener Viren mit ihren Ausgangsformen." *Biologisches
Zentralblatt* 65: 187–222.

Frisch, Karl von. 1938. "Zur Psychologie des Fisch-Schwarmes." *Die Naturwissenschaften* 26: 601–606.

———. 1943. "Versuche über die Lenkung des Bienenschwarmes durch
Duftstoffe." *Die Naturwissenschaften* 31: 445–460.

———. 1946. "Die Tänze der Bienen." *Österreichischer Biologischer Anzeiger* 1: 1–48.

———. 1973. *Lebenserinnerungen eines Biologen.* Berlin: Springer.

Gaffron, Hans, and Kurt Wohl. 1936. "Zur Theorie der Assimilation." *Die
Naturwissenschaften* 24: 81–90.

Gasman, Daniel. 1971. *The Scientific Origins of National Socialism: Social
Darwinism in Ernst Haeckel and the German Monist League.* London:
Macdonald.

Geitler, Lothar. 1939. "Die Entstehung der polyploiden Somakerne der Het-
eropteren durch Chromosomenteilung ohne Kernteilung." *Chromosoma* 1:
1–22.

Gellermann, Günther W. 1986. *Der Krieg, der nicht stattfand.* Koblenz: Bernard and Graefe.

Gerlach, Joachim. 1941. "Über den Verbleib natürlich radioaktiver Stoffe im
Organismus nach parenteraler Zuführung." *Die Naturwissenschaften* 29:
300.

Gerlach, Joachim, M. Wolf, and Hans Joachim Born. 1942. "Zur Methodik
der Kreislaufbestimmung beim Menschen." *Archiv für experimentelle
Pathologie und Pharmakologie* 199: 83–88.

"Die Geschichte der Kaiser-Wilhelm-Gesellschaft und Max-Planck-Gesellschaft 1945–1949." Unpublished Festschrift for Otto Hahn on his seventieth
birthday, March 8, 1949. Arch. MPG.

Geuter, Ulfried. 1984. *Die Professionalisierung der deutschen Psychologie im
Nationalsozialismus.* Frankfurt: Suhrkamp.

Gilbert, Scott F. 1991. "Cellular politics: Ernest Everett Just, Richard B. Goldschmidt, and the attempt to reconcile embryology and genetics," in R.
Rainger, K. R. Benson, and J. Maienschein, ed., *The American Development of Biology,* 2d ed. New Brunswick: Rutgers University Press, pp. 317,
326.

Gilbert, Scott F., and Lauri Saxén. 1993. "Speman's organizer: models and
molecules." *Mechanisms of Development* 41: 73–89.

Glum, Friedrich. 1936. "Die Kaiser Wilhelm-Gesellschaft zur Förderung der

Wissenschaften," in Max Planck, ed., *25 Jahre Kaiser-Wilhelm-Gesell-schaft zur Förderung der Wissenschaften.* Berlin: Springer, pp. 1–21.

———. 1964. *Zwischen Wissenschaft, Wirtschaft und Politik: Erlebtes und Erdachtes in vier Reichen.* Bonn: H. Bouvier.

Goetsch, Wilhelm, and R. Güger. 1940. "Die Pilze der Blattschneider-Amei-sen und ihre Vernichtung." *Die Naturwissenschaften* 28: 764–765.

Goetsch, Wilhelm, and Rose Stoppel. 1940. "Die Pilze der Blattschneider-ameisen." *Biologisches Zentralblatt* 60: 393–398.

Goldschmidt, Richard. 1927a. *Die Lehre von der Vererbung.* Berlin: J. Springer.

———. 1927b. *Physiologische Theorie der Vererbung.* Berlin: Springer.

———. 1928. *Der Mendelismus,* 2d ed. Berlin: Parey.

———. 1940. *The Material Basis of Evolution.* New Haven: Yale University Press.

———. 1956a. *In and Out of the Ivory Tower.* Seattle: University of Wash-ington Press.

———. 1956b. *Portraits from Memory: Recollections of a Zoologist.* Seattle: University of Washington Press.

Goldstein, Avram. 1987. "Otto Krayer: Biographical memoirs." *Washington D.C.: The National Academy Press* 57: 151–225.

Gottschewski, Georg. 1939. "Eine Analyse bestimmter Drosophila pseu-doobscura—Rassen- und Artkreuzungen." *Zeitschrift für induktive Ab-stammungs- und Vererbungslehre* 78: 338–398.

Goudsmit, Samuel A. 1947. *Alsos.* New York: H. Schuman.

Gould, James L. 1976. "The dance-language controversy." *Quarterly Review of Biology* 51: 211–244.

Grabowski, Ulrich. 1938. "Experimentelle Untersuchungen über das angeb-liche Lernvermögen von Paramecium." *Zeitschrift für Tierpsychologie* 2: 265–282.

Graham, Loren R. 1974. *Science and Philosophy in the Soviet Union.* New York: Vintage.

Granin, Daniil. 1988. *Der Genetiker. Das leben des Nikolai Timofejew-Res-sowski, genannt Ur.* Cologne: Pahl-Rugenstein.

Groß, Walter. 1937. "Drei Jahre rassenpolitische Aufklärungsarbeit." *Volk und Rasse* 11: 331–337.

Grüneberg, Hans, and Werner Ulrich. 1950. *Moderne Biologie—Festschrift zum 60. Geburtstag von Hans Nachtsheim.* Berlin: F. W. Peters.

Gulland, J. Masson. 1938. "Nucleic acids." *Journal of the Chemical Society,* 1722–1734.

———. 1944. "Some aspects of the chemistry of nucleotides." *Journal of the Chemical Society,* 208–217.

Gumbel, Ernil J. 1938. *Freie Wissenschaft: Ein Sammelbuch aus der deutschen Emigration.* Strasbourg: S. Brant.

Günther, Hans F. K. 1922. *Rassenkunde des deutschen Volkes.* Munich: J. F. Lehmann.

Gütt, Arthur, Ernst Rüdin, and Falk Ruttke. 1936. *Gesetz zur Verhütung erbkranken Nachwuchses vom 14.7.1933 nebst Ausführungsbestimmungen,* 2d ed. Munich: J. F. Lehmann.

Guttenberg, Hermann von. 1942. "Über die Bildung und Aktivierung des Wuchsstoffes in den höheren Pflanzen." *Die Naturwissenschaften* 30: 109–112.

Haecker, Valentin. 1918. *Entwicklungsgeschichtliche Eigenschaftsanalyse (Phänogenetik): Gemeinsame Aufgaben der Enticklungsgeschichte, Vererbungs- und Rassenlehre.* Jena: G. Fischer.

Haevecker, Herbert. 1951. "40 Jahre Kaiser-Wilhelm-Gesellschaft: Im Auftrag der MPG nach den Archiven der Generalverwaltung bearbeitet." *Jahrbuch 1951 der MPG* (Göttingen), 7–59.

Hamburger, Viktor. 1934. "The effect of wing bud extirpation on the development of the central nervous system in chick embryos." *Journal of Experimental Zoology* 68: 449–494.

———. 1939. "The development and innervation of transplanted limb primordia of chick embryos." *Journal of Experimental Zoology* 80: 347–385.

———. 1979/1980. "S. Ramón y Cajal, R. G. Harrison, and the beginnings of neuroembryology." *Perspectives in Biology and Medicine* 23: 600–616.

———. 1980. "Evolutionary theory in Germany: A comment," in E. Mayr and W. Provine, eds., *The Evolutionary Synthesis: Perspectives on the Unification of Biology.* Cambridge, Mass.: Harvard University Press.

———. 1985. "Hans Spemann, Nobel laureate 1935." *Trends in Neuroscience* 8: 385–387.

———. 1988. *The Heritage of Experimental Embryology: Hans Spemann and the Organizer.* Oxford: Oxford University Press.

———. 1990. "The rise of experimental neuroembryology—A personal reassessment." *International Journal of Developmental Neuroscience* 8: 121–131.

Hämmerling, Joachim. 1939. "Über die Bedingungen der Kernteilung und der Zytenbildung bei Acetabularia mediterranea." *Biologisches Zentralblatt* 59: 158–193.

Hansen, Friedrich. 1990. "Zur Geschichte der deutschen biologischen Waffen." *1999* 1: 53–81.

Harder, Richard. 1946. "Über photoperiodisch bedingte Organ- und Gestaltbildung bei den Pflanzen." *Die Naturwissenschaften* 33: 41–49.

Harder, Richard, and O. Bode. 1943. "Über die Wirkung von Zwischenbelichtungen währen der Dunkelperiode auf das Blühen, die Verlaubung, und die Blattsukkulenz bie der Kurztagspflanze Kalanchoe blossfeldiana." *Planta* 33: 469–504.

Harder, Richard, and Hans von Witsch. 1942. "Über Massenkultur von Diatomeen." *Bericht der deutschen botanischen Gesellschaft* 60: 146–152.

Harris, Sheldon H. 1992. "Japanese biological warfare experiments and other atrocities in Manchuria, 1932–1945: A preliminary statement," in C. Roland, H. Friedlander, and B. Müller-Hill, eds., *Medical Science without Compassion: Past and Present.* Hamburg: Hamburger Stiftung für Sozialgeschichte des 20. Jahrhunderts, pp. 117–167.

Harrison, Ross G. 1937. "Embryology and its relations." *Science* 85: 369–374.

Hartmann, Max. 1944. "Befruchtungsstoffe bei Fischen (Regenbogenforelle)." *Die Naturwissenschaften* 32: 231.

Hartmann, Max, and O. Schartau. 1939. "Untersuchungen über die Befruchtungsstoffe der Seeigel I." *Biologisches Zentralblatt* 59: 571–587.

Hartshorne, E. Y. 1937. *The German Universities and National Socialism.* London: Unwin Brothers.

Hartwig, Hermann. 1940. "Metamorphose-Reaktionen auf einen lokalisierten Hormonreiz." *Biologisches Zentralblatt* 60: 473–478.

Hartwig, Hermann, and Eckhard Rotmann. 1940. "Experimentelle Untersuchungen an einem Massenauftreten von neotenen Triton taeniatus." *Archiv für Entwicklungsmechanik* 140: 195–251.

Harwood, Jonathan. 1985. "The erratic career of cytoplasmic inheritance." *Trends in Genetics* 1: 298–300.

———. 1987. "National styles in science: Genetics in Germany and the United States between the world wars." *Isis* 78: 390–414.

———. 1993. *Styles of Scientific Thought: The German Genetics Community, 1900–1933.* Chicago: University of Chicago Press.

Hayes, William. 1972. "Sexuelle Differenzierung bei Bakterien," in J. Cairns, G. S. Stent, J. D. Watson, eds., *Phagen und die Entwicklung der Molekularbiologie.* Berlin: Akademie-Verlag, pp. 198–211.

Heberer, Gerhard. 1933. *50 Jahre Chromosomentheorie der Vererbung.* Tübingen: Akademische Verlagsbuchhandlung F. F. Heine.

———, ed. 1937. "Neuere Funde zur Urgeschichte des Menschen und ihre Bedeutung für Rassenkunde und Weltanschauung." *Volk und Rasse* 12: 422–427, 435–444.

———. 1942. "Makro- und Mikrophylogenie." *Der Biologe* 11: 169–180.

———, ed. 1943a. *Die Evolution der Organismen: Ergebnisse und Probleme der Abstammungslehre.* Jena: G. Fischer.

———. 1943b. *Rassengeschichtliche Forschungen im indogermanischen Urheimatgebiet.* Jena: G. Fischer.

———. 1968. *Der gerechtfertigte Naeckel. Einblicke in seine Schriften aus*

Anlaß des Erscheinens seines Hauptwerkes "Generelle Morphologie der Organismen" vor 100 Jahren. Stuttgart: G. Fischer.

———. 1972. *Der Ursprung des Menschen.* Stuttgart: G. Fischer.

Heberer, Gerhard, and F. Schwanitz. 1960. *Hundert Jahre Evolutionsforschung.* Stuttgart: G. Fischer.

Heiber, Helmut. 1958. "Der Generalplan Ost." *Vierteljahreshefte für Zeitgeschichte* 6: 281–325.

———. 1968. *Reichsführer!... Briefe von und an Himmler.* Stuttgart: DVA.

Heindorf, Horst, and Heinz Schwabe. 1968. "Arnold Japha (1877–1943)." *Wissenschaftliche Zeitschrift der Universität Halle* 17: 125–142.

Heinroth, Katharina. 1971. *Oskar Heinroth, Vater der Verhaltensforschung 1871–1945.* Stuttgart: Wissenschaftliche Verlagsgesellschaft.

———. 1978. "Oskar Heinroth, Karl von Frisch, Otto Koehler, Konrand Lorenz und Nikolaas Tinbergen," in K. Faßmann, ed., *Die Großen der Weltgeschichte.* Zurich: Kindler, pp. 428–445.

Heisenberg, Werner. 1953. "Notwendigkeit wissenschaftlicher Forschung" (*Münchner Allgemeine,* December 12, 1949), in H. Eickemeyer, ed., *Abschlußbericht des Deutschen Forschungsrates über seine Tätigkeit von seiner Gründung am 9.3.1949 bis zum 15.8.1951.* Munich: Oldenbourg.

———. 1956. "Über die Arbeiten zur technischen Ausnutzung der Atomenergie in Deutschland." *Die Naturwissenschaften* 33: 325–329.

Heitz, Emil, and Hans Bauer. 1933. "Beweise für die Chromosomennatur der Kernschleifen in den Knäuelkernen von Bibio hortulanus L." *Zeitschrift für Zellforschung* 17: 67–82.

Henke, Karl. 1946. "Über die verschiedenen Zellteilungsvorgänge in der Entwicklung des beschuppten Flügelepithels der Mehlmotte Ephestia kühniella Z." *Biologisches Zentralblatt* 65: 120–135.

Henke, Karl, E. v. Finck, and S. Y. 1941. Ma. "Über sensible Perioden für die Auslösung von Hitzemodifikationen bei Drosophila und die Beziehungen zwischen Modifikationen und Mutationen." *Zeitschrift für induktive Abstammungs- und Vererbungslehre* 79: 267–316.

Henning, Eckhart. 1987. *Otto Warburg.* Berlinische Lebensbilder I. Berlin: Colloquim.

Herbst, Curt. 1941. "Hans Driesch als experimenteller und theoretischer Biologe." *Archiv für Entwicklungsmechanik* 141: 111–153.

Herbst, Ludolf, and Constantin Goschler, eds. 1989. *Wiedergutmachung in der Bundesrepublik Deutschland.* Munich: Oldenbourg.

Herre, Wolf. 1943. "Domestikation und Stammesgeschichte," in G. Heberer, ed., *Die Evolution der Organismen: Ergebnisse und Probleme der Abstammungslehre.* Jena: G. Fischer.

Hertwig, Paula. 1942. "Neue Mutationen und Koppelungsgruppen bei der

Hausmaus." *Zeitschrift für induktive Abstammungs- und Vererbungslehre* 80: 220–246.

Hintsches, Eugen. 1983. "Vor 50 Jahren—Exodus der Wissenschaftler aus Deutschland." *Max-Planck-Spiegel* 5: 44–53.

Hippius, Rudolf, I. G. Feldmann, K. Jellinek, and K. Leider. 1943. *Volkstum, Gesinnung und Charakter. Bericht über psychologische Untersuchungen an Posener deutsch-polnischen Mischlingen und Polen, Sommer 1942.* Stuttgart: Prag, W. Kohlhammer.

Hoffmann, Walter. 1937. "Die Winterfestigkeit keimgestimmter Gersten." *Der Züchter* 9: 281–284.

Höflechner, Walter. 1988. *Die Baumeister des künftigen Glücks: Fragment einer Geschichte des Hochschulwesens in Österreich vom Ausgang des 19. Jahrhunderts bis in das Jahr 1938.* Graz: Akademische Druck- und Verlagsanstalt.

Höfler, Karl. 1939. "Kappenplasmolyse und Ionenantagonismus." *Protoplasma* 33: 545–578.

———. 1943. "Über Fettspeicherung und Zuckerpermeabilität einiger Diatomeen." *Protoplasma* 38: 71–104.

Höfler, Karl, H. Migsch, and W. Rottenburg. 1941. "Über die Austrocknungsresistenz landwirtschaftlicher Kulturpflanzen." *Forschungsdienst* 12: 50–61.

Höhne, Heinz. 1988. *Der Orden unter dem Totenkopf. Die Geschichte der SS.* Munich: Bertelsmann.

Holst, Erich von. 1939. "Entwurf eines Systems der lokomotorischen Periodenbildung bei Fischen: Ein kritischer Beitrag zum Gestaltproblem." *Zeitschrift für vergleichende Physiologie* 26: 481–528.

Holtfreter, Johannes. 1931. "Über die Aufzucht isolierter Teile des Amphibienkeimes II." *Roux' Archiv für Entwicklungsmechanik* 124: 404–446.

———. 1933. "Eigenschaften und Verbreitung induzierender Stoffe." *Die Naturwissenschaften* 21: 766–770.

———. 1938. "Differenzierungspotenzen isolierter Teile der Urodelengastrula." *Archiv für Entwicklungsmechanik* 138: 522–656.

———. 1939a. "Gewebeaffinität, ein Mittel der embryonalen Formbildung." *Archiv für experimentelle Zellforschung* 23: 657–738.

———. 1939b. "Studien zur Ermittlung der Gestaltungsfaktoren in der Organentwicklung der Amphibien I: Dynamisches Verhalten isolierter Furchungszellen und Entwicklungsmechanik der Entodermanlage." *Archiv für Entwicklungsmechanik* 139: 110–190.

———. 1943a. "A study of the mechanics of gastrulation, Part I." *Journal of Experimental Zoology* 94: 261–318.

———. 1943b. "Properties and functions of the surface coat in amphibian embryos." *Journal of Experimental Zoology* 93: 251–323.

————. 1944. "A study of the mechanics of gastrulation, Part II." *Journal of Experimental Zoology* 95: 171–212.

Horder, T. J., and P. J. Weindling. 1985. "Hans Spemann and the organizer," in Horder, Witkowski, and Wylie 1985.

Horder, T. J., J. A. Witkowski, and C. C. Wylie, eds. 1985. *A History of Embryology*. Cambridge: Cambridge University Press.

Höxtermann, Ekkehard. 1991. "Wilhelm Ruhland und seine Leipziger 'Schüler' in Berlin." *NTM-Schriftenreihe Geschichte Naturwissenschaft Technik* (Leipzig) 28: 95–107.

Hunt, Linda. 1985. "U.S. coverup of Nazi scientists." *Bulletin of the Atomic Scientists*, 16–24.

Huxley, Julian. 1949. *Heredity—East and West: Lysenko and World Science*. New York: H. Schuman.

Irving, David. 1967. *The Virus House*. London: Kimber.

Jacob, François. 1988. *Die innere Statue*. Zurich: Ammann.

Jacob, François, and Jacques Monod. 1961. "Genetic regulatory mechanisms in the synthesis of proteins." *Journal of Molecular Biology* 3: 318–356.

Jahn, Ilse, Rolf Löther, and Konrad Senglaub. 1982. *Geschichte der Biologie: Theorien, Methoden, Institutionen, Kurzbiographien*. Jena: VEB Fischer.

Jahrbuch der Kaiser-Wilhelm-Gesellschaft zur Förderung der Wissenschaften, 1939–1942.

Jahrbuch der Max-Planck-Gesellschaft zur Förderung der Wissenschaften, 1951–1956.

Joravsky, David. 1970. *The Lysenko Affair*. Repr. Chicago: University of Chicago Press.

Jordan, Pascual. 1938. "Zur Frage einer spezifischen Anziehung zwischen Genmolekülen." *Physikalische Zeitschrift* 39: 711–714.

————. 1940. "Die Stellung der Quantenphysik zu den aktuellen Problemen der Biologie." *Archiv für die gesamte Virusforschung* 1: 1–20.

————. 1944. "Zum Problem der Eiweißautokatalysen." *Die Naturwissenschaften* 32: 20–26.

Judson, Horace Freeland. 1979. *The Eighth Day of Creation: Makers of the Revolution in Biology*. New York: Simon and Schuster.

Just, Günther. 1939. "Die erbbiologischen Grundlagen der Leistung." *Die Naturwissenschaften* 27: 154–161, 170–176.

Kalikow, Theodora J. 1980. "Die ethologische Theorie von Konrad Lorenz: Erklärung und Ideologie, 1938–1943," in H. Mehrtens and S. Richter, eds., *Naturwissenschaft, Technik und NS-Ideologie*. Frankfurt: Suhrkamp, pp. 189–214.

————. 1983. "Konrad Lonrenz's ethological theory: Explanations and ideology." *Journal of the History of Biology* 16: 39–73.

Kappert, Hans. 1978. *Vier Jahrzehnte miterlebte Genetik*. Berlin: Paul Parey.

Kappert, Hans, and Wilhelm Rudorf. 1958. *Handbuch der Pflanzenzüchtung.* 2d ed. Berlin: P. Parey.

Karlson, Peter. 1990. *Adolf Butenandt. Biochemiker, Hormonforscher, Wissenschaftspolitiker.* Stuttgart: Wissenschaftliche Verlags-Gesellschaft.

Kater, Michael. 1966. "Das Ahnenerbe. Die Forschungs-und Lehrgemeinschaft in der SS-Organisationsgeschichte von 1933–1945." Ph.D. diss., University of Heidelberg.

———. 1989. *Doctors under Hitler.* Chapel Hill and London: University of North Carolina Press.

Kaudewitz, Fritz. 1957. *Grundlagen der Vererbungslehre.* Munich: Lehnen.

Kaufmann, C., H. Aurel Müller, A. Butenandt, and H. Friedrich-Freksa. 1949. "Experimentelle Beiträge zur Bedeutung des Follikelhormons für die Carcinomentstehung." *Zeitschrift für Krebsforschung* 56: 482–542.

Kempner, Robert M. W. 1987. *SS im Kreuzverhör.* Nördlingen: Hamburger Stiftung für Sozialgeschichte des 20. Jahrhunderts.

Kirsche, W. 1986. *Oskar Vogt 1870–1959.* Berlin.

Klee, Ernst. 1983. *"Euthanasie" im NS-Staat—Die "Vernichtung lebensunwerten Lebens."* Frankfurt: Fischer.

Knaape, Hans-Hinrich. 1988. "Die medizinische Forschung an geistig behinderten Kindern in Brandenburg-Görden in der Zeit des Faschismus," in *Das Schicksal der Medizin im Faschismus* (Internationales wissenschaftliches Symposium europäischer Sektionen der IPPNW Erfurt/Weimar 17.-20.11.1988).

Knapp, Edgar, A. Reuss, O. Risse, and H. Schreiber. 1939. "Quantitative Analyse der mutationsauslösenden Wirkung monochromatischen UV-Lichtes." *Die Naturwissenschaften* 27: 304.

Koch, Gerhard. 1993. *Humangenetik und Neuro-Psychiatrie in meiner Zeit (1932–1978). Jahre der Entstehung.* Erlangen: Palm and Enke.

Koehler, Otto. 1934. "Die biologische Gestaltung der Völker durch Fortpflanzung, Vererbung und Auslese." *Der Biologe* 3: 192–202.

———. 1937. "Können Tauben 'zählen'?" *Zeitschrift für Tierpsychologie* 1: 38–48.

———. 1943. "'Zähl'-Versuche an einem Kolkraben und Vergleichsversuche am Menschen." *Zeitschrift für Tierpsychologie* 5: 575–712.

———. 1963. "Konrad Lorenz 60 Jahre." *Zeitschrift für Tierpsychologie* 20: 385–409.

Kosswig, Curt. 1929. "Das Gen in fremder Erbmasse." *Der Züchter* 1: –152–157.

Kramer, Erich. 1957. "Eletrophoretische Untersuchungen an Mutanten des Tabakmosaikvirus." *Zeitschrift für Naturforschung* 12b: 609–614.

Kuckuck, Hermann. 1934. *Von der Wildpflanze zur Kulturpflanze: Die Be-*

deutung der natürlichen und künstlichen Zuchtwahl für die Entstehung neuer Pflanzenrassen. Berlin: Metzner.

———. 1939. *Pflanzenzüchtung.* Berlin: de Gruyter.

———. 1962. "Genzentrentheorie in heutiger Sicht." *Eucarpia,* May, pp. 177–196.

———. 1988. *Wandel und Beständigkeit im Leben eines Pflanzenzüchters.* Hamburg.

Kühn, Alfred. 1957. "Karl Henke." *Die Naturwissenschaften* 44: 25.

Kühn, Alfred, and Erich Becker. 1942. "Quantitative Beziehungen zwischen zugeführtem Kynurenin und Augenpigment bei Ephestia kühniella Z." *Biologisches Zentralblatt* 62: 303–317.

Laibach, Friedrich. 1941. "Wuchstoffbildung mit Kohlenhydratstoffwechsel." *Berichte der deutschen botanischen Gesellschaft* 59: 257–271.

Lang, Anton. 1987. "Elisabeth Schiemann: Life and career of a woman scientist in Berlin." *Englera* 7: 17–28.

Lange, Otto. 1989. "Michael Evenari alias Walter Schwarz, 1904–1989." *Botanica Acta* 102: 261–340.

Lehmann, Ernst. 1933a. *Biologie im Leben der Gegenwart.* Munich: J. F. Lehmann.

———. 1933b. "Werbeaktion: Erbbiologie—Rassenhygiene." *Der Biologe* 2: 343–344.

———. 1934a. *Biologischer Wille—Wege und Ziele biologischer Arbeit im neuen Reich.* Munich: J. F. Lehmann.

———. 1934b. "Der Einfluß der Biologie auf unser Weltbild," in *Deutschland in der Zeitenwende* (public lectures at the University of Tübingen, summer semester 1933). Stuttgart: Kohlhammer.

———. 1935. "Die Biologie an der Zeitenwende." *Der Biologe* 4: 375–381.

———. 1936. *Wege und Ziele einer deutschen Biologie.* Munich: J. H. Lehmann.

———. 1937. "Biologie und Weltanschauung." *Der Biologe* 6: 337–341.

———. 1941. "Zur Genetik der Entwicklung in der Gattung Epilobium III." *Jahrbuch für wissenschaftliche Botanik* 89: 637–686.

———. 1942a. "Mendel und seine Entdeckung." *Süddeutsche Apotheker-Zeitung,* July 25.

———. 1942b. "Zur Genetik der Entwicklung in der Gattung Epilobium IVB." *Jahrbuch für wissenschaftliche Botanik* 90: 49–98.

———. 1946. *Irrweg der Biologie.* Stuttgart: Hatje.

Lehmann, Ernst, G. Hinderer, H. Graze, and G. Schlenker. 1936. "Versuche zur Klärung der reziproken Verschiedenheiten von Epibolium-Bastarden (I-III)." *Jahrbuch für wissenschaftliche Botanik* 82: 657–668.

Lehmensick, Rudolf. 1940. "Deutsche Wissenschaftler als Kolonialpioniere." *Deutschlands Erneuerung,* November 559–565.

Levene, P. A., and E. S. London. 1929. "The structure of thymonucleic acid." *Journal of Biological Chemistry* 83: 793–802.

Levi-Montalcini, Rita. 1987. "The nerve growth factor thirty-five years later." *Science* 237: 1154–1162.

———. 1988. *In Praise of Imperfection.* New York: Basic Books.

Linskens, H. F. 1962. "Die Situation der Biologie in Holland." *Bericht der Oberhessischen Gesellschaft für Natur- und Heilkunde zu Gießen, Naturwissenschaftliche Abteilungen* 32: 13–37.

Lorenz, Konrad. 1935. "Der Kumpan in der Umwelt des Vogels." *Journal für Ornithologie* 83: 137–213, 289–413. Eng. trans. in Lorenz 1970, vol. 1.

———. 1937a. "Biologische Fragestellungen in der Tierpsychologie." *Zeitschrift für Tierpsychologie* 1: 24–32.

———. 1937b. "Über den Begriff der Instinkthandlung." *Folia biotheoretica,* 17–50.

———. 1937c. "Über die Bildung des Instinktbegriffs." *Die Naturwissenschaften* 25: 289–300, 304–318, 324–331. Eng. trans. in Lorenz 1970, vol. 1.

———. 1939. "Vergleichende Verhaltensforschung." *Zoologischer Anzeiger,* Supp. 12, 69–102.

———. 1940a. "Durch Domestikation verursachte Störungen arteigenen Verhaltens." *Zeitschrift für angewandte Psychologie und Charakterkunde* 59: 2–81.

———. 1940b. "Nochmals: Systematk und Entwicklungsgedanke im Unterricht." *Der Biologe,* 24–36.

———. 1941. "Oskar Heinroth 70 Jahre." *Der Biologe,* 45–47.

———. 1943a. "Die angeborenen Formen möglicher Erfahrung." *Zeitschrift für Tierpsychologie* 5: 235–409.

———. 1943b. "Psychologie und Stammesgeschichte," in G. Heberer, ed., *Die Evolution der Organismen: Ergebnisse und Probleme der Abstammungslehre.* Jena: G. Fischer.

———. 1963. *On Aggression,* trans. Marjorie Kerr Wilson. New York: Harcourt, Brace, and World.

———. 1965. *Über tierisches und menschliches Verhalten. Aus dem Werdegang der Verhaltenslehre,* vols. 1 and 2. Munich: Piper.

———. 1970. *Studies in Animal and Human Behavior,* trans. Robert Martin, 2 vols. Cambridge, Mass.: Harvard University Press.

———. 1973. *Die Rückseite des Spiegels.* Munich and Zurich: Piper.

———. 1978. *Vergleichende Verhaltensforschung. Grundlagen der Ethologie.* Vienna: Springer.

———. 1979. *Das sogenannte Böse. Zur Naturgeschichte der Aggression,* 6th ed. Munich. *On Aggression,* trans. Marjorie Kerr Wilson, New York: Harcourt, Brace, and World, 1963.

———. 1980. *Die acht Todsünden der zivilisierten Menschheit*, 11th ed. Munich: Piper. *Civilized Man's Eight Deadly Sins*, trans. Marjorie Kerr Wilson, New York: Harcourt, Brace, Jovanovich, 1973.

———. 1981. *The Foundations of Ethology*, trans. Konrad Lorenz and Robert Warren Kickert. New York: Springer-Verlag.

Lorenz, Konrad, and Franz Kreuzer. 1988. *Leben ist Lernen*, 4th ed. Munich: R. Piper.

Loring, Hubert S. 1938. "The diffusion constants and approximate molecular weights of tobacco mosaic virus nucleic acid and yeast nucleic acid." *Journal of Biological Chemistry* 128: lxi–lxii.

Macrakis, Kristie I. 1989. "Scientific Research in National Socialist Germany: The Survival of the Kaiser Wilhelm Gesellschaft." Ph.D. diss. Harvard University.

———. 1993. *Surviving the Swastika: Scientific Research in Nazi Germany*. Oxford: Oxford University Press.

Maienschein, Jane. 1991. "Epistemic styles in German and American embryology." *Science in Context* 4: 407–427.

Manning, Aubrey. 1979. *An Introduction to Animal Behavior*, 3d ed. Reading, Mass.: Addison-Wesley.

Marburger Hochschultage 1946. Frankfurt, 1947.

Marquardt, Hans. 1957. *Natürliche und künstliche Erbänderungen—Probleme der Mutationsforschung*. Hamburg: Rowohlt.

Martius, Heinrich, and Friedrich Kröning. 1936. "Meerschweinversuche zur Frage der Röntgenstrahlenwirkung auf die Keimdrüsen." *Deutsche medizinische Wochenschrift* 26: 1046–1054.

May, Eduard. 1941. *Am Abgrund des Relativismus*. Berlin: G. Luttke.

Mayer, Arno J. 1988. *Why Did the Heavens Not Darken? The "Final Solution" in History*. New York: Pantheon Books.

Mayr, Ernst. 1963. *Animal Species and Evolution*. Cambridge, Mass.: Harvard University Press.

Mayr, Ernst, and William Provine. 1980. *The Evolutionary Synthesis. Perspectives of the Unification of Biology*. Cambridge, Mass.: Harvard University Press.

———. 1982. *The Growth of Biological Thought*. Cambridge, Mass.: Harvard University Press.

Medvedev, Zhores A. 1969. *The Rise and Fall of T. D. Lysenko*, trans. I. Michael Lerner. New York: Doubleday.

Mehrtens, Herbert, and Steffen Richter, eds. 1980. *Naturwissenschaft, Technik und NS-Ideologie: Beiträge zur Wissenschaftsgeschichte des Dritten Reiches*. Frankfurt: Suhrkamp.

Melchers, Georg. 1942. "Über einige Mutationen des Tabakmosaikvirus und eine 'Parallelmutation' des Tomatenmosaikvirus." *Die Naturwissenschaften* 30: 48–49.

———. 1958. "Die Bedeutung der Virusforschung für die moderne Genetik," in *Arbeitsgemeinschaft für Forschung des Landes Nordrhein-Westfalen,* vol. 77, Cologne: Westdeutscher Verlag.

———. 1960a. "Die Bedeutung der Virusforschung für die allgemeine Biologie." *Berichte der deutschen Botanischen Gesellschaft* 73: 6–15.

———. 1960b. "Warum interessiert den Biologen das Tabakmosaikvirus." *Jahrbuch der MPG,* pp. 90–113.

———. 1968. "Techniques for the quantitative study of mutation in plant viruses." *Theoretical and Applied Genetics* 38: 275–279.

———. 1987. "Ein Botaniker auf dem Wege in die Allgemeine Biologie auch in Zeiten moralischer und materieller Zerstörung und Fritz von Wettstein, 1895–1945." *Berichte der deutschen botanischen Gesellschaft* 100: 373–405.

Melchers, Georg, Gerhard Schramm, H. Trurnit, and H. Friedrich-Freksa. 1940. "Die biologische, chemische und eletronenmikroskopische Untersuchung eines Mosaikvirus aus Tomaten." *Biologisches Zentralblatt* 60: 527.

Meyer, Konrad. 1939. "Unsere Aufgabe." *Forschungsdienst* 8: 463–467.

———. 1941. "Unsere Forschungsarbeit im Kriege." Main report on the wartime meeting of the Forschungsdieust, 1941. *Forschungsdienst* 11: 253–286.

———. 1942. "Planung und Aufbau in den besetzten Ostgebieten," in *Jahrbuch der Kaiser-Wilehlm-Gesellschaft zur Förderung der Wissenschaften, 1942,* pp. 250–275.

Michaelis, Anthony R., and Roswitha Schmidt. 1983. *Wissenschaft in Deutschland: Niedergang und neuer Aufstieg.* Stuttgart: Wissenschaftliche Verlagsgesellschaft.

Michaelis, Peter. 1938. "Über die Konstanz des Plasmons." *Zeitschrift für induktive Abstammungs- und Vererbungslehre* 74: 435–459.

———. 1942. "Experimentelle Untersuchungen über die geographische Verbreitung von Plasmon-Unterschieden und der auf diese Unterschiede empfindliche Gene, sowie deren theoretische Bedeutung für das Kein-Plasma-Problem." *Biologisches Zentralblatt* 62: 170–186.

Michaelis, Peter, and H. Ross. 1944. "Untersuchungen an reziprok verschiedenen Artbastarden bei Epilobium II. Über Abänderungen an reziprok verschiedenen und reziprok gleichen Epilobium-Arbastarden." *Flora* 37: 24–56.

Mitscherlich, Alexander. 1963. "Eugenik—Notwendigkeit und Gefahr." *Fortschr. Med.* 81: 714–715.

Mitscherlich, Alexander, and Fred Mielke, eds. 1978. *Medizin ohne Men-*

schlichkeit. Dokumente des Nürnberger Ärzteprozesses, 2d ed. Frankfurt: Fischer.

Möglich, F., R. Rompe, and N. W. Timoféeff-Ressovsky. 1944. "Über die Indeterminiertheit und die Verstärkererscheinungen in der Biologie." *Die Naturwissenschaften* 32: 316–324.

Mothes, Kurt. 1940. "Zur Biosynthese der Säureamide Asparagin und Glutamin." *Planta* 30: 726–756.

Mothes, Kurt, and K. Hieke. 1943. "Die Tabakwurzel als Bildungsstätte des Nikotins." *Die Naturwissenschaften* 31: 17–18.

Mothes, Kurt, and D. Kretschmer. 1946. "Über die Alkaloidsynthese in isolierten Lupinenwurzeln." *Die Naturwissenschaften* 33: 26.

Möwus, Franz. 1940. "Carotinoid-Derivate als geschlechtsbestimmende Stoffe von Algen." *Biologisches Zentralblatt* 60: 143–166.

Müller, Rolf-Dieter. 1980. "Die deutschen Gaskriegsvorbereitungen 1919–1945." *Militärgeschichtliche Mitteilungen* 1: 25–54.

Müller-Hill, Benno. 1984. *Tödliche Wissenschaft: Die Aussonderung von Juden, Zigeunern und Geisteskranken 1933–1945.* Reinbek: Rowohlt. *Murderous Science: Elimination by Selection of Jews, Gypsies, and Others, Germany, 1933–1945,* trans. George R. Fraser, Oxford: Oxford University Press, 1988.

————. 1987. "Genetics after Auschwitz." *Holocaust and Genocide Studies* 2: 3–20.

————. 1988. "Daidalos, Faust und Mengele: Wissenschaftler als Verbrecher." Lecture at the International Scientific Symposium of the IPPNW, Erfurt, November 17–20.

————. 1993. "The shadow of genetic injustice." *Nature* 362: 491–492.

Mussgnug, Dorothee. 1988. *Die vertriebenen Heidelberger Dozenten. Zur Geschichte der Ruprecht-Karls-Universität nach 1933.* Heidelberg: Winter.

Nachmansohn, David. 1979. *German-Jewish Pioneers in Science, 1900–1933: Highlights in Atomic Physics, Chemistry, and Biochemistry.* New York: Springer.

Nachtsheim, Hans, and Hans Stengel. 1977. *Vom Wildtier zum Haustier,* 3d. ed. Berlin: Parey.

Nachtsheim, Hans. 1938. "Erbkranke Kaninchen (Schüttellähmung, eine erbliche Nervenkrankheit)." Publications of the Reich Office for Educational Films concerning university film no. C240.

————. 1940a. "Allgemeine Grundlagen der Rassenbildung," in G. Just, ed., *Handbuch der Erbbiologie des Menschen.* Berlin: Springer, pp. 552–583.

————. 1940b. "Krampfbereitschaft und Erbbild des Epileptikers." *Der Erbarzt* 8.

————. 1942. "Krampfbereitschaft und Genotypus III: Das Verhalten epileptischer und nichtepileptischer Kaninchen im Cardiazolkrampf." *Zeitschrift für menschliche Vererbungs- und Konstitutionslehre* 26: 22–74.

———. 1948. "Bourgeoise Biologie?" *Der Tagesspiegel,* August 5.

———. 1949. "Weshalb ich die Humboldt-Hochschule verlasse." *Der Tagesspiegel,* February 19.

———. 1951a. *Ein halbes Jahrhundert Genetik.* Berlin: Duncker and Humblo.

———. 1951b. "Neue Schuld der Wissenschaft." *Der Tagesspiegel,* June 23.

———. 1952. *Für und wider die Sterilisation aus eugenischer Indikation.* Stuttgart: G. Fischer.

———. 1953. Foreword to "Wissenschaft und Freiheit." International Conference, Hamburg, June 23–26. Published by Der Kongreß für die Freiheit der Kultur.

———. 1956. "Lyssenkos Ende." *Deutsche Kommentare,* April 14.

———. 1962. "Das Gesetz zur Verhütung erbkranken Nachwuchses aus dem Jahre 1933 in heutiger Sicht." *Ärztliche Mitteilungen* 59: 1640–1644.

———. 1963. "Warum Eugenik?" *Fortschr. Med.* 81: 711–713.

Nakamura, O., and S. Toivonen. 1978. *Organizer—A Milestone a Half-Century from Spemann.* Elsevier: Biomedical Press.

Needham, Joseph. 1941. *The Nazi Attack on International Science.* London: Watts and Co.

Nipperdey, Thomas, and Ludwig Schmugge. 1970. *50 Jahre Forschungsförderung in Deutschland. Ein Abriß der Geschichte der Deutschen Forschungsgemeinschaft 1920–1970.* Berlin: Deutsche Forschungsgemeinschaft.

Noack, Kurt. 1943. "Über den biologischen Abbau des Chlorophylls." *Biochemische Zeitschrift* 316: 166–187.

Noack, Kurt, and E. Timm. 1942. "Vergleichende Untersuchung der Proteine in den Chloroplasten und im Cytoplasma des Spinatblattes." *Die Naturwissenschaften* 30: 453.

Nolte, Ernst. 1986. "Vergangenheit die nicht vergehen will." *Frankfurter Allgemeine Zeitung,* June 6.

———. 1992. *Martin Heidegger: Politik und Geschichte im Leben und Denken.* Berlin: Propyläen.

Notgemeinschaft deutscher Wissenschaftler im Ausland, ed. 1936. *List of Displaced German Scholars.* London.

Nyiszli, Miklós. 1992. *Im Jenseits der Menschlichkeit: Ein Gerichtsmediziner in Auschwitz.* Berlin: Dietz Verlag GmbH.

Oehlkers, Friedrich. 1943. "Die Auslösung von Chromosomenmutationen in der Meiosis durch Einwirkung von Chemikalien." *Zeitschrift für induktive Abstammungs- und Vererbungslehre* 81: 313–341.

———. 1946. "Weitere Versuche zur Mutationsauslösung durch Chemikalien." *Biologisches Zentralblatt* 65: 176—186.

Olby, Robert. 1970. "The macromolecular concept and the origins of molecular biology." *Journal of Chemical Education* 47: 168–174.

———. 1974. *The Path to the Double Helix.* London: MacMillan.

Oppenheimer, J. M. 1955. "Methods and techniques," in B. J. Willier, P. A. Weiss, and V. Hamburger, eds., *Analysis of Development.* Philadelphia: Saunders Co.

———. 1970. "Cells and organizers." *American Zoologist* 10: 75–88.

Oppenheimer, J. M., and B. H. Willier. 1974. *Foundations of Experimental Embryology,* 2d ed. New York: Macmillan.

Osietzki, Maria. 1982. "Wissenschaftsorganisation und Restauration. Der Aufbau außeruniversitärer Forschungseinrichtungen und die Gründung des westdeutschen Staates 1945–1952." Ph.D. diss., University of Cologne.

Pash, Boris T. 1969. *The Alsos Mission.* New York: Award Books.

Pätau, Klaus. 1941. "Eine statistische Bemerkung zu Möwus' Arbeit: Die Analyse von 42 erblichen Eigenschaften der Chlamydomonas-engametos-Gruppe III. Teil." *Zeitschrift für induktive Abstammungs- und Vererbungslehre* 79: 317–319.

Paul, Diane. 1992. "Eugenic anxieties, social realities, and political choices." *Journal of Social Research* 59: 663–683.

Paul, Diane, and Costas B. Krimbas. 1992. "Nikolai V. Timoféeff-Ressovsky." *Scientific American* 266: 86–92.

Pauling, Linus, and Max Delbrück. 1940. "The nature of the intermolecular forces in biological processes." *Science* 92: 77–79.

Pfankuch, E., G. A. Kausche, and H. Stubbe. 1940. "Über die Entstehung, die biologische und physikalisch-chemische Charakterisierung von Röntgen- und Gamma-Strahlen induzierten 'Mutationen' des Tabakmosaikvirusproteins." *Biochemische Zeitschrift* 304: 238–258.

Pirschle, Karl. 1942a. "Quantitative Untersuchungen über Wachstum und 'Ertrag' autopolyploider Pflanzen." *Zeitschrift für induktive Abstammungs- und Vererbungslehre* 80: 126–156.

———. 1942b. "Weitere Untersuchungen über Wachstum und 'Ertrag' von Autopolyploiden (2n, 3n, 4n) und ihren Bastarden." *Zeitschrift für induktive Abstammungs- und Vererbungslehre* 80: 247–270.

Planck, Max, ed. 1936. *25 Jahre Kaiser-Wilhelm-Gesellschaft zur Förderung der Wissenschaften.* Berlin: Springer.

Plarre, Werner. 1987. "A contribution to the history of the science of heredity in Berlin." *Englera* 7: 147–217.

Portugal, Franklin H., and Jack S. Cohen. 1979. *A Century of DNA: A History of the Discovery of the Structure and Function of the Genetic Substance.* Cambridge, Mass.: MIT Press.

Proctor, Robert. 1988. *Racial Hygiene. Medicine under the Nazis.* Cambridge, Mass.: Harvard University Press.

Rajewsky, Boris. 1939. "Bericht über die Schneeberger Untersuchungen." *Zeitschrift für Krebsforschung* 49: 315–340.

Rajewsky, Boris, A. Schraub, and G. Kahlau. 1943. "Experimentelle Geschwulsterzeugung durch Einatmung von Radiumemanation." *Die Naturwissenschaften* 31: 170–171.

Rajewsky, Boris, A. Schraub, and E. Schraub. 1942a. "Über die toxische Dosis bei Einatmung von Ra-Emanation." *Die Naturwissenschaften* 30: 489–492.

———. 1942b. "Zur Frage der Toleranz-Dosis bei der Einatmung von Ra-Em." *Die Naturwissenschaften* 30: 733–734.

Rehm, S. 1992. "Heinz Brücher." *Angewandte Botanik* 66: 1–2.

Reitlinger, Gerald. 1956. *Die SS: Tragödie einer deutschen Epoche.* Vienna and Munich: Desch.

Renner, Otto. 1946. "Artbildung in der Gattung Oenothera." *Die Naturwissenschaften* 33: 211–218.

Richter, Steffen. 1980. "Die 'Deutsche Physik,'" in H. Mehrtens and S. Richter, eds., *Naturwissenschaft, Technik und NS-Ideologie.* Frankfurt: Suhrkamp, pp. 116–141.

Riehl, Nikolaus, Nikolai W. Timoféeff-Ressovsky, and Karl Günter Zimmer. 1941. "Mechanismus der Wirkung ionisierter Strahlen auf biologische Elementareinheiten." *Die Naturwissenschaften* 29: 625–639.

Roggenbau, C. H. 1937. "Albrecht Langelüdekke, Über die differentialdiagnostische Bedeutung der Cardiazolkrämpfe." *Zentralblatt für die gesamte Neurologie und Psychiatrie* 84: 85.

Römer, Theodor. 1941–1943. "Ausgangsmaterial für die Resistenzzüchtung bei Getreide." *Zeitschrift für Pflanzenzüchtung* 24: 304–332.

———. 1936. "Die Bedeutung des Gesetzes der Parallelvariationen für die Pflanzenzüchtung." *Nova Acta Leopoldina* 4: 350–365.

Römer, Theodor, and Wilhelm Rudorf, eds. 1938–1944. *Handbuch der Pflanzenzüchtung,* 5 vols. Berlin: Parey.

Rose, Steven. 1987. "Biotechnology at war." *New Scientist,* March 19, 33–37.

Rudorf, Wilhelm. 1937a. *Die politischen Aufgaben der deutschen Pflanzenzüchtung.* Berlin.

———. 1937b. "Entwicklungstendenzen in der Pflanzenzüchtung." *Forschungsdienst* 4: 153–162.

———. 1938. "Keimstimmung und Photoperiode in ihrer Bedeutung für die Kälteresistenz." *Der Züchter* 10: 238–246.

Rudorf, Wilhelm, and O. Schröck. 1941. "Neuere Beobachtungen über den Photoperiodismus." *Zeitschrift für Pflanzenzüchtung* 24: 129–133.

Ruhenstroth-Bauer, Gerhard, and Hans Nachtsheim. 1944. "Die Bedeutung des Sauerstoffmangels für die Auslösung des epileptischen Anfalls." *Klinische Wochenschrift* 23: 18–21.

Ruska, Helmut. 1941. "Über ein neues bei der bakteriophagen Lyse auftretendes Formelement." *Die Naturwissenschaften* 29: 367–368.

Saller, Karl. 1964. *Leitfaden der Anthropologie*, 2d ed. Stuttgart: Fischer.

Sander, Klaus. 1986. "Hans Spemann und die Entdeckung des Organisatoreffekts." *MPG-Spiegel* 2: 39–45.

———. 1990. "From germ plasm theory to synergetic pattern formation: A century of developmental modelling in retrospect." *Verhandlungen der deutschen zoologischen Gesellschaft* 83: 133–177.

———. 1991. "Landmarks in developmental biology." *Roux's Archive of Developmental Biology* 200: 1–3

Sapp, Jan. 1987. *Beyond the Gene*. Oxford: Oxford University Press.

Scheibe, Arnold. 1956. "Planung und Aufbau des 'Deutsch-Bulgarischen Instituts für landwirtschaftliche Forschungen (Kaiser-Wilhelm-Institut)' in Sofia," in Boris Rejewsky and Georg Schreiber, eds., *Aus der deutschen Forschung der letzten Dezennien*. Stuttgart, pp. 326–337.

Schiemann, Elisabeth. 1943. "Entstehung der Kulturpflanzen." *Ergebnisse der Biologie* 19: 409–552.

———. 1944. "Artkreuzungen bei Fragaria III. Die vesca Bastarde." *Flora* 37: 166–192.

———. 1960. "Erinnerung an meine Berliner Universitätsjahre," in *Gedenkschrift der westdeutschen Rektorenkonferenz und der Freien Universität Berlin zur 150. Wiederkehr des Gründungsjahres der Friedrich Wilhelm Universität zu Berlin*. Berlin: Walter de Gruyter.

Schmacke, Norbert, and Hans-Georg Güse. 1984. *Zwangssterilisiert, verleugnet, vergessen: Zur Geschichte der nationalsozialistischen Rassenhygiene am Beispiel Bremen*. Bremen: Brockkamp.

Schmelcher, Robert. 1965. "Freiwillige Sterilisationen straffrei—Urteilbegründung des Bundesgerichtshofes im Falle Chefarzt Dr. Dohrn." *Deutsche medizinische Wochenschrift* 90: 448–450.

Schmidt, Gerhard, and P. A. Levene. 1938. "The effect of nucleophosphatase on 'native' and depolymerized thymonucleic acid." *Science* 88: 172–173.

Schmidt, Martin. 1942. "Beiträge zur Züchtung frostwiderstandsfähiger Obstsorten." *Der Züchter* 14: 1–19.

Schmidt, Wilhelm J. 1941. "Einiges über optische Anisotropie und Feinbau von Chromatin und Chromosomen." *Chromosoma* 2: 86–110.

Schnarrenberger, Claus. 1987. "Botany at the Kaiser Wilhelm Institutes." *Englera* 7: 105–146.

Schönefeldt, Maria. 1935. "Entwicklungsgeschichtliche Untersuchungen bei

Neurospora tetrasperma und Neurospora sitophila." *Zeitschrift für induktive Abstammungs- und Vererbungslehre* 69: 193–209.

Schramm, Gerhard. 1941. "Über die enzymatische Abspaltung der Nucleinsäure aus dem Tabakmosaikvirus." *Bericht der deutschen chemischen Gesellschaft* 74: 532–536.

———. 1943. "Uber die Spaltung der Tabakmosaikvirus in niedermolekulare Proteine und die Rückbildung hochmolekularen Proteins aus den Spaltstücken." *Die Naturwissenschaften* 31: 94–96.

———. 1947. "Über die Spaltung der Tabakmosaikvirus und die Wiedervereinigung des Spaltstückes zu höhermolekularen Proteinen." *Zeitschrift für Naturforschung* 2b: 112–121.

———. 1948. "Zur Chemie des Mutationsvorganges beim Tabakmosaikvirus." *Zeitschrift für Naturforschung* 3b: 320–327.

———. 1951. "Makromolekulare Struktur der Nucleinsäuren," in "2. Colloquium der Deutschen Gesellschaft für physiologische Chemie," April pp. 69–80.

———. 1954. "Die Struktur des Tabakmosaikvirus und seiner Mutanten," *Advances in Enzymology and Related Subjects of Biochemistry.*

Schramm, Gerhard, and Heinz Dannenberg. 1944. "Über die Ultraviolettabsorption des Tabakmosaikvirus." *Bericht der deutschen chemischen Gesellschaft* 77: 53–60.

Schramm, Gerhard, and Hans Müller. 1940. "Zur Chemie des Tabakmosaikvirus: Über die Einwirkung von Keten und Phenylisocyanat auf das Virusprotein." *Zeitschrift für physiologische Chemie* 266: 43–55.

———. 1942. "Über die Bedeutung der Aminogruppen für die Vererbungsfähigkeit des Tabakmosaikvirus." *Zeitschrift für physiologische Chemie* 274: 267–275.

Schramm, Gerhard, and L. Rebensburg. 1942. "Zur vergleichenden Charakterisierung einiger Mutanten des Tabakmosaikvirus." *Die Naturwissenschaften* 30: 48–51.

Schreiber, Georg. 1951. "Die Kaiser-Wilhelm-Gesellschaft im Reichetat und Reichgeschehen." *Jahrbuch MPG,* 60–107.

Schrödinger, Erwin. 1944. *What Is Life?* Cambridge: Cambridge University Press.

Schubert, Gerhard. 1942. "Anwendungen der Neutronenstrahlen und der künstlichen Radioaktivität in der Medizin." *Strahlentherapie* 71: 599–626.

———. 1947. "Erfahrungen und Ergebnisse von Untersuchungen mit künstlichen radioaktiven Indikatoren." *Strahlentherapie* 76: 389–406.

Schubert, Gerhard, and Wolfgang Riezler. 1947. "Zur biologischen Wirkung injizierter künstlich radioaktiver Substanzen." *Strahlenbiologie* 71: 407–416.

Schwanitz, Franz. 1940. "Polyploidie und Pflanzenzüchtung." *Die Naturwissenschaften* 28: 353–361.

Schwemmle, Julius. 1941. "Weitere Untersuchungen an Eu-Oenotheren über die genetische Bedeutung des Plasmas und der Plastiden." *Zeitschrift für induktive Abstammungs- und Vererbungslehre* 79: 321–335.

———. 1944. "Plastiden und Genmanifestationen." *Flora* 37: 61–72.

Seidel, Friedrich. 1954. "Geschichtliche Linien und Problematik der Entwicklungsphysiologie." *Verhandlungen deutscher Naturforscher und Ärzte* 98: 93–104.

Seidel, Friedrich, E. Bock, and G. Krause. 1940. "Die Organisation des Insekteneies (Reaktionsablauf, Induktionsvorgänge, Eitypen)." *Die Naturwissenschaften* 28: 433–446.

Sell-Beleites, Ilse, and Alexander Catsch. 1942. "Mutationsauslösung durch ultraviolettes Licht bei Drosophila." *Zeitschrift für induktive Abstammungs- und Vererbungslehre* 80: 551–557.

Seybold, August. 1943. "Zur Kenntuis der herbstlichen Laubblattfärbung." *Botanisches Archiv* 44: 551–569.

Seybold, August, Karl Egle, and W. Hülsbruch. 1941. "Chlorophyll- und Carotinoidbestimmungen von Süßwasseralgen." *Botanisches Archive* 42: 239–253.

———. 1943. "Zur Kenntnis der herbstlichen Laubblattfärbung." *Botanisches Archive* 44: 551–568.

Seyfarth, Ernst-August, and Henryk Pierzchala. 1992. "Sonderaktion Krakau 1939. Die Verfolgung von polnischen Biowissenschaftlern und Hilfe durch Karl von Frisch." *Biologie in unserer Zeit* 22: 218–225.

Smith, Bradley F., and Agnes F. Peterson, eds. 1974. *Heinrich Himmler. Geheimreden 1933 bis 1945.* Frankfurt: Ullstein.

Spemann, Hans. 1942. "Über das Verhalten embryonalen Gewebes im erwachsenen Organismus." *Archiv für Entwicklungsmechanik* 141: 693—769.

Spemann, Hans, and Hildge Mangold. 1924. "Über die Induktion von Embryonalanlagen durch Implantation artfremder Organisatoren." *Archiv für mikroskopische Anatomie und Entwicklungsmechanik* 100: 599–638.

Spengler, Oswald. 1920. *Der Untergang des Abendlandes.* Munich: Beck.

Stein, Emmy. 1922. "Untersuchungen über die Radiomorphosen von Antirrhinum." *Zeitschrift für induktive Abstammungs- und Vererbungslehre* 29: 1–15.

———. 1926. "Über den Einfluß von Radiumbestrahlung auf Antirrhinum." *Zeitschrift für induktive Abstammungs- und Vererbungslehre* 43: 1–15.

———. 1930. "Weitere Mitteilungen über die durch Radiumbestrahlung induzierten Gewebeentartungen im Antirrhinum (Photocarcinome) und ihr erbliches Verhalten." *Biologisches Zentralblatt* 50: 129–155.

————. 1935. "Weitere Analysen der Gruppe A von den durch Radiumbestrahlung veränderten Erbanlagen bei Antirrhinum." *Zeitschrift für induktive Abstammungs- und Vererbungslehre* 69: 303–326.

Stent, Gunther S. 1963. *Molecular Biology of Bacterial Viruses.* San Francisco: W. H. Freeman.

————. 1967. "That was the molecular biology that was." *Science* 160: 390–395.

Stent, Gunther S., and Richard Calender. 1978. *Molecular Genetics,* 2d ed. San Francisco: Freeman and Co.

Stern, Curt. 1931. "Zytologisch-genetische Untersuchungen als Beweise für die Morgan'sche Theorie des Faktorenaustausches." *Biologisches Zentralblatt* 51: 547–587.

Stieve, H. 1944. "Paracyclische Ovulationen." *Zentralblatt für Gynäkologie* 68: 257–271.

Stocker, Otto, S. Rehm, and H. Schmidt. 1943. "Der Wasser- und Assimilationshaushalt dürreresistenter und dürreempfindlicher Sorten landwirtschaftlicher Kulturpflanzen." *Jahrbuch für wissenschaftliche Botanik* 91: 1–53.

Stokar, Walter v. 1938. "Die Herkunft des Getreides, besonders des Weizens." *Deutsche Apothekerzeitung* 64: 970–973.

Straub, Josef. 1939. "Zytogenetik." *Fortschritte der Botanik* 8: 261–279.

————. 1941. "Ergebnisse und Probleme der Polyploidenforschung." *Forschungsdienst* 12: 318–324.

Strauß, Herbert A., and Werner Röder. 1983. *International Biographical Dictionary of Central European Emigrés, 1933–1945.* vol. 2, *The Arts, Sciences, and Literatures.* Munich: K. G. Saur.

Strugger, Siegfried. 1940. "Die Kultur von Didymium migripes aus Myxamöben mit vitalgefärbtem Plasma und Zellkernen." *Zeitschrift für wissenschaftliche Mikroskopie* 57: 415–419.

Stubbe, Hans. 1934. "Erwin Baur (Nachruf)." *Zeitschrift für induktive Abstammungs- und Vererbungslehre* 66: v–ix.

————. 1935. "Erbkrankheiten bei Pflanzen." *Der Erbarzt* 8: 65–69.

————. 1937. "Der gegenwärtige Stand der experimentellen Erzeugung von Mutationen durch Einwirkung von Chemikalien." *Angewandte Chemie* 50: 241–246.

————. 1940. "Neuere Forschungen zur experimentellen Erzeugung von Mutationen." *Biologisches Zentralblatt* 60: 113–129.

Stubbe, Hans, and Helmut Döring. 1938. "Untersuchungen über experimentelle Auslösung von Mutationen bei Antirrhinum majus VII." *Zeitschrift für induktive Abstammungs- und Vererbungslehre* 75: 341–357.

Stubbe, Hans, and Karl Pirschle. 1940. "Über einen monogen bedingten Fall von Meiosis bei Antirrhinum majus." *Berichte der deutschen botanischen Gesellschaft* 58: 546–558.

Stuckart, Wilhelm, and Hans Globke. 1936. *Kommentar zur deutschen Rassengesetzgebung,* vol. 1. Munich and Berlin: J. H. Lehmann.

Studnitz, G. v., H. K. Loevenich, and H. J. Neumann. 1943. "Über die Löslichkeit und Trennbarkeit der Farbsubstanzen." *Zeitschrift für vergleichende Physik* 30: 74–83.

Süss, R., V. Kinzel, and J. Scribner. 1970. *Krebs. Experimente und Denkmodelle—eine elementare Einführung in Probleme der experimentellen Tumorforschung.* Berlin: Springer.

Thienemann, August. 1939. "Grundzüge einer allgemeinen Ökologie." *Archiv für Hydrobiologie* 35: 267–285.

———. 1941. "Vom Wesen der Ökologie." *Biologia generalis* 15: 312–331.

Thomae, Hans. 1977. *Psychologie in der modernen Gesellschaft.* Hamburg.

Timoféeff-Ressovsky, Nikolai W. 1935. "Experimentelle Untersuchungen der arblichen Belastung von Populationen." *Der Erbarzt* 8: 117–118.

Timoféeff-Ressovsky, Nikolai W., and Max Delbrück. 1936. "Strahlengenetische Versuche über sichtbare Mutationen und die Mutabilität einzelner Gene bei Drosophila melangoster." *Zeitschrift für induktive Abstammungs- und Vererbungslehre* 71: 322–334.

Timoféeff-Ressovsky, Nikolai W., and Elena A. Timoféeff-Ressovsky. 1941a. "Populationsgenetische Versuche an Drosophila I." *Zeitschrift für induktive Abstammungs- und Vererbungslehre* 78: 28–34.

———. 1941b. "Populationsgenetische Versuche an Drosophila II." *Zeitschrift für induktive Abstammungs- und Vererbungslehre* 78: 35–43.

———. 1941c. "Populationsgenetische Versuche an Drosophila III." *Zeitschrift für induktive Abstammungs- und Vererbungslehre* 78: 44–49.

Timoféeff-Ressovsky, Nikolai W., and Karl Günter Zimmer. 1944. "Strahlengenetik." *Strahlentherapie* 74: 183–211.

Timoféeff-Ressovsky, Nikolai W., Karl Günter Zimmer, and Max Delbrück. 1935. "Über die Natur der Genmutation und der Genstruktur." *Nachrichten der Gesellschaft für Wissenschaften in Göttingen,* Fachgr. VI, N.F. 1, no. 13: 189–245.

Trinkaus, J. P. 1966. "Morphogenetic cell movements," in M. Locke, ed., *Major Problems in Developmental Biology.* New York: Academic Press. 1966.

Troll, Wilhelm. 1935/1936. "Die Wiedergeburt der Morphologie aus dem Geiste deutscher Wissenschaft." *Zeitschrift für die gesamte Naturwissenschaft* 1: 349–356.

———. 1941. *Gestalt und Urbild. Gesammelte Aufsätze zu Grundfragen der organischen Morphologie.* Leipzig: Akademische Verlagsgesellschaft.

Twitty, V. C. 1966. *Of Scientists and Salamanders*. San Francisco: W. H. Freeman.

Ubisch, Gerta von. 1919. "II. Beitrag zu einer Faktorenanalyse von Gerste." *Zeitschrift für induktive Abstammungs- und Vererbungslehre* 20: 65–117.

———. 1934. "Das Fertilitätsproblem im Pflanzenreich." *Zeitschrift für induktive Abstammungs- und Vererbungslehre* 67: 225–241.

Vierhaus, Rudolf, and Bernhard von Brocke, eds. 1990. *Forschung im Spannungsfeld von Politk and Gesellschaft: Geschichte und Struktur der Kaiser-Wilhelm-/Max-Planck-Gesellschaft*. Stuttgart: DVA.

Vogel, Christian. 1988. "Gibt es eine natürliche Moral? Oder: Wie widernatürlich ist unsere Ethik?" in Heinrich Meier, ed., *Die Herausforderung der Evolutionsbiologie*. Munich: Piper, pp. 193–219.

Vogt, Cécile, and Oskar Vogt. 1936. "Bericht über die wissenschaftliche Arbeit des Kaiser-Wilhelm-Instituts für Hirnforschung in Berlin-Buch," in Max Planck, ed., *25 Jahre Kaiser Wilhelm Gesellschaft zur Förderung der Wissenschaften*, vol. 2. Berlin: Springer, pp. 391–394.

Walk, Joseph, ed. 1981. *Das Sonderrecht für die Juden im NS-Staat: Eine Sammlung der gesetzllichen Maßnahmen und Richtlinien—Inhalt und Bedeutung*. Heidelberg and Karlsruhe: C. F. Müller, 1981.

Walker, Mark. 1989. *German National Socialism and the Quest for Nuclear Power, 1939–1949*. Cambridge: Cambridge University Press.

Wallenfels, Kurt. 1953. Review of "Physiologische Chemie. Ein Lehr- und Handbuch für Ärzte, Biologen, und Chemiker." *Angewandte Chemie* 65: 67–68.

Waterson, A. P., and Lise Wilkinson. 1978. *An Introduction to the History of Virology*. Cambridge: Cambridge University Press.

Watson, James D. 1968. *The Double Helix*. London: Weidenfeld and Nicholson.

Wehler, Hans-Ulrich. 1988. *Entsorgung der deutschen Vergangenheit? Ein polemischer Essay zum 'Historikerstreit.'* Munich: Beck.

Weindling, Paul. 1989. *Health, Race, and German Politics between National Unification and Nazism, 1870–1945*. Cambridge: Cambridge University Press.

Weinert, Hans. 1938. *Entstehung der Menschenrassen*. Stuttgart: F. Enke.

Weingart, Peter, and Matthias Winterhager. 1984. *Die Vermessung der Forschung—Theorie und Praxis der Wissenschaftsindikatoren*. Frankfurt: Campus.

Weingart, Peter, Jürgen Kroll, and Kurt Bayertz. 1988. *Rasse, Blut und Gene: Geschichte der Eugenik und Rassenhygiene in Deutschland*. Frankfurt: Suhrkamp.

Weinreich, Max. 1946. *Hitler's Professors: The Part of Scholarship in Germany's Crime against the Jewish people*. New York: YIVO.

Weinzierl, H., and B. Lötsch. 1988. "Konrad Lorenz—eine Legende wird 85." *Natur* 11: 28–33.

Weiss, Sheila. 1990. "The race hygiene movement in Germany, 1904–1945," in Mark B. Adams, ed., *The Wellborn Science*. New York: Oxford University Press, pp. 8–68.

Wettstein, Fritz von. 1937. "Die genetische und entwicklungsphysiologische Bedeutung des Cytoplasmas." *Zeitschrift für induktive Abstammungs- und Vererbungslehre* 73: 349–366.

———. 1940. "Experimentelle Untersuchungen zum Artbildungsproblem II. Zur Frage des Polyploidie als Artbildungsfaktor." *Berichte der deutschen botanischen Gesellschaft* 58: 374–388.

———. 1941. "Was ist aus der neueren Vererbungsforschung für die Pflanzenzüchtung zu verwerten?" *Forschungsdienst* 14: 116–130.

———. 1943. "Warum hat der diploide Zustand bei den Organismen den größeren Selektionswert." *Die Naturwissenschaften* 31: 574–577.

Wettstein, Fritz von, and Karl Pirschle. 1940. "Klimakammern mit konstanten Bedingungen für die Kultur höherer Pflanzen." *Die Naturwissenschaften* 28: 537–543.

Wettstein, Fritz von, and Josef Straub. 1942. "Experimentelle Untersuchungen zum Artbildungsproblem III." *Zeitschrift für induktive Abstammungs- und Vererbungslehre* 80: 271–280.

Wissenschaft und Freiheit. 1954. Edited by Der Kongress für die Freiheit der Kultur. Berlin: Grunewald.

Wistrich, Robert. 1982. *Who's Who in Nazi Germany.* New York: Macmillan.

Wolf, P. M., and Hans Joachim Born. 1941. "Über die Verteilung natürlich radioaktiver Substanzen im Organismus nach parenteraler Zufuhr." *Strahlentherapie* 70: 342–348.

Wolf, P. M., G. H. Radu, and Alexander Catsch. 1944. "Über die Verteilung natürlich-radioaktiver Substanzen im Organismus nach parenteraler Zufuhr IV: Versuch mit Uran X an Kaninchen." *Strahlentherapie* 75: 452–456.

Wollman, Elie L. 1972. "Bakterienkonjugation," in J. Cairns, G. S. Stent, and J. D. Watson, eds., *Phagen und die Entwicklung der Molekularbiologie.* Berlin: Akademie-Verlag, pp. 212–220.

Wolters, Gereon. 1991. "Die Natur der Erkenntnis. Ein Thema der Philosophie oder der Biologie?" in Helmut Bachmaier and Ernst Peter Fischer, eds., *Glanz und Elend der zwei Kulturen: Über die Verträglichkeit der Natur- und Geisteswissenschaften.* Konstanz: Universitätsverlag Konstanz, pp. 141–155.

Wuketits, Franz M. 1990. *Konrad Lorenz: Leben und Werk eines großen Naturforschers.* Munich: Piper.

Wülker, Heinz. 1935. "Untersuchungen über Tetradenaufspaltung bei Neurospora sitophila Shear et Dodge." *Zeitschrift für induktive Abstammungs- und Vererbungslehre* 69: 210–248.

Wuttke-Groneberg, Walter. 1983. "Von Heidelberg nach Dachau," in Gerhard Baader and Ulrich Schultz, eds., *Medizin und Nationalsozialismus. Tabuisierte Vergangenheit—Ungebrochene Tradition?* 2d ed. Berlin: Verlagsgesellschaft Gesundheit, pp. 113–138.

Zacharov, I., and J. Surikov. 1991. "Genetik in Acht und Bann." *Wissenschaft in der UdSSR* 3: 110–117.

Zachau, H. G. 1989. "40 Jahre Forschung in der Bundesrepublik—Rückkehr in die internationale Forschergemeinschaft." Talk given at the Symposium der Wissenschaftsorganisationen in Bonn on October 26.

Zarnitz, Marie Luise. 1968. *Molekulare und physikalische Biologie: Bericht zur Situation eines interdisziplinären Forschungsgebietes in der Bundesrepublik Deutschland.* Göttingen: Vandenhoeck and Ruprecht.

Zierold, Kurt. 1968. *Forschungsförderung in drei Epochen: Deutsche Forschungsgemeinschaft, Geschichte, Arbeitsweise, Kommentar.* Wiesbaden: Steiner.

Zimmer, Karl Günter. 1938a. "Aussichten der praktischen Verwendung der Bestrahlung mit schnellen Neutronen." *Fortschritte auf dem Gebiet der Röntgenstrahlen* 58: 77–79.

———. 1938b. "Versuche mit schnellen Neutronen I." *Strahlentherapie* 63: 517–527.

———. 1940. "Dosimetrische und strahlenbiologische Versuche mit schnellen Neutronen III." *Strahlentherapie* 68: 74–78.

———. 1960. "Studien zur quantitativen Strahlenbiologie." *Akademie der Wissenschaften und der Literatur in Mainz. Mathematisch-Naturwissenschaftliche Klasse,* no. 3.

Zimmer, Karl Günter, and Nikolai W. Timoféeff-Ressovsky. 1937. "Auslösung von Mutationen bei Drosophila melanogaster durch Alpha-Teilchen nach Emanationseinatmung." *Strahlentherapie* 55: 77.

———. 1942. "Über einige physikalische Vorgänge bei der Auslösung von Genmutationen durch Strahlung." *Zeitschrift für induktive Abstammungs- und Vererbungslehre* 80: 353–372.

Zippelius, Hanna-Maria. 1992. *Die vermessene Theorie: Eine kritische Auseinandersetzung mit der Instinkttheorie von Konrad Lorenz und verhaltenskundlicher Praxis.* Braunschweig: Vieweg.

Index